Robotics Technology

by

James W. Masterson
Department of Technology
Eastern Kentucky University
Richmond, KY

Robert L. Towers
Department of Technology
Eastern Kentucky University
Richmond, KY

Stephen W. Fardo
Department of Technology
Eastern Kentucky University
Richmond, KY

South Holland, Illinois
THE GOODHEART-WILLCOX COMPANY, INC.
Publishers

Library of Congress Catalog Card Number 93-43463
International Standard Book Number 1-56637-046-9

1 2 3 4 5 6 7 8 9 10 96 99 98 97 96

Library of Congress Cataloging in Publication Data

Masterson, James W.
Robotics Technology / by James W. Masterson, Robert L. Towers, and Stephen W. Fardo
p. cm.
Includes index.
ISBN 1-56637-046-9
1. Robotics. I. Towers, Robert L. II. Fardo, Stephen W. III. Title.
TJ211.M3672 1996
670.42'72—dc20 93-43463
CIP

About the Authors

James W. Masterson is currently a professor in the Department of Technology at Eastern Kentucky University. Dr. Masterson has developed instructional materials and taught classes in the areas of robotics, material handling, design for manufacturability, and automated manufacturing. In addition, he has worked for IBM's Department of Advanced Technical Education and has consulted for several industries concerning manufacturing problems. Dr. Masterson, who has 27 years teaching experience, received his doctorate in Industrial Education from the University of Missouri.

Robert L. Towers is a licensed professional engineer and is an assistant professor in the Department of Technology at Eastern Kentucky University. He has worked as an engineering consultant and has held engineering positions with several firms in the telecommunications field. Dr. Towers received his doctorate in Industrial Systems Design from the University of Kentucky. He has coauthored a number of textbooks and articles.

Stephen W. Fardo has more than twenty years' experience teaching the technical curriculum. Dr. Fardo has developed technical curricula for several state vocational-technical education programs and is the author of numerous textbooks and laboratory manuals. He is a member of the National Association of Industrial Technology and of the Society of Manufacturing Engineers. He received his doctorate in Vocational-Technical Education from the University of Kentucky and is currently a professor in the Department of Technology at Eastern Kentucky University.

This robotic system is used to assemble the components of automobile engines. (Hirata Corporation of America)

Introduction

The development of the computer has created what some experts have called the Second Industrial Revolution. Many consider robots to be the prime movers of this revolution. Their use in our society is increasing and will continue to increase. The need to know more about robots has become more important.

This text is a comprehensive approach to learning the technical aspects of robotics. It is divided into four units, covering the broad areas of robotic principles, power supplies and movement systems, sensing and end-of-arm tooling, and control systems.

Unit I is devoted to the basic principles of robotic technology. Chapter 1 introduces industrial robotics with a discussion of some of the historical events leading to its development. The industrial robot is defined and the place of industrial robots in the area of automation is established. Chapter 2, one of the key chapters in the book, prepares a solid foundation for understanding the characteristics and fundamentals of robotics. Chapter 3 provides an overview of the languages and techniques used to program industrial robots. Chapter 4 includes numerous photographs that provide insight into the many applications for industrial robots. Chapter 5 deals with some of the major factors to consider when using robotics in an industrial environment.

Unit II deals with robotic power supplies and movement systems. Chapter 6 provides an overview of electromechanical systems used with robots. Alternating current and direct current systems are discussed in detail. Fluid power systems, which include hydraulic and pneumatic power, are covered in Chapter 7. These robot power sources are used for numerous applications. Chapter 8, "Maintaining Robotic Systems," provides an overview of maintenance procedures. The benefits of preventive maintenance are discussed and a general plan is included for implementing preventive maintenance.

Unit III deals with robotic sensing systems and end-of-arm tooling. Chapter 9 discusses various types of sensors commonly used by a robot to gain information about its external environment. Some of these include tactile sensors, limit switches, proximity sensors, and photoelectric sensors. Chapter 10 provides information about various end effectors and tools used to move workpieces from one location to another within the robot's work envelope.

Unit IV covers robotic control systems. The basics of digital electronics are discussed in Chapter 11. This chapter also includes information on the microcomputers and microprocessors used. Chapter 12 deals with the use of PLCs to program and control robots and other automated machines. Chapter 13 explains how the robot controller communicates with peripheral equipment found in robotic workcells. Vision systems are also covered in this chapter. Chapter 14, "The Future of Robotics," discusses the factory of the future, robots outside the factory, artificial intelligence and expert systems, and how training in robotics can be acquired.

This textbook is designed for vocational/technical schools, college/university technical programs, industrial training programs, and for technical high school programs. It would also make an excellent reference book.

The authors would like to thank the many companies that provided photographs and technical information during preparation of the manuscript. With their cooperation, much up-to-date material was provided.

James W. Masterson
Robert L. Towers
Stephen W. Fardo

Contents

PRINCIPLES OF ROBOTICS

Since the first industrial robot was installed at a U.S. automotive plant in 1961, robotics technology has become an important factor in most types of manufacturing. Robots are widely used for applications that require extreme precision, for repetitive and tedious tasks, and for work that would be unpleasant or dangerous for humans. Robots are vital components of flexible manufacturing systems, which can be quickly reconfigured to meet changing production requirements.

POWER SUPPLIES AND MOVEMENT SYSTEMS

In automated applications, three kinds of motion are used: rotary, linear, and reciprocating. These motions can be produced by either electrical or fluid (hydraulic or pneumatic) power operating motors, relays, solenoids, actuators, or cylinders. The basic mechanical unit of a robot, the manipulator, has several moving joints and performs the actual work function of the machine.

SENSING AND END-OF-ARM TOOLING

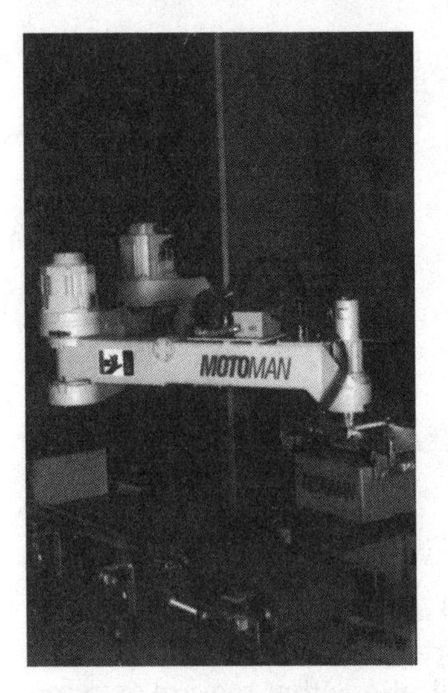

The control of industrial robots often depends upon a sensing system which uses devices called transducers to convert light, heat, or mechanical energy into electrical energy. The signal output of the transducer is used to affect the operation of the robot's end effector (end-of-arm tooling). End effectors, attached to the wrist of a manipulator, can grasp, lift, transport, maneuver, or perform operations on a workpiece.

CONTROL SYSTEMS

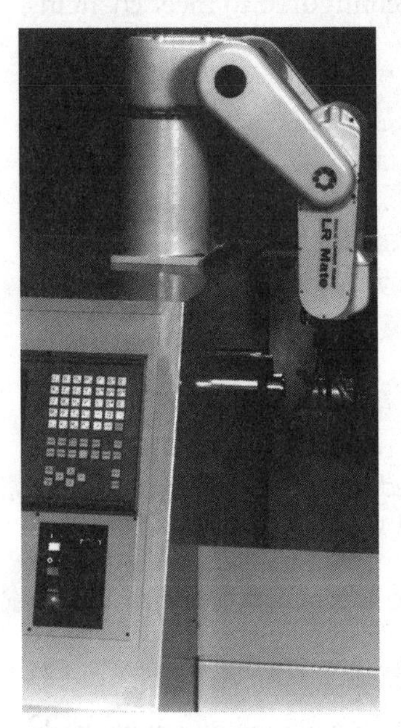

The heart of a robotic control system is a microprocessor (computer chip) linked to input/output and monitoring devices. The control system has a series of instructions, called a program, stored in its memory. The program supplies the commands that control motors, hydraulic systems, or pneumatic systems to activate the robot's motion control mechanism. This mechanism is typically an actuator, a device that converts power into robot movement.

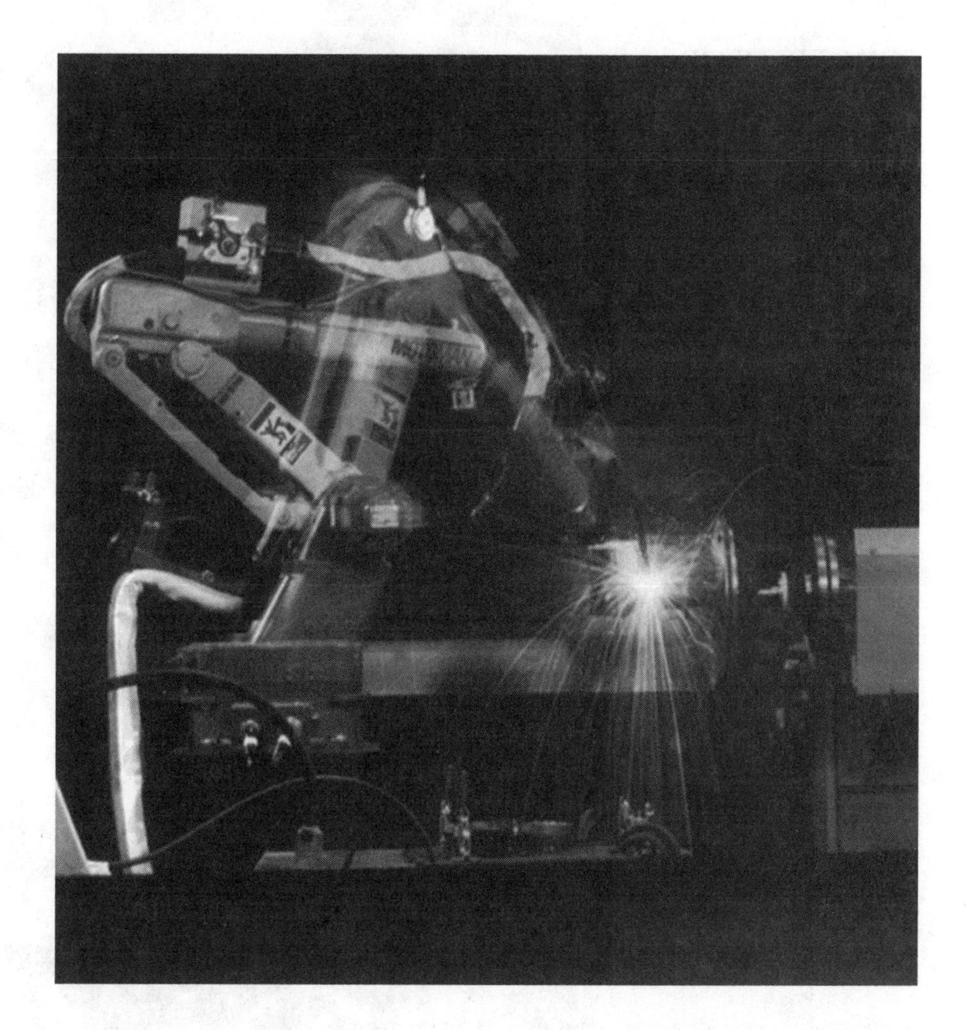

Welding is one of the most common industrial applications for robots. (Motoman, Inc.)

PRINCIPLES OF ROBOTICS

Since the first industrial robot was installed at a U.S. automotive plant in 1961, robotics technology has become an important factor in most types of manufacturing. Robots are widely used for applications that require extreme precision, for repetitive and tedious tasks, and for work that would be unpleasant or dangerous for humans. Robots are vital components of flexible manufacturing systems, which can be quickly reconfigured to meet changing production requirements.

Hirata Corporation of America

1 Introduction to Industrial Robotics

Overview

For centuries, the idea of robots has captured people's imaginations. Who first conceived of them, and how did they become so popular? In this chapter, the origins of the robot will be discussed. The way in which robots began to be used in industry will also be covered. Not everyone defines a robot in the same way; this chapter attempts to sort out the differences. It also covers the different types of automation.

Early Robots

Robots have long played important roles in the movies, in books, and on television. R2-D2 and C-3PO of the *Star Wars*© movies present robots in a favorable, comic light. Some real robots like those in Figures 1-1, 1-2, and 1-3 resemble R2-D2 and look nonthreatening and humanlike. However, most books and movies in the past have shown robots as a threat to humankind, rather than a help. To fully understand why many people regard robots as a threat, you have to go back to the robot's origins.

At first, the term used for what we now consider a robot was ***automaton,*** a human-made object that moved automatically. The first useful automatons were clockworks introduced during the Middle Ages to automatically keep track of time, Figure 1-4. As these clocks became more complex, systems of gears and pulleys enabled the workings and figures attached to the clocks to move in lifelike ways. Many advanced automatons were built to entertain royalty and the nobility, for they appealed to people's imaginations. The English scientist, Roger Bacon (ca. 1220-1292) even had ideas for a flying machine, a diving bell, a mechanical chariot, and mechanical birds.

As people became more knowledgeable about physiology, some believed that even the human body was merely a complex automaton separate from the mind and soul. The seventeenth-century French philosopher René Descartes (1596-1650) strongly supported this viewpoint, and people had visions of building truly lifelike automatons. However, since the machines would be mindless and soulless, would they wreak havoc on

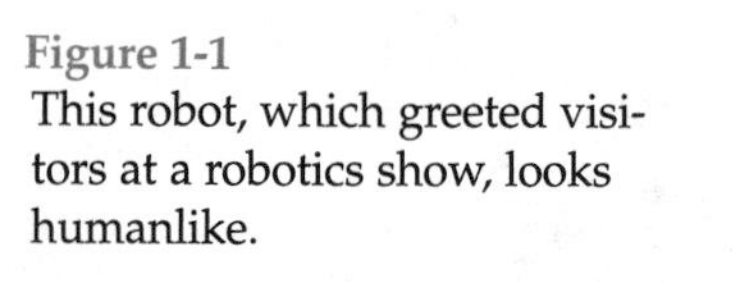

Figure 1-1
This robot, which greeted visitors at a robotics show, looks humanlike.

Figure 1-2
A humanlike robot used in educational activities. (General Robotics Corporation)

Figure 1-3
These service robots are used for building security and possess humanlike features. They have limited intelligence. (Denning Mobile Robotics, Inc.)

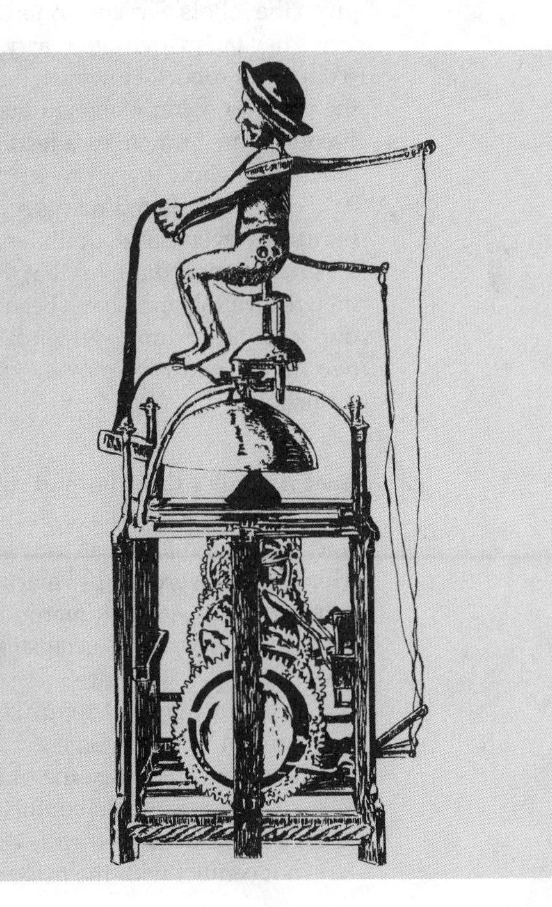

Figure 1-4
Artists' sketches of an early clockworks showing the addition of lifelike figures and movements.

humans if they were to go out of control? The story is told that Descartes built a female automaton and took it on a sea voyage. The captain of the ship, thinking it was the work of the devil, promptly threw it overboard.

Robots in Literature

In 1818, Mary Shelley's novel *Frankenstein* did little to ease people's fears of technology. Dr. Frankenstein's monster, an artificial man, was unhappy when his creator neglected him, so he took a horrible revenge.

Then, in 1920, the Czech dramatist ***Karel Capek*** (1890-1938) wrote a play titled *R.U.R.* ("Rossum's Universal Robots"). ***Robot*** is from the Czech word *robota,* which means "forced labor." The play was so popular that the word robot began to be used instead of automaton in virtually every language. As a result, a robot is commonly thought of today as any manufactured structure that can perform functions normally done by human beings. Robots are often visualized as having humanlike forms.

In Capek's play, Rossum develops a formula for making mechanical robots, whose only function is work. Work is what they exist for, and they are more perfect than humans in many ways. They are extremely strong and dedicated to their tasks. Some are even quite intelligent. All they lack is a soul and emotions.

Then, a young woman in the story objects to the inhumane treatment given the robots. She confronts the factory manager, who finds her very attractive. They fall in love and marry. Over the next ten years, the factory floods the world with robots. However, the young woman persuades the physiologist at the plant to secretly change the formula for making the robots so that they become more human. As a result of this change, some robots develop interests other than work.

The new, altered robots organize the lesser robots and turn them against humanity, eventually annihilating everyone except the company's founder. When the robots discover that their parts are wearing out and need replacing, they realize this may have been a mistake. The man they have spared cannot duplicate his formula, which has been lost. However, at the end of the play, two humanized robots evolve into different sexes; robot-kind is saved from extinction.

Isaac Asimov's three laws of robotics

Not all writers have portrayed robots in a negative manner. In 1939, when ***Isaac Asimov*** was only nineteen, he began to write science fiction in which robots were simply machines that could be built with safety measures in mind. In his story "Runaround," which was published in the March, 1942, issue of the magazine *Astounding,* Asimov established his three fundamental laws of robotics. They are:

1. A robot may not injure a human being or, through inaction, allow a human being to come to harm.
2. A robot must obey the orders given it by human beings, except where such orders would conflict with the first law.
3. A robot must protect its own existence as long as such protection does not conflict with the first and second laws.

At the time, Asimov did not have a clear idea of how robots would actually be built. For instance, he described them as having "positronic," rather than electronic, brains. His three laws of robotics, however, have been taken more seriously, and his term ***robotics,*** which means the use of robots, has become part of the language. Asimov's version of the robot is now the more generally accepted one, and Capek's version has essentially faded into the past.

The Advent of Computers

By the end of World War II, computers began to be built. To many people they seemed to be "thinking machines." Self-guided control systems had been developed during the war, and scientists combined them with computers. The idea of building humanlike machines came back into popularity. Both scientists and laypeople began to consider the consequences of merging computers with another new development, artificial intelligence. ***Artificial intelligence*** is the ability of a computer program to make decisions based on known information. Could a computer be installed into a structure resembling a human body and become a "true" robot?

The First Industrial Robots

The invention of the first industrial robot was now close at hand. After the war, North American industry benefited from state-of-the-art equipment and technologies, while the other industrialized countries lay in shambles. By 1950, industry in the United States was at its zenith. The demand for goods both at home and abroad was overwhelming, and the need for increased production through automated systems became apparent.

In 1954, ***George C. Devol, Jr.,*** Figure 1-5, patented the first industrial robot, but faced many problems trying to finance his invention or sell the idea to corporations in the United States. Then, in 1956, Devol met ***Joseph Engelberger,*** Figure 1-6. Engelberger, a young engineer, was impressed with Devol's idea. He tried to persuade his employer, Aircraft Products, to help develop an industrial robot. However, it was not until Aircraft Products was acquired by Consolidated Diesel Electric Corporation that the much-needed capital was made available.

Unimation, Inc., a subsidiary, was formed in 1958 to develop Devol's invention, and Engelberger became president. In 1961, the first Unimation robot, called the ***Unimate,*** was sold to General Motors. The name Unimate stood for "universal automation." Like many other early robots used in industry, the Unimate was not called a robot because that word had too many negative connotations. The first Unimate was used in a die-casting operation. It had to be first guided through the desired sequence of steps, which were recorded. Later, the recording was played back and the robot automatically performed the required task.

At about the same time that Unimation was being formed, Versatran, a company that built the manipulators used on atomic energy projects, became interested in robots. In 1979, Versatran was purchased by Prab Conveyor Company, and together they formed Prab Robots, Inc. Prab developed its own line of industrial robots in the late 1960s.

Figure 1-5
George C. Devol, Jr., has been called the "grandfather of industrial robots."

Figure 1-6
Joseph F. Engelberger, who became the president of Unimation, is often called the "father of industrial robots."

Other major robot manufacturers involved in early development for commercial use were DeVilbiss, Asea, and Cincinnati Milacron. DeVilbiss became one of the leading producers of finishing robots and, in 1982, introduced a new arc-welding robot. Asea was one of the early developers of electric-powered anthropomorphic robot units. ***Anthropomorphic*** means "humanlike in form." Cincinnati Milacron also entered the robot market with an anthropomorphic unit, this one powered by hydraulics.

During the late 1970s and early 1980s, robots moved into assembly operations. In 1978, engineers at Unimation introduced a smaller robot called PUMA (Programmable Universal Machine). PUMAs were designed to handle small parts used in the assembly of motors and instruments.

Eventually other American companies, such as IBM, Bendix, and GE, also entered the robotics business. These companies offered foreign-built but American-packaged robots through various licensing agreements with Japanese, German, and Italian companies.

Japan Enters the Market

The early success of robotics technology in the United States had not gone unnoticed by the Japanese. In 1966, many Japanese companies sent representatives to this country to see what was happening. In 1967, Joseph Engelberger was invited to tour Japan. He lectured in Tokyo to an audience of 700 engineers and executives, and robotics technology grew rapidly in that country.

Japan's first industrial robot was developed in 1969, after the first Versatran robots had been exhibited and sold there. The Japanese copied the Versatrans. According to Engelberger, Japanese companies did not resist technology as American companies had, and Japan was able to enter the market quickly. Japan's Industrial Robot Association (JIRA) was founded in 1971, three years before the Robot Institute of America (now called the Robotic Industries Association), and six years before the British Robot Association (BRA). In 1978, the SCARA (Selective Compliance Assembly Robot Arm) was developed at Yamanashi University. Today, Japan is the world's largest user of robots.

What is an Industrial Robot?

On the surface, it seems easy to define what is meant by the term "industrial robot." This is what Webster's dictionary says:

> *"An automatic apparatus or device that performs functions ordinarily ascribed to human beings, or operates with what appears to be human intelligence."*

However, Webster's definition is rather broad and does not make specific reference to industrial robots.

The Japanese use a wide range of classifications, from simple "arms" to what they call "intelligent" robots. These classifications are:

- Δ ***Manual manipulator.*** A manipulator worked by a human operator.
- Δ ***Fixed-sequence robot.*** A manipulator that performs successive steps of a given operation repetitively, according to a predetermined sequence, condition, and position. Its intructions cannot be easily changed.
- Δ ***Variable-sequence robot.*** A manipulator that is similar to the fixed-sequence robot, but its sequence of movement can be changed easily.

- Δ ***Playback robot.*** A manipulator that can reproduce operations originally executed under human control. An operator initially feeds in the instructions relating to sequence of movement, conditions, and positions. These are then stored in memory.
- Δ ***NC (numerically controlled) robot.*** A manipulator that can perform the sequence of movement, conditions, and positions of a given task which are communicated by means of numerical data.
- Δ ***Intelligent robot.*** A robot that can itself detect changes in the work environment by means of sensory perception (visual and/or tactile). Then, using its decision-making capability, it can proceed with the appropriate operations.

Japan lists types of automated machinery which are not considered robots in the United States, raising questions as to how many true robots are being used by Japanese industry, compared to those used here.

The most widely accepted definition for the ***industrial robot*** in the United States has been published by the ***Robotic Industries Association (RIA):***

> *"A programmable, multifunction manipulator designed to move materials, parts, tools, or special devices through programmed motions for the performance of a variety of tasks."*

This definition contains several important points:

- Δ The robot is a machine.
- Δ The robot is ***programmable***; therefore, it can be given new instructions to meet new requirements.
- Δ The robot has a multifunction manipulator arm, which means it may be used in different ways, even within the same program.
- Δ The robot is flexible, enabling it to perform a variety of operations to meet special needs.

Figure 1-7 shows robots that fulfill the requirements of the RIA definition. The robotic system consists of a motor-driven multifunction manipulator arm, an electronic memory system containing the program that controls manipulator movement, and a microcomputer for reprogramming the robot for new tasks. Figure 1-8 describes three generations of manufacturing robots. It lists several applications used today and possible applications for the future. Many of these applications will be discussed in Chapter 4.

Because of a robot's programmability, it can function in many different jobs. In the future, the majority of these jobs will be outside the manufacturing area.

Types of Automation

Today's industries use various types of automation to manufacture parts and products. Two common classifications are hard automation and flexible automation.

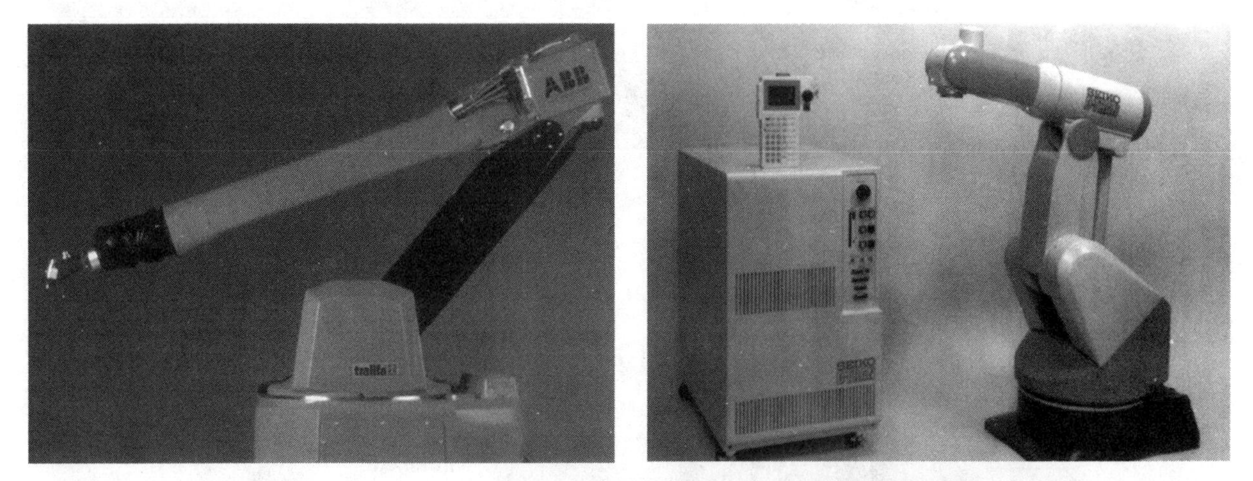

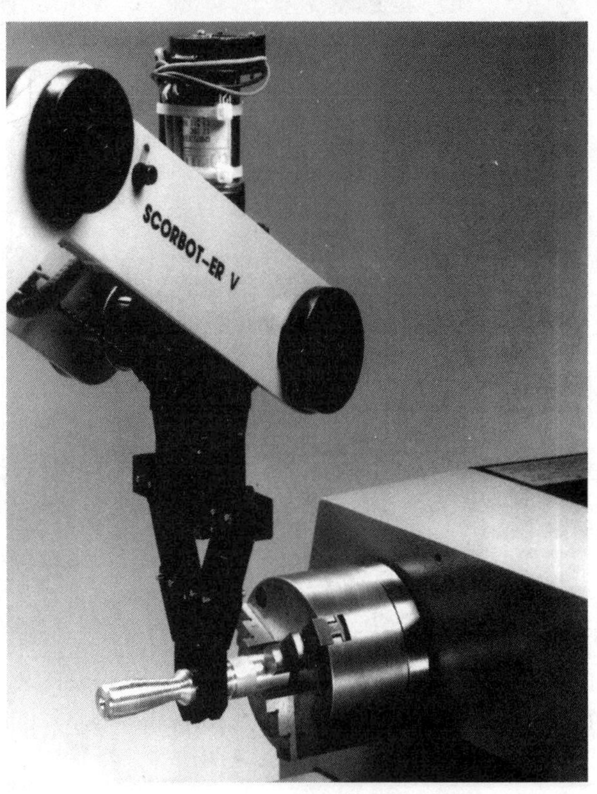

Figure 1-7
These industrial robots meet the RIA definition. (ABB Graco, Seiko, Eshed Robotec, Inc.)

Hard Automation

Hard automation refers to machinery that has been specifically designed and built to perform one particular task within an assembly line. After the item that the machine was designed to make is no longer needed, the machine must be discarded. This kind of automation can be very costly. The demand for new products and new models of existing products means shorter and shorter product life spans. Today, new hard automation equipment is cost-effective only where volume is high and the machine will be used for a long time.

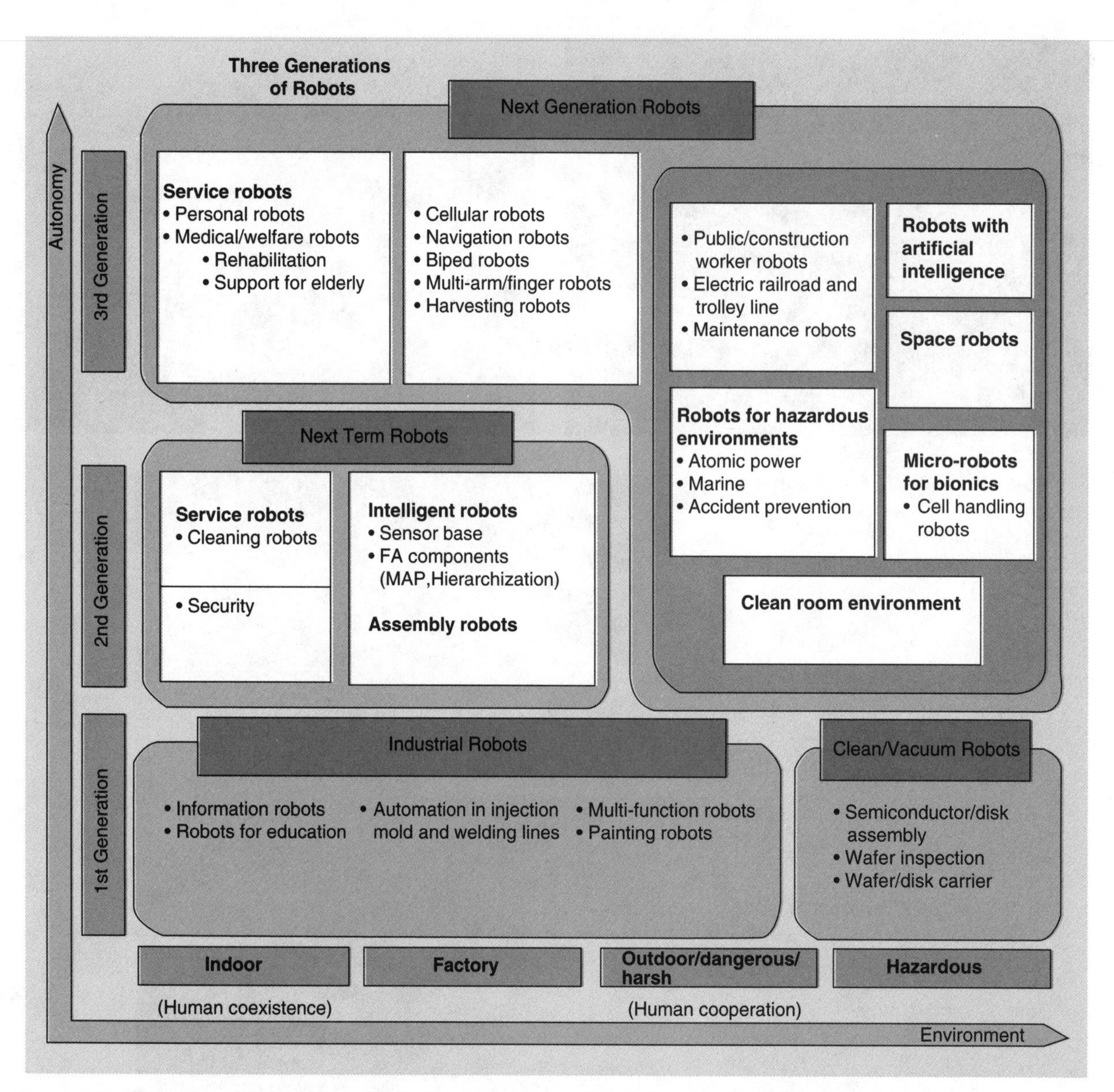

Figure 1-8 Three generations of industrial robots show increasing ability to accomplish more difficult tasks. (Motoman, Inc.)

Flexible Automation

Flexible automation includes machines that can perform different tasks. Robots belong in this category. As new products or new models are needed, the flexible machine can be reprogrammed to make the parts required. This flexibility saves money because the equipment does not have to be discarded or rebuilt. In addition, it takes much less time to reprogram the same machine than to install a new one. Figure 1-9 shows a work cell for flexible automation.

Robots are the most flexible type of automation available today. Not only can they be reprogrammed easily, quickly, and economically, but they can also be moved from one location in a plant to another.

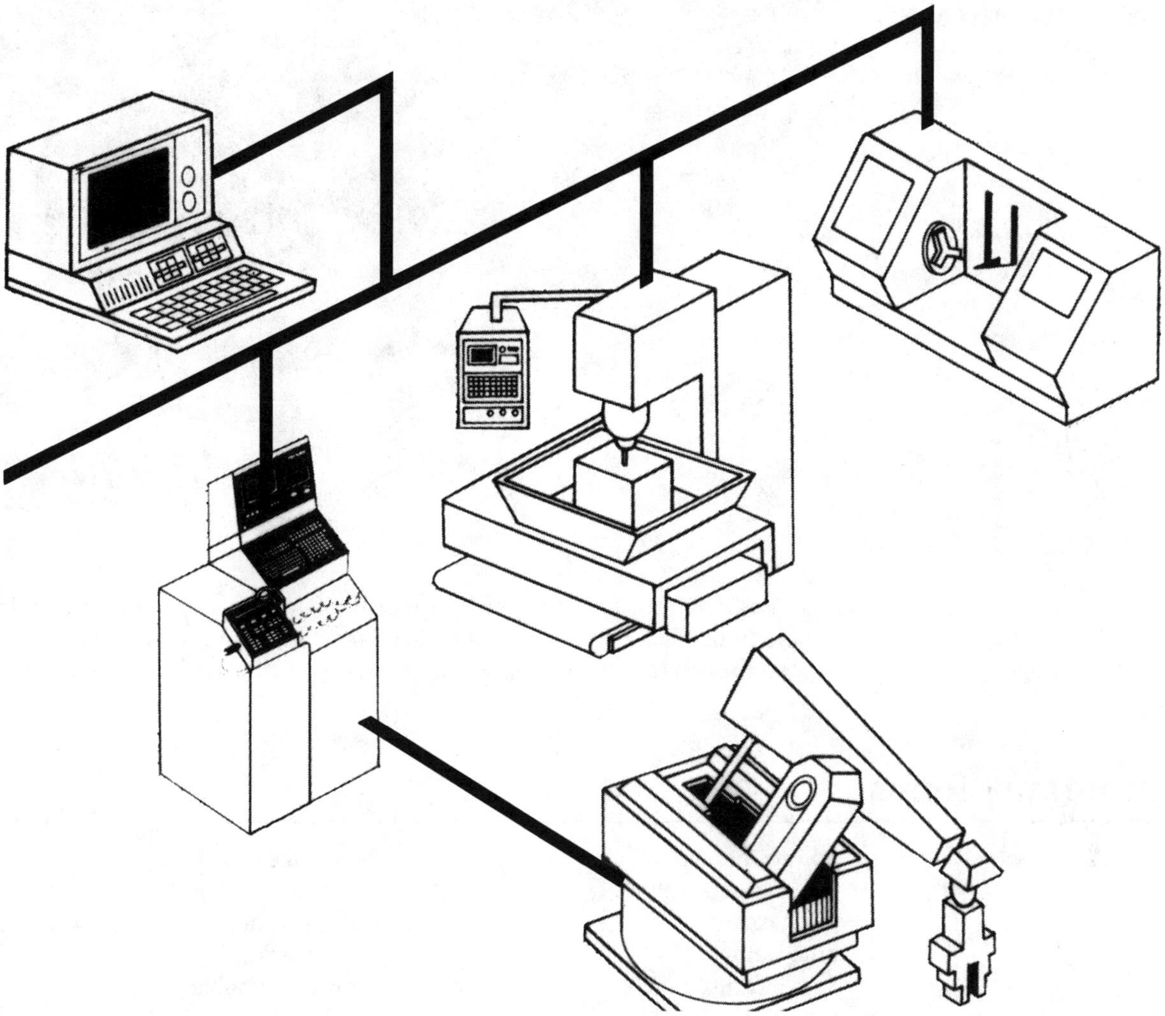

Figure 1-9 A flexible manufacturing workcell can be programmed for more than one task. (NUM Corp.)

Are Robots a Threat?

We have come a long way since Karel Capek wrote his play in 1920. However, some of his ideas are still being considered. Robots are more productive than humans. They can perform work more cheaply and have taken jobs away from humans. For these reasons, workers and labor unions often see robots as a threat. At the same time, robots are relieving men and women from repetitive work that is boring, sometimes dangerous, and often unpleasant, Figure 1-10.

Figure 1-10
This remote-controlled machine can be sent into hazardous or otherwise inaccessible locations to inspect objects or parts. (Visual Inspection Technologies, Inc.)

Perhaps it is best to keep the words of Joseph Engelberger in mind: "Nobody needs a robot. . . . There isn't anything that a robot can do that a willing human being can't do better."

Important Terms

anthropomorphic
artificial intelligence
Isaac Asimov
automaton
Karel Capek
George C. Devol, Jr.
Joseph F. Engelberger
fixed-sequence robot
flexible automation
hard automation
industrial robot
intelligent robot
manual manipulator
NC (numerically controlled) robot
playback robot
programmable
robot
robotics
Robotic Industries Association (RIA)
Unimate
variable-sequence robot

Review Questions

Write your answers on a separate sheet of paper.

1. Who were George Devol and Joseph Engelberger?
2. In what year was the first industrial robot patented?

3. Robots are considered to be a key element in automated manufacturing. Define the term "industrial robot" and discuss the key factors in that definition.
4. Some believe that the use of automation and robots will replace human workers. Others believe that if automation and robots are not incorporated into manufacturing, we will lose a far greater number of jobs. What statements can you give in defense of both views?
5. Who was the Czech dramatist who coined the word robot? What does the Czech word "robota" mean?
6. What are some key negative points concerning robots that have been passed on from *R.U.R.* ("Rossum's Universal Robots"), *Frankenstein,* and other literature?
7. List Isaac Asimov's three laws of robotics and discuss their significance.
8. Analyze the six Japanese classifications of robots. Which of them best fit the Robotic Industries Association's definition for an industrial robot?
9. Distinguish between hard automation and flexible automation. List some of the advantages and disadvantages of each.
10. Summarize the historical development of robotics in the United States.
11. Some robots are considered anthropomorphic. What does the term "anthropomorphic" mean?

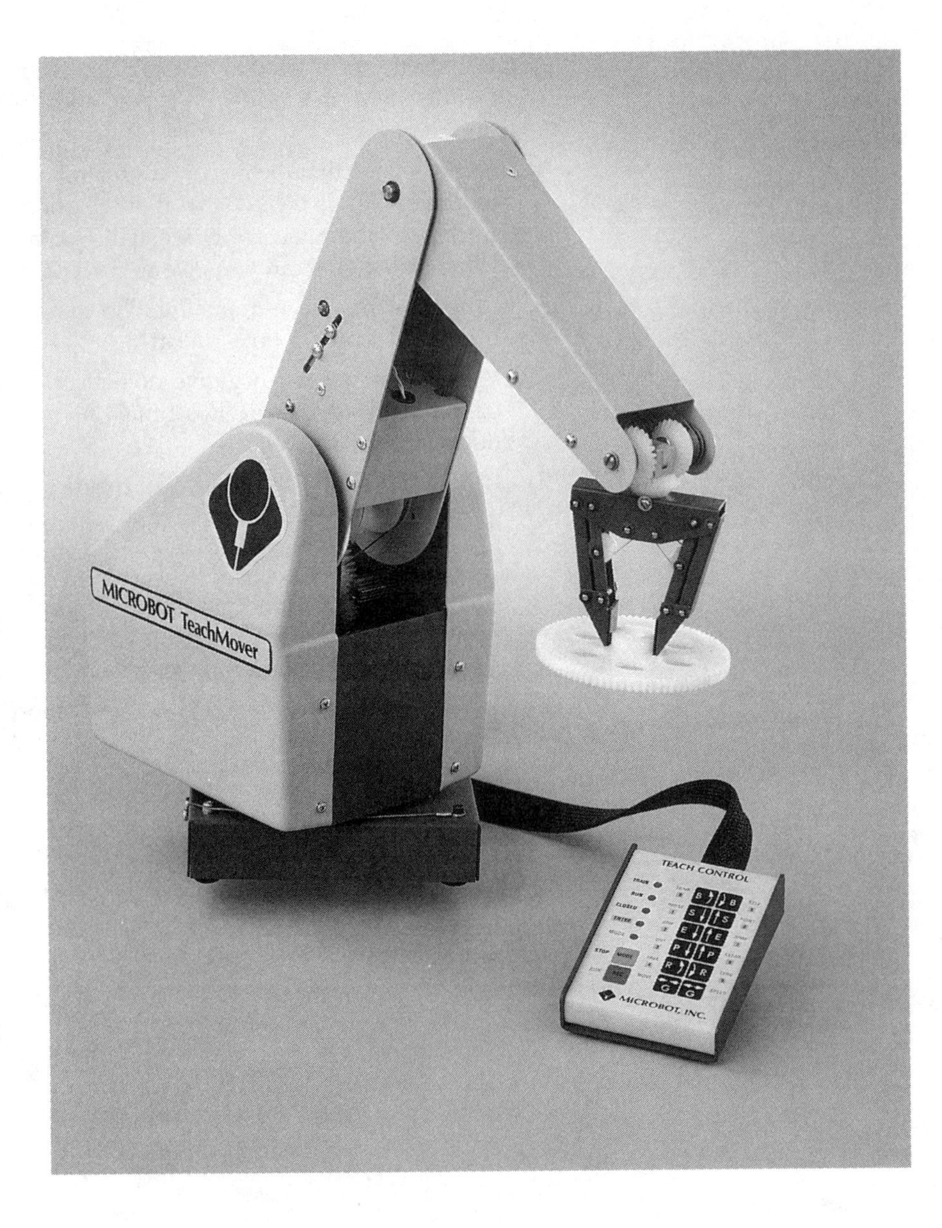

Table-top robots are often used to teach students the fundamentals of robotics. (Microbot, Inc.)

2 Fundamentals of Robotics

Overview

Even the most complex robotic system can be broken down into a few basic subsystems, which are covered in this chapter. They provide an overview of how a robot works. More detail about selected subsystems will be given in later chapters. Freedom of motion and the resulting shape of the robot's work area are also covered.

The Parts of a Robot

Robots come in many shapes and sizes. The industrial robots illustrated in Figures 2-1 and 2-2 resemble an inverted human arm mounted on a base. Industrial robots consist of a number of subsystems that work together. These subsystems, labeled in Figure 2-1, are the controller, the manipulator, the power supply, the end effector, and a means for programming. Figure 2-3 is a diagram of the relationships among these five major systems.

The ***controller*** is the part of a robot that coordinates all movements of the mechanical system. It also receives input from the immediate environment through various kinds of sensors. The ***manipulator*** consists of segments that may be jointed and that move about, allowing the robot to do work. In Figure 2-2, for example, the manipulator is the robot's "arm." The ***power supply*** may convert ac line voltage to the dc voltages required by the robot's internal circuits, or it may be a pump or compressor providing hydraulic or pneumatic power. The ***end effector*** is the robot's hand. The area which the robot's end effector can reach is called its ***work envelope.*** The means for programming is used to record movements into the robot's memory.

The Controller

The heart of the robot's controller is generally a microprocessor (computer chip) linked to input/output and monitoring devices. The commands issued by the controller activate the motion control mechanism, consisting of various controllers, amplifiers, and actuators. An ***actuator*** is a motor or valve that converts power into robot movement.

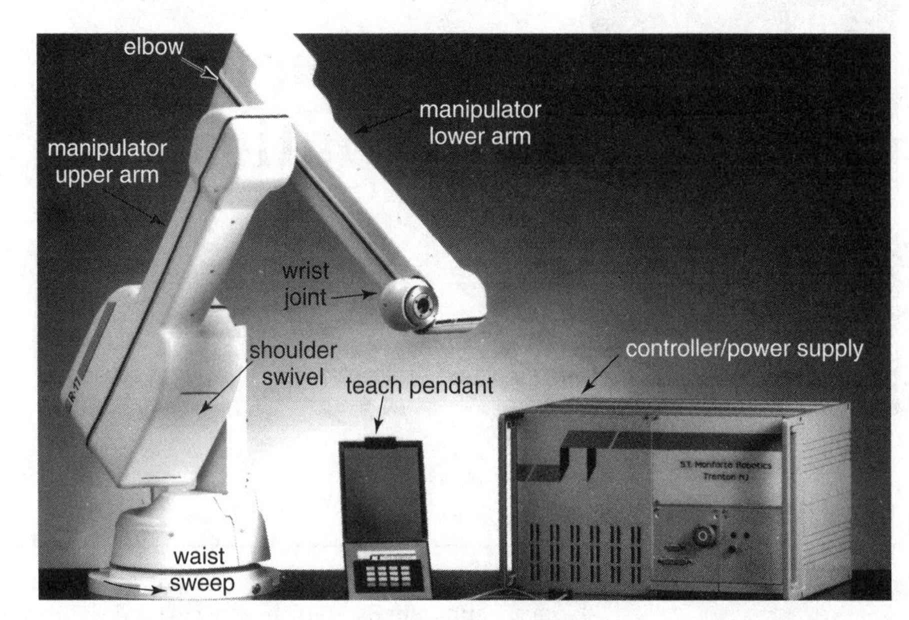

Figure 2-1
This robot clearly illustrates the systems of a typical industrial robot. This electric robot can be used in a variety of industrial applications.
(S.T. Monforte Robotics, Inc.)

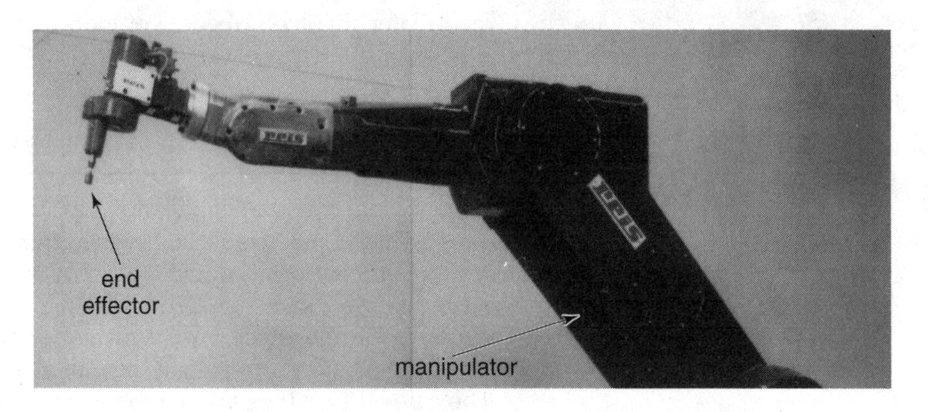

Figure 2-2
This robot has been designed expressly for use in precise path-oriented tasks such as deburring, milling, sanding, gluing, bonding, cutting, and assembly.
(Reis Machines, Inc.)

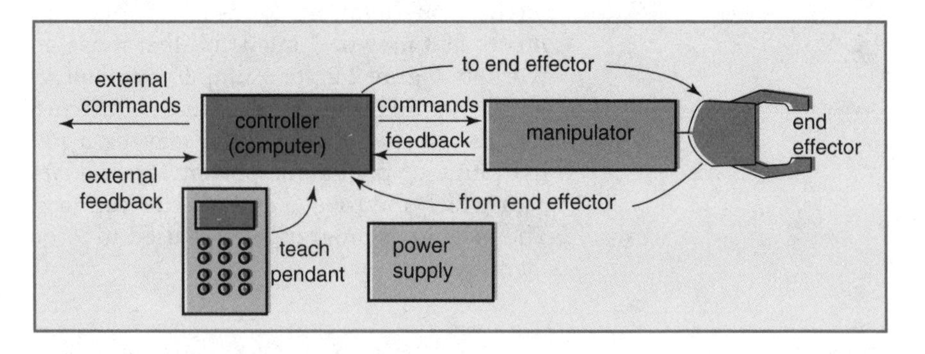

Figure 2-3
The relationships among the five major systems that make up an industrial robot are shown in this diagram.

This movement is initiated by a series of instructions, called a ***program,*** stored in the controller's memory. Figure 2-4 shows a Kawasaki controller/power supply and teach pendant.

Figure 2-4
A controller/power supply with a teach pendant. (Kawasaki)

The controller has three levels of ***hierarchical control.*** A hierarchical arrangement is one in which a given level is lower (more elemental) than the one above it and is dependent on the level above it for its instructions. Figure 2-5 illustrates the relationships among the basic levels of hierarchical control. The three levels are:

Level I: Actuator Control. This is the most elementary level, at which the separate movements of the robot along various planes, such as the X, Y, and Z axes are controlled. These movements will be explained in detail in the section of this chapter that covers robot configurations.

Level II: Path Control. The path control (intermediate) level coordinates the separate movements along the planes determined in Level I into the desired ***trajectory*** (path).

Level III: Main Control. The primary function of this highest control level is to interpret the written instructions from the human programmer as to the tasks

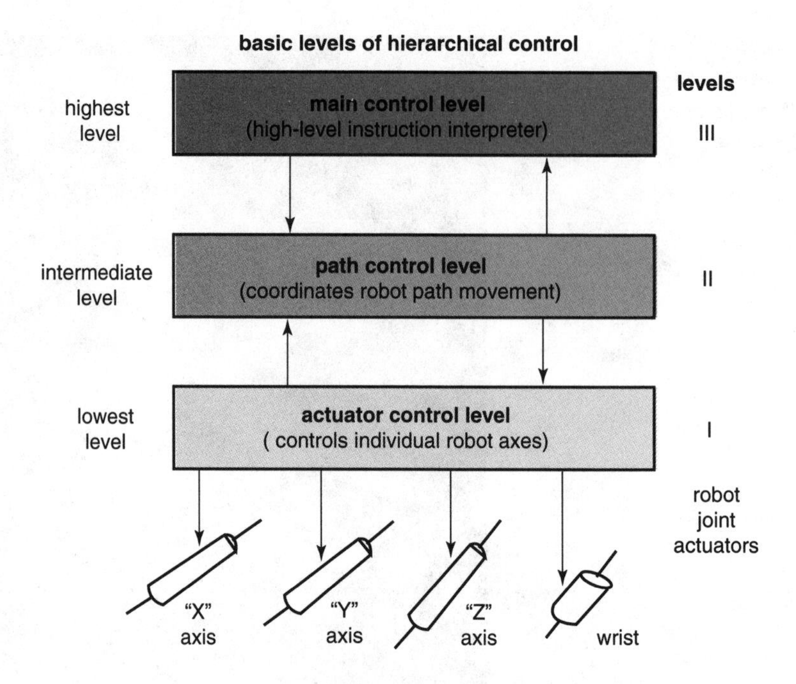

Figure 2-5
The three basic levels of hierarchical control.

required. The instructions are then combined with various environmental signals and translated by the controller into the more elementary instructions that Level II can understand.

Robots can be classified according to the type of control system used. The ***nonservo robot*** is a nonintelligent robot. ***Servo robots*** are classified as either intelligent or highly intelligent. The primary difference between an intelligent and highly intelligent robot is its level of awareness of its environment.

Nonservo robots

Nonservo robots are the simplest robots, according to the definition used in the United States. (See Figure 2-6 for a comparison of U.S. and Japanese classifications.) Nonservo robots are often referred to as "limited sequence," "pick-and-place," or "fixed-stop" robots.

Figure 2-6
Differing definitions of robot types by Japanese and U.S. industry are shown in this table.

Comparison Of Robots By Definition

Item No.	*Japanese Definition*	*U.S. Definition* *Nonsophisticated*	*U.S. Definition* *Sophisticated*
1	Manual Manipulator		
2	Fixed Sequence		
3	Variable Sequence	Nonservo	
4	Playback		Servo—Intelligent
5	Numerically Controlled		Servo—Intelligent
6	Intelligent		Servo—Highly Intelligent

The nonservo robot is an ***open-loop system.*** That is, no feedback mechanism is used to compare programmed positions to actual positions. A good example of an open-loop system is the operating cycle of a modern washing machine. Figure 2-7 is a block diagram of the steps performed by a typical washing machine. At the beginning of the operation, the dirty clothes and the detergent are placed in the machine's tub. The timer/control is set for the proper cleaning cycle, and the machine is activated by the start button. The machine fills with water and begins to go through the various washing, rinsing, and spinning cycles. The machine finally stops after the set sequence is completed. The washing machine is considered an open-loop system for two reasons. First, because the clothes are never examined by sensors during the washing cycle to see if they are clean. Second, the length of the cycle is not automatically adjusted to compensate for the amount of dirt remaining in the clothes. The cycle and its time span are determined by the fixed sequence of the timer/control.

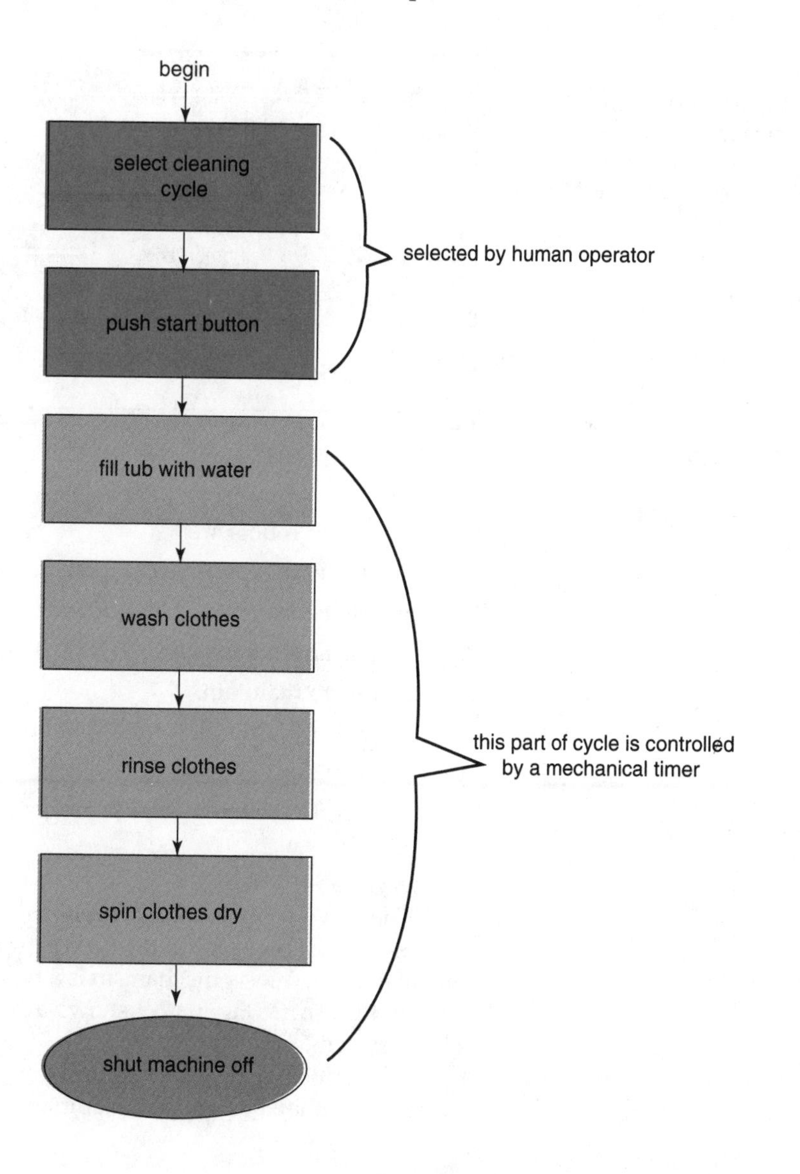

Figure 2-7
In this block diagram depicting the sequence of steps performed by a washing machine, no feedback is used. In such an open-loop control system, the condition of the clothes during the washing operation is not monitored and used to alter the process.

Study the diagram in Figure 2-8, which represents a three-axis pneumatic (air-controlled) robot. The three axes allow movement along certain planes. For the sake of simplicity, only one axis is shown. At the beginning of the cycle, the controller sends a signal to the control valve of the manipulator. As the valve opens, air passes into the air cylinder, causing the rod in the cylinder to move. As long as the valve remains open, this rod continues to move until it is restrained by the end stop. After the rod reaches the limit of its travel, a limit switch tells the controller to close the control valve, and the controller sends the valve a signal to close. The controller then moves to the next step in the program and initiates the necessary signals. If the signals go to the end effector, for example, they might cause the gripper to close in order to grasp an object. The process is repeated until all the steps in the program have been completed.

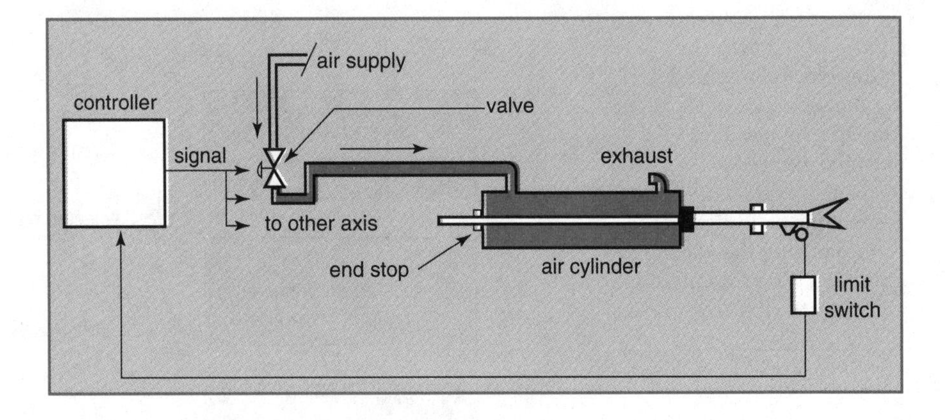

Figure 2-8
In a nonservo system, movement is regulated by such devices as a limit switch, which signals the controller when it is activated.

Nonservo robots are:

- Δ Relatively inexpensive compared to servo robots.
- Δ Simple to understand and operate.
- Δ Precise and reliable.
- Δ Simple to maintain.
- Δ Capable of fairly high speeds of operation.
- Δ Small in size.
- Δ Limited to relatively simple programs.

The servo robot

The servo robot is a ***closed-loop system*** because it allows for feedback. The feedback signal sent to the servo amplifier affects the output of the system. In a sense, a servomechanism is a type of control system that detects and corrects for errors. Figure 2-9 shows a block diagram of a servo-controlled robotic system.

The principle of servo control can be compared to many tasks performed by human beings. One example is cutting a circle from a piece of stock

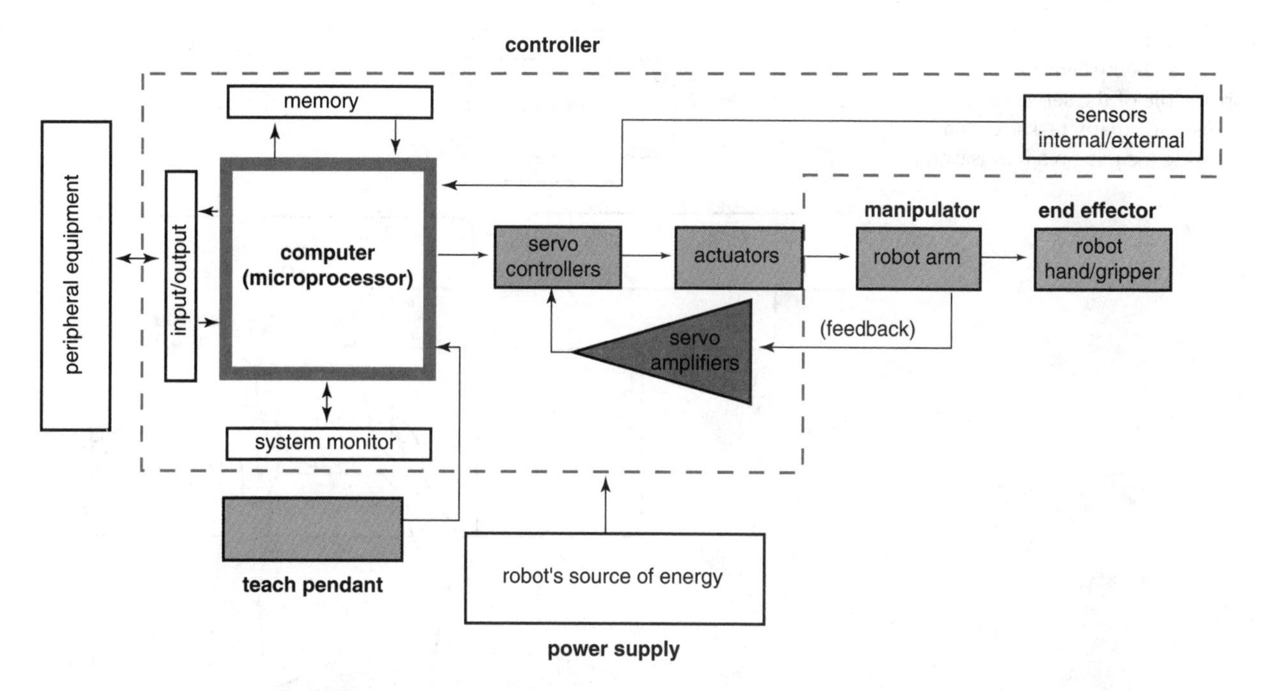

Figure 2-9
A servo-controlled industrial robotic system such as the one depicted in this block diagram might be classified as "intelligent" or "highly intelligent," depending on the level of sensory data it can interpret.

on a power bandsaw, shown in Figure 2-10. The machine operator's eye studies the position of the stock to be cut in relation to the cutting edge of the blade. The eye transmits a signal to the brain. The brain compares the actual position to the desired position. The brain then sends a signal to the arms to move the stock beneath the cutting edge of the blade. The eye is used as a feedback sensing device, while the brain compares desired locations with actual locations. The brain sends signals to the arms to make necessary adjustments. This process is repeated as the operator follows the scribed line during the sawing operation.

The diagram in Figure 2-11 helps to explain the operation of a six-axis hydraulic robot. Only one axis is shown in detail. When the cycle begins, the controller searches the robot's programming for the desired locations along each axis. By means of feedback signals, the controller determines the actual locations on the various axes of the manipulator. The desired locations and actual locations are compared. ***Error signals*** are fed back to the servo amplifier. The greater the error, the higher the intensity of the signal. These error signals are amplified (increased) by the servo amplifier and applied to the servo control valve on the appropriate axis. The valve opens in proportion to the level of the command signal received. The opened valve admits fluid to the proper actuator to move the various segments of the manipulator.

New signals are generated as the manipulator moves. When there are no more error signals, the servo control valves close, shutting off the flow of fluid. The manipulator comes to rest at the desired position. The controller then

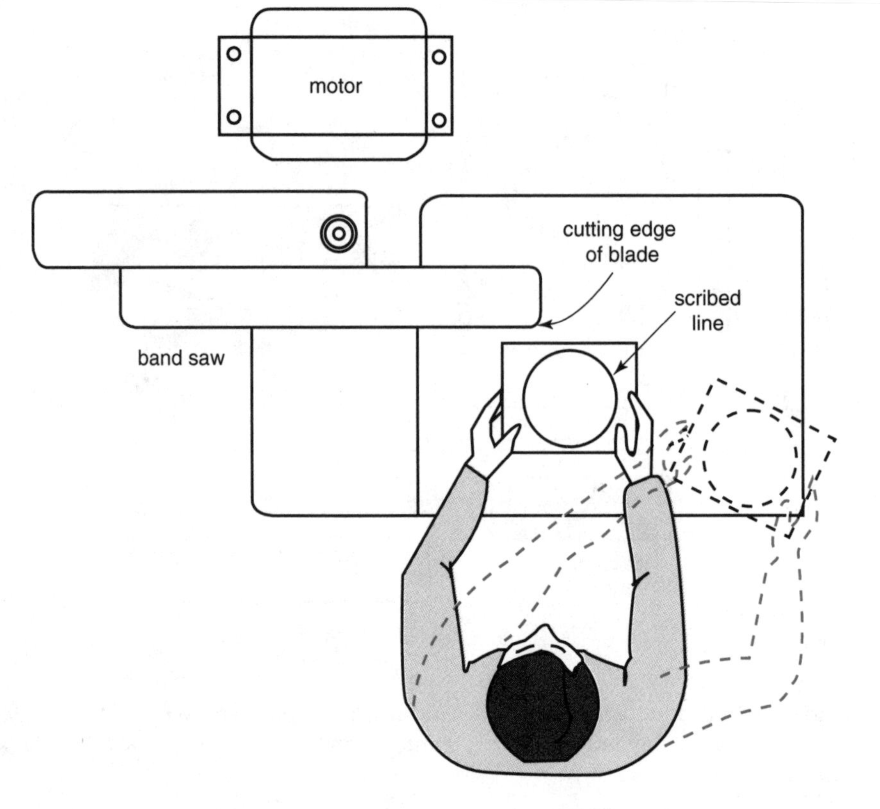

Figure 2-10
Human beings employ the principle of the servomechanism for many tasks, such as cutting a circle on a bandsaw.

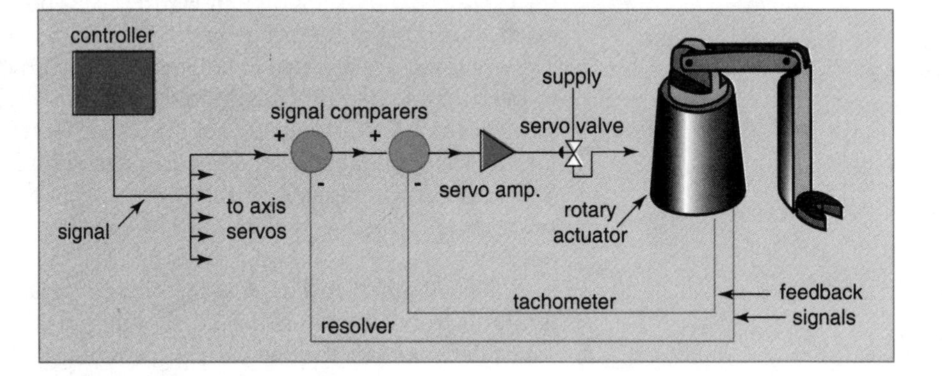

Figure 2-11
Feedback signals from the tachometer and resolver sensing systems allow the system to make corrections whenever the actual speed or position of the robot does not agree with the values contained in the robot's program.

addresses the next instruction in the program. It may be to move to another location, or to operate some peripheral equipment. The process is repeated until all steps of the program are completed. The output of a ***tachometer,*** a speed measuring device, is used to control acceleration and deceleration of the manipulator's movements.

Servo robots:

Δ Are relatively expensive to buy and cost more than nonservo robots to operate and maintain.

- Δ Use a sophisticated, closed-loop controller.
- Δ Have a wide range of capabilities.
- Δ Can perform multiple point-to-point transfer as well as transfer along a controlled, continuous path.
- Δ Can respond to very sophisticated programming.
- Δ Use a manipulator arm that can be programmed to avoid obstructions within the work envelope.

The Manipulator

The manipulator must move materials, parts, tools, or special devices through various motions to provide useful work. The manipulator is the arm of the robot, and is made up of a series of segments and joints much like those found in the human arm. Joints connect two segments together and allow them to move relative to one another. The joints provide either linear (straight line) or rotary (circular) movement. Figure 2-12 shows simple linear and rotary joints commonly found in manipulator arms.

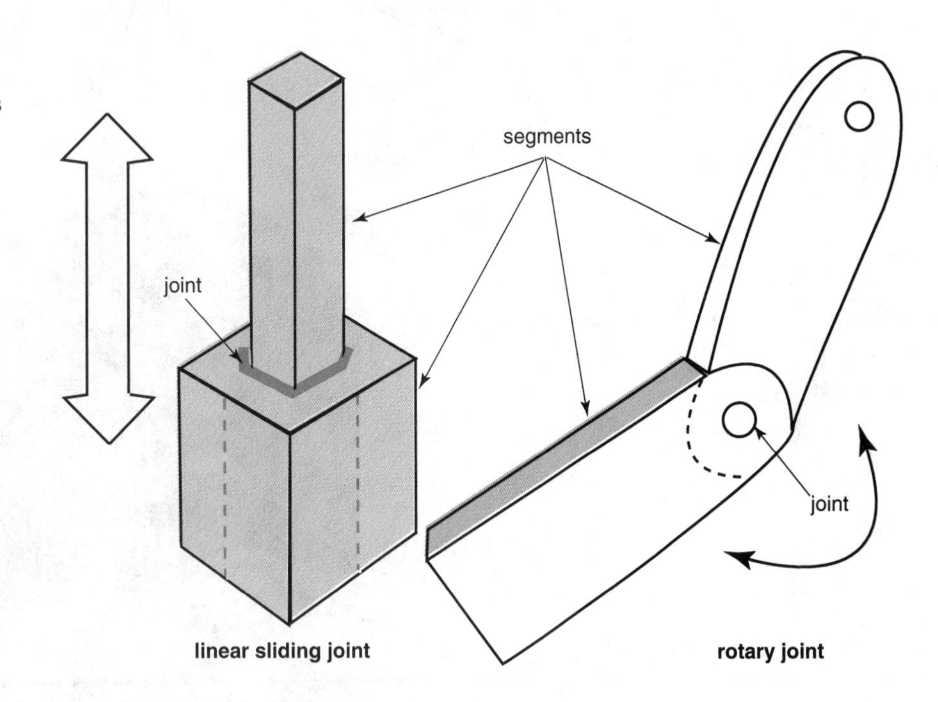

Figure 2-12
Both linear and rotary joints are commonly found in robots.

The muscles of the human body supply the driving force that moves the various body joints. Similarly, a robot uses actuators to move its arm along programmed paths and then to hold its joints rigid once the correct position is reached.

There are two basic types of motion provided by actuators. ***Linear actuators*** provide motion along a straight line. They either extend or retract their attached loads. ***Angular actuators*** provide rotation. They move

their loads in an arc or circle. Rotary motion can be converted into linear motion using a lead screw or other mechanical means of conversion.

Figure 2-13A shows a linear actuator; Figure 2-13B shows an angular (rotary) actuator. These types of actuators are also used outside the robot to move workpieces and provide other kinds of motion within the work envelope.

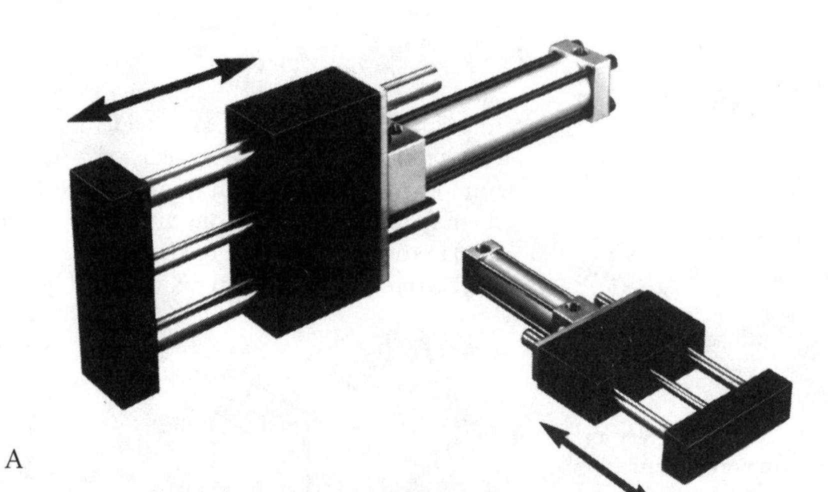

A

B

Figure 2-13
A—Linear actuators provide straight-line movement. B—Rotational movement around an axis is provided by the angular (rotary) actuator. Actuators can be powered by electric motors, pneumatic (air) cylinders, or hydraulic (oil) cylinders. (Phd, Inc.)

Types of actuator drive

One common method of classifying robots is by the type of drive required by the actuators. Electrical actuators use electric power. Pneumatic actuators use pneumatic (air) power. Hydraulic actuators, shown in Figure 2-14, use hydraulic (fluid) power.

Figure 2-14
A large hydraulic actuator provides up-and-down motion to the manipulator arm of this industrial robot.

Electric drive. Three types of motors are commonly used for ***electric drive*** actuators: ac servo motors, dc servo motors, and stepper motors. Both ac and dc servo motors have built-in methods for controlling exact position. Many newer robots use dc servo motors rather than hydraulic or pneumatic ones. Dc servo motors are commonly found on small and medium-size robots. Ac servo motors are found in heavy-duty robots because of their high torque capabilities, Figure 2-15. A stepper motor is a type of incrementally controlled dc motor. Stepper motors are rarely used in commercial industrial robots, but are commonly found in educational robots, Figure 2-16.

Conventional electric-drive motors have several advantages. They are quiet and simple and can be used in clean-air environments. Electric robots generally require less floor space, and their energy source is readily available. However, the conventionally geared drive causes problems of backlash, friction, compliance, and wear. These problems cause inaccuracy, poor dynamic response, need for regular maintenance, poor torque control capability, and limited maximum speed on longer moves. Loads that are heavy enough to stall (stop) the motor can cause damage. Conventional electric motors also

Figure 2-15
The arrows indicate the ac servomotors on this heavy duty industrial robot. (Kawasaki)

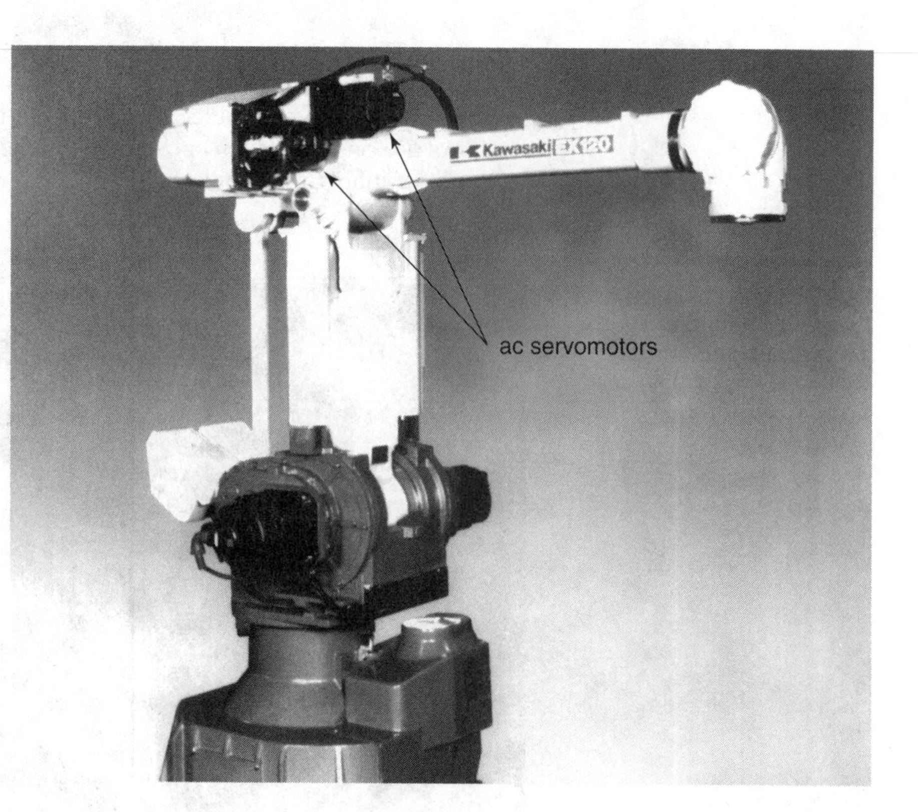

Figure 2-16
Dc stepper motors are used on this tabletop educational robot. Note the three-fingered gripper. (D & M Computing, Inc.)

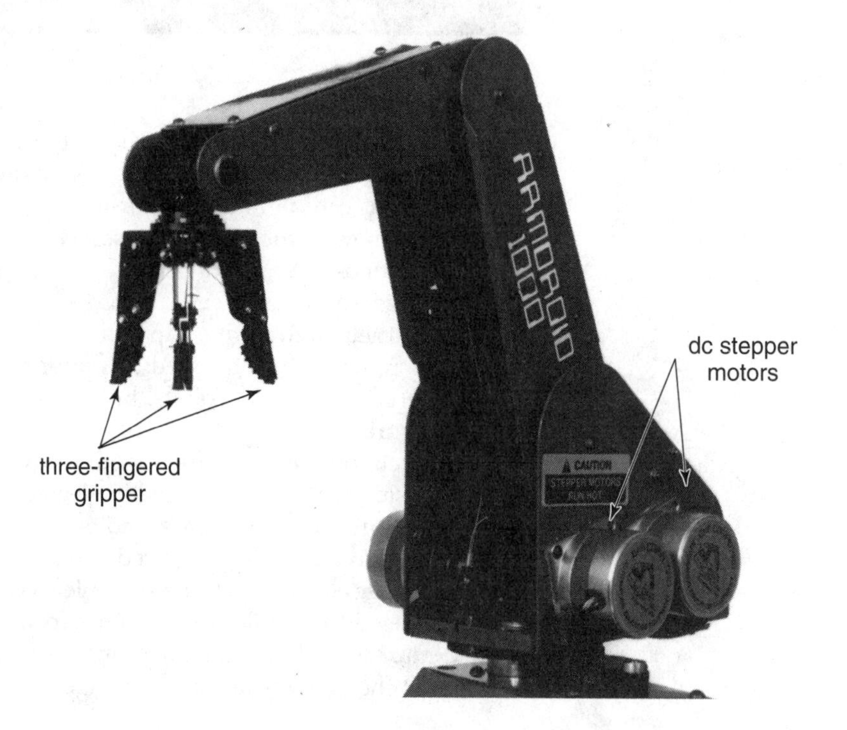

have poor output power compared to their weight. This means that a larger, heavier motor must be mounted on the robot arm when a large amount of torque is needed.

The rotary motion of most electric drive motors must be geared down (reduced) to provide the speed or torque required by the manipulator. However, manufacturers are beginning to offer robots that use ***direct-drive electric motors,*** which eliminate some of these problems. These high-torque motors drive the arm directly, without the need for reducer gears. The prototype of a direct-drive arm was developed by scientists at Carnegie-Mellon University in 1981.

The basic construction of a direct-drive motor is shown in Figure 2-17 Coupling the motor with the arm segment to be manipulated eliminates backlash, reduces friction, and increases the mechanical stiffness of the drive mechanism. Figure 2-18 shows a direct-drive robot arm. Compare its design to the non-direct-drive robot in Figure 2-15. Simplicity of design results when direct-drive motors are used. Maintenance requirements are also reduced. Direct-drive robots provide higher speeds, greater flexibility, and greater accuracy than non-direct-drive types.

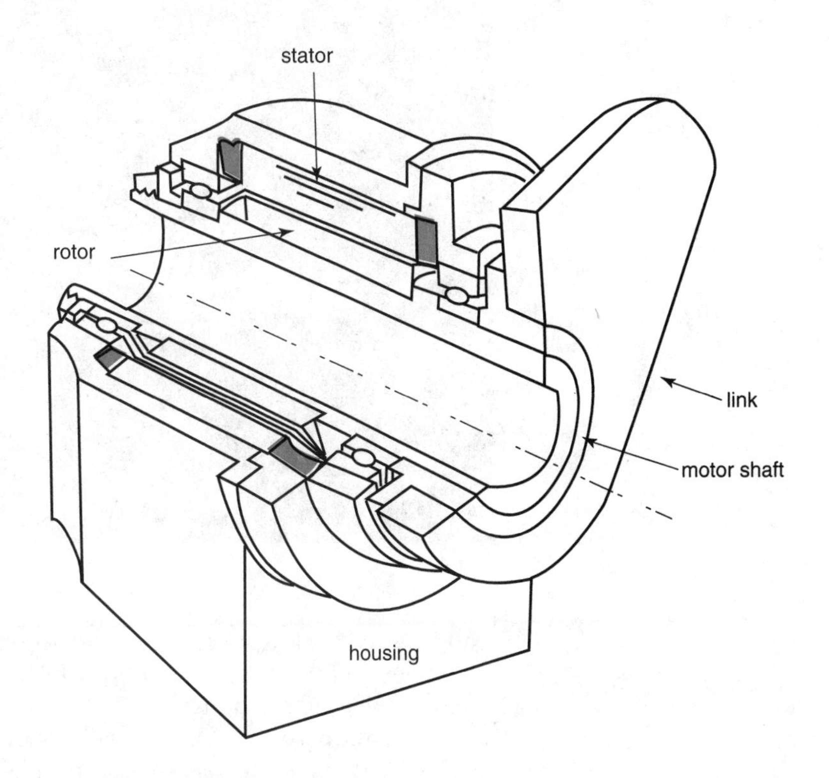

Figure 2-17
This drawing shows the basic construction of the type of dc motor commonly used in direct-drive robots.

Applications currently being performed by electric direct-drive robots are mechanical assembly, electronic assembly, and material handling. These robots will increasingly meet the demands of advanced, high-speed, precision applications. Such applications include laser cutting of sheet metal, which

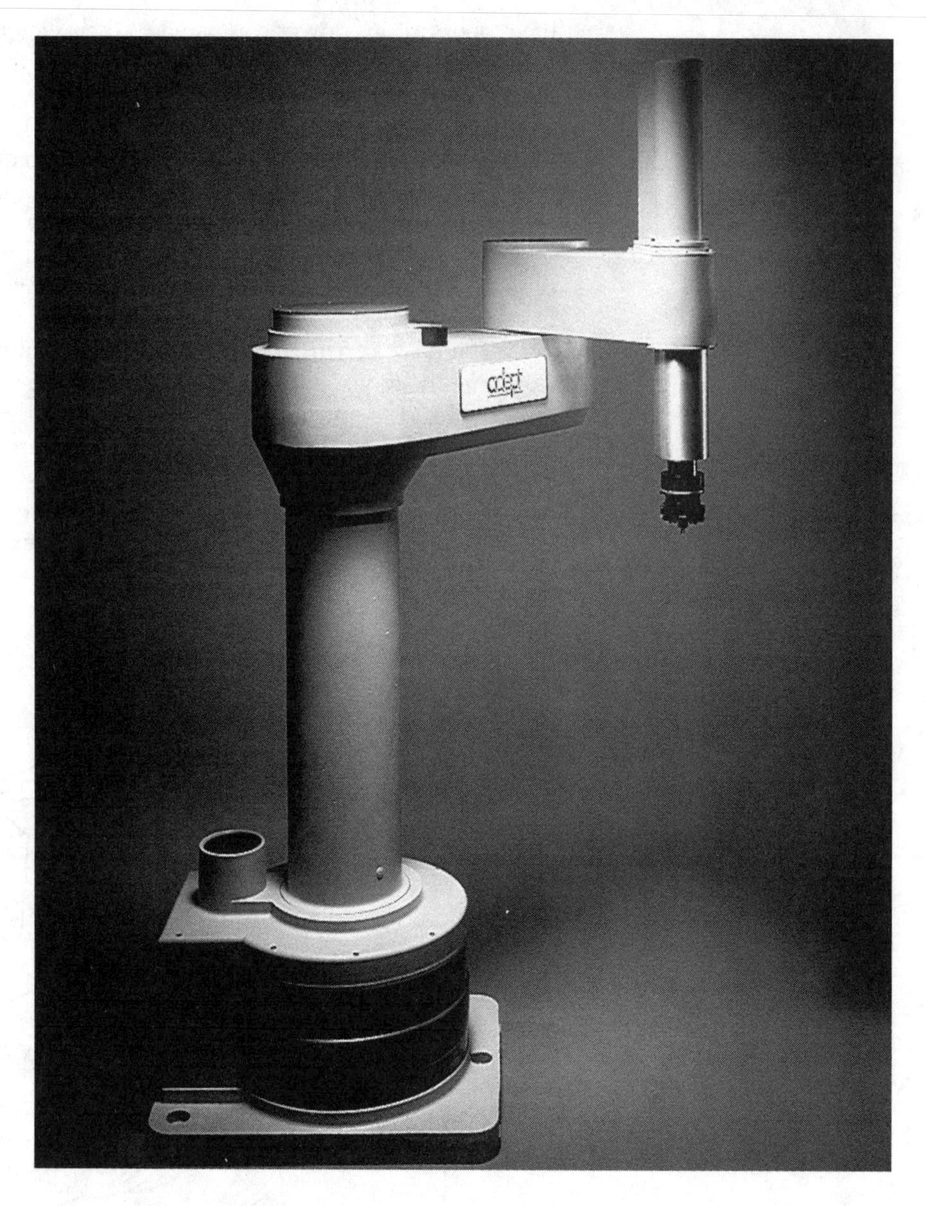

Figure 2-18
Note the simplified mechanical design of this direct-drive robot as compared to the non-direct drive robot shown in Figure 2-15 (Adept Technology, Inc.)

requires speeds of over one meter per second and tolerances of less than 0.1 millimeter, and fiber optic assembly, which requires accuracy within 0.1 to 0.5 microns (millionths of a meter).

Hydraulic drive. Many earlier robots were driven by ***hydraulic drive*** systems. A ***hydraulic system*** consists of an electric pump connected to a reservoir tank, control valves, and a hydraulic actuator. Hydraulic systems provide both linear and rotary motion using a much simpler arrangement than non-direct-drive electrical systems, Figure 2-19. The storage tank supplies large amounts of instant power not available from electric systems.

Hydraulic-drive units have several advantages. They provide precise motion control over a wide range of speeds. They can handle heavy loads on

Figure 2-19
A large hydraulically operated robot being used for demonstration at a manufacturing trade show.

the end of the manipulator arm, can be used around high explosives, and are not easily damaged when quickly stopped while carrying a heavy load. However, they are expensive to purchase and maintain, and are not energy efficient. They are also noisier than electric units. Because of hydraulic fluid leaks, they are not recommended for clean environments.

Pneumatic drive. *Pneumatic systems* make use of air-driven actuators. Since air is also a fluid, many of the same principles that apply to hydraulic systems are applicable to pneumatic systems. Pneumatic and hydraulic motors and cylinders are very similar. Since most industrial plants have a compressed air system running throughout assembly areas, air is an economical and readily available energy source. This makes the installation of pneumatic robots easier and less costly than that of hydraulic robots. For lightweight pick-and-place applications that require both speed and accuracy, a pneumatic robot, like the one in Figure 2-20, is potentially a good choice.

Figure 2-20
This twin-arm assembly robot is controlled by a pneumatic servo system. Note the dual grippers, which are placing components on electronic circuit boards. (Rexroth-Pneumatic)

Pneumatic-drive units work at high speeds and are most useful for small-to-medium loads. They are economical to operate and maintain, and can be used in explosive atmospheres. However, since air is compressible, precise placement and positioning is more difficult to control. It is also difficult to keep the air as clean and dry as needed by the control system. Pneumatic robots are noisy and vibrate as the air cylinders and motors stop.

Non-direct electric, pneumatic, and hydraulic systems are compared in Figure 2-21. A more detailed discussion of movement systems is provided in Unit II.

Figure 2-21
Characteristics of robots with three types of drives are compared on this chart.

	Non-direct	Pneumatic	Hydraulic
Cost	poor	good	poor
Cleanliness	good	good	poor
Controllability	good	poor	good
Noise	good	poor	poor
Weight	poor	good	good
Strength	fair	poor	good
Speed	poor	good	good
Stiffness	good	poor	good

Power Supply

The robot power supply provides the energy to drive the controller and actuators. The three basic types of power supplies are electrical, hydraulic, and pneumatic.

The most common energy source available where industrial robots are used is electricity. The second most common is compressed air, and the least common is hydraulic power. These primary sources of energy must be converted into the form and amount required by the type of robot being used. The electronic part of the control unit and any electric motor actuators require electrical power. A robot containing hydraulic actuators requires the conversion of electrical power into hydraulic energy through the use of an electric motor-driven hydraulic pump. A robot with pneumatic actuators requires compressed air, which is usually supplied by a compressor driven by an electric motor. Robotic power supplies are discussed in more detail in Unit II.

End Effector

End effector is the technical term for the end-of-arm tooling on the robot. It is often referred to as the hand or gripper. An end effector is better-defined as a device attached to the wrist of the manipulator for the purpose of grasping, lifting, transporting, maneuvering, or performing operations on a workpiece. Chapter 10 is devoted to end effectors and end-of-arm tooling.

Means for Programming

A robot may be programmed using any of several different methods. The ***teach pendant***, also called a teach box or hand-held programmer, is one commonly used device. It teaches a robot the movements required to perform a useful task. The human operator uses a teach pendant to move the robot through the series of points that describe its desired path. The points are recorded by the controller for later use. The teach pendants shown in Figures 2-1 and 2-4 are typical of those found with most industrial robots. Teaching methods are covered in Chapter 3.

Degrees of Freedom

Although robots have a certain amount of dexterity, theirs is nothing compared to human dexterity. The movements of the human hand are controlled by 35 muscles. Fifteen of these muscles are located in the forearm. The arrangement of muscles in the hand provides great strength to the fingers and thumb for grasping objects. Each finger can act alone or together with the thumb. This enables the hand to do many intricate and delicate tasks. In addition, the human hand has 27 *bones.* Figure 2-22 shows the bones found in the hand and wrist. This bone, joint, and muscle arrangement gives the hand its dexterity.

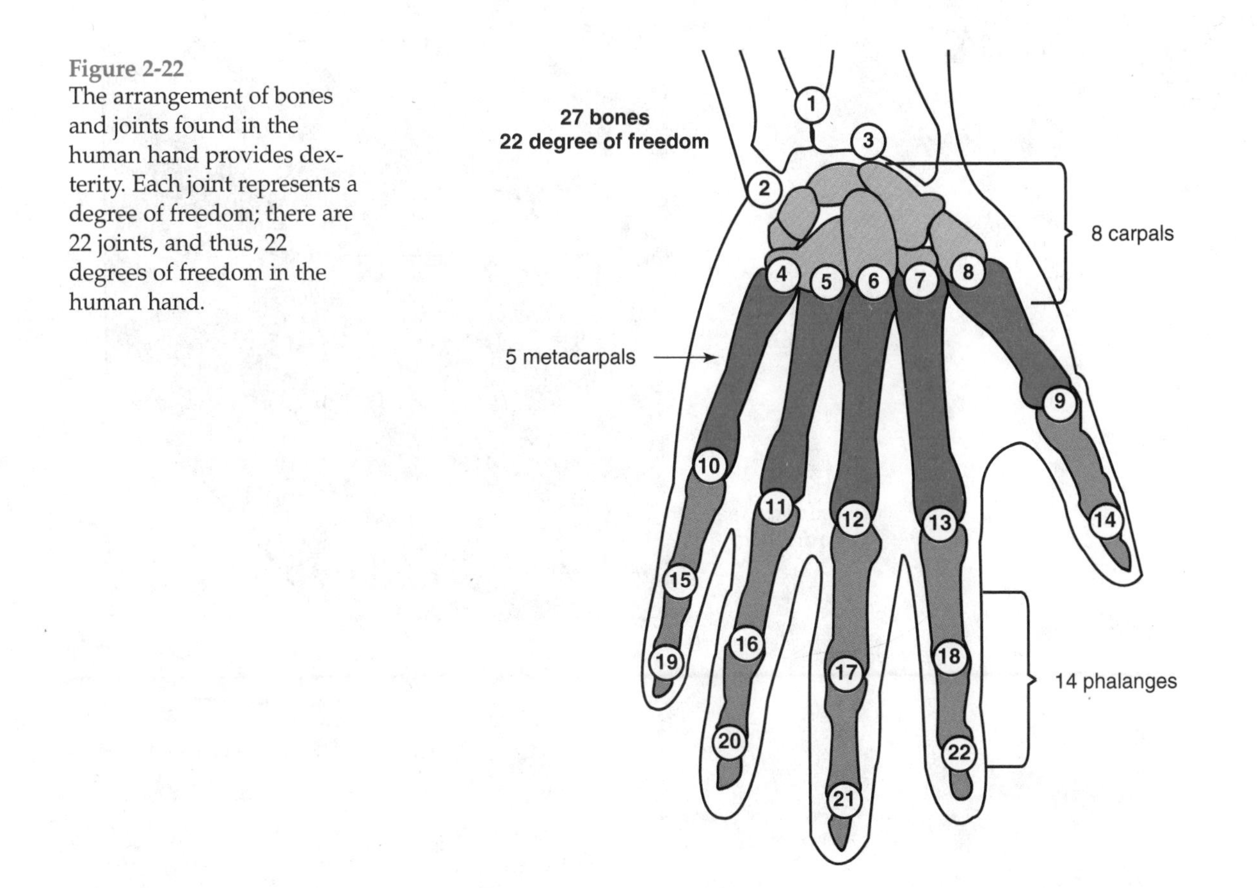

Figure 2-22
The arrangement of bones and joints found in the human hand provides dexterity. Each joint represents a degree of freedom; there are 22 joints, and thus, 22 degrees of freedom in the human hand.

Degrees of freedom is a term used to describe a robot's dexterity, or freedom of motion. For each degree of freedom, a joint is required. A robot requires six degrees of freedom to be completely versatile. Its movements are clumsier than those of a human hand, which has 22 degrees of freedom.

The number of degrees of freedom defines the robot's configuration. For example, many simple applications require movement along three axes — X, Y, and Z. See Figure 2-23. These tasks would require three joints, or three degrees of freedom. The three degrees of freedom in the robot arm are the rotational traverse, the radial traverse, and the vertical traverse. The ***rotational traverse*** is movement about a vertical axis. This is the side-to-side swivel of the robot's arm about its base. The ***radial traverse*** is the extension and retraction of the arm, creating in-and-out motion relative to the base. The ***vertical traverse*** provides up-and-down motion.

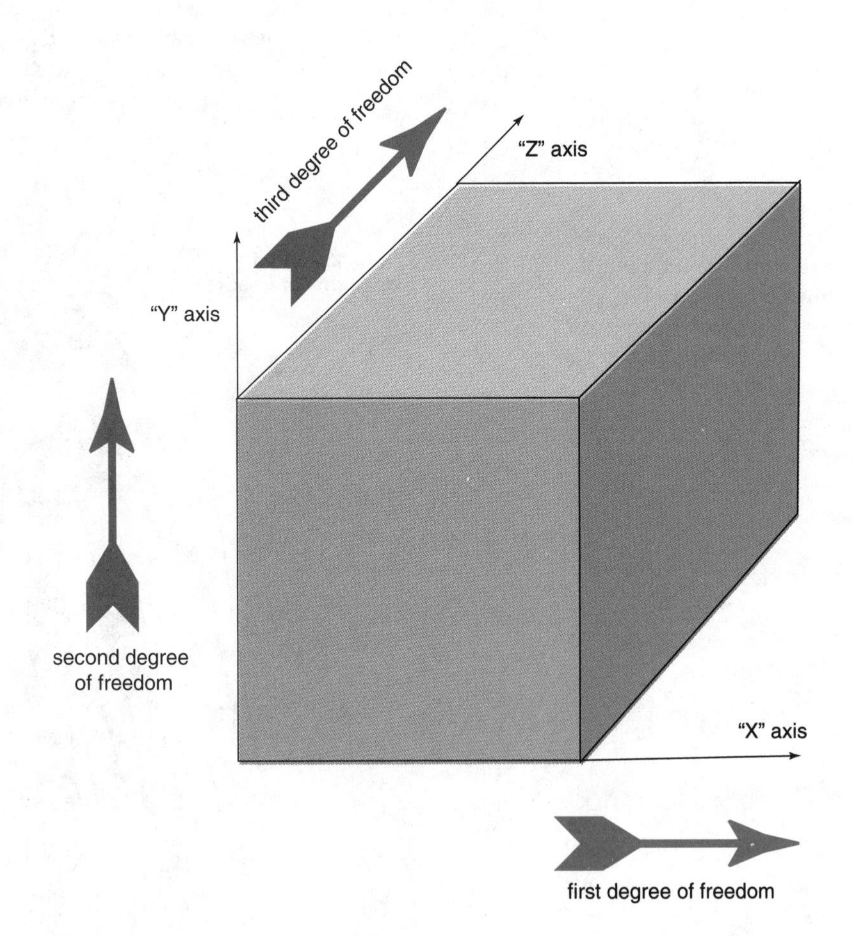

Figure 2-23
The three basic degrees of freedom associated with movement along the X, Y, and Z axes of the Cartesian coordinate system.

For applications that require more freedom, additional degrees can be obtained from the wrist, which gives the end effector its flexibility. See Figure 2-24. The three degrees in the wrist bear aeronautical names: pitch, yaw, and roll. The ***pitch***, or bend, is the up-and-down movement of the wrist. The ***yaw*** is the side-to-side movement, and the ***roll***, or swivel, involves rotation.

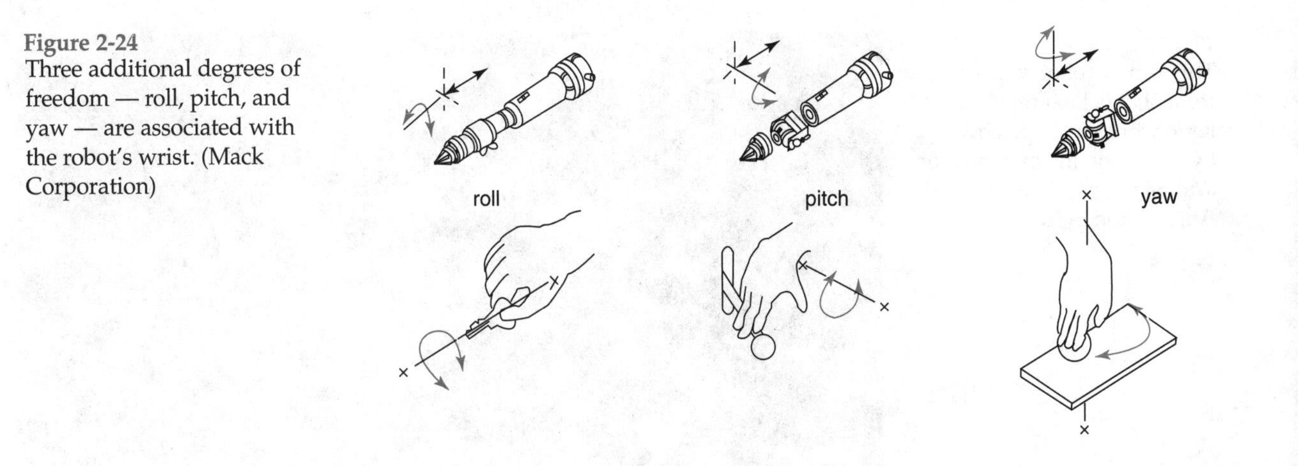

Figure 2-24
Three additional degrees of freedom — roll, pitch, and yaw — are associated with the robot's wrist. (Mack Corporation)

A robot requires a total of six degrees of freedom to locate and orient its hand at any point in its work envelope, Figure 2-25. Although six degrees of freedom are required for maximum flexibility, most applications require only three to five. The more degrees of freedom required, the more complex must be the robot's motions and controller design. Some industrial robots have seven or eight degrees of freedom. These additional degrees are achieved by mounting the robot on a track or moving base, as shown in Figure 2-26. The track-mounted robot shown in Figure 2-27 has a total of seven. This addition also increases the robot's reach.

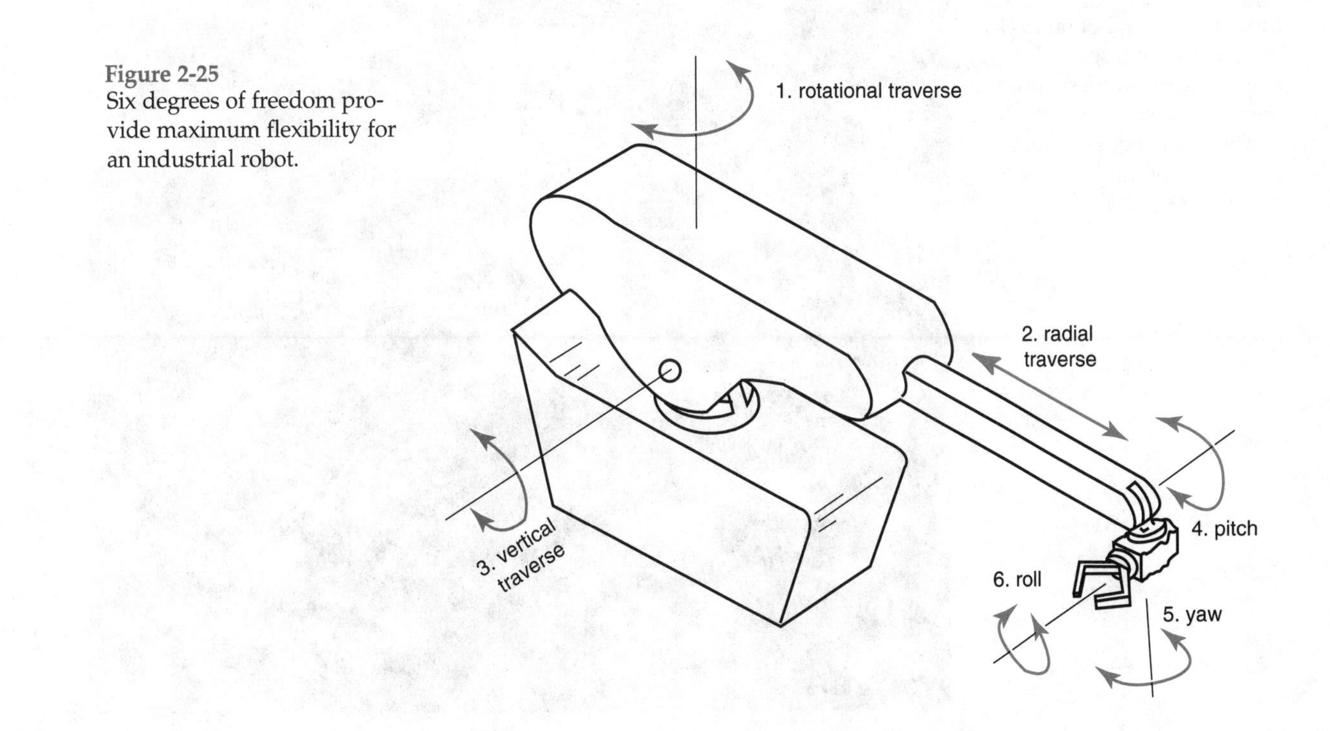

Figure 2-25
Six degrees of freedom provide maximum flexibility for an industrial robot.

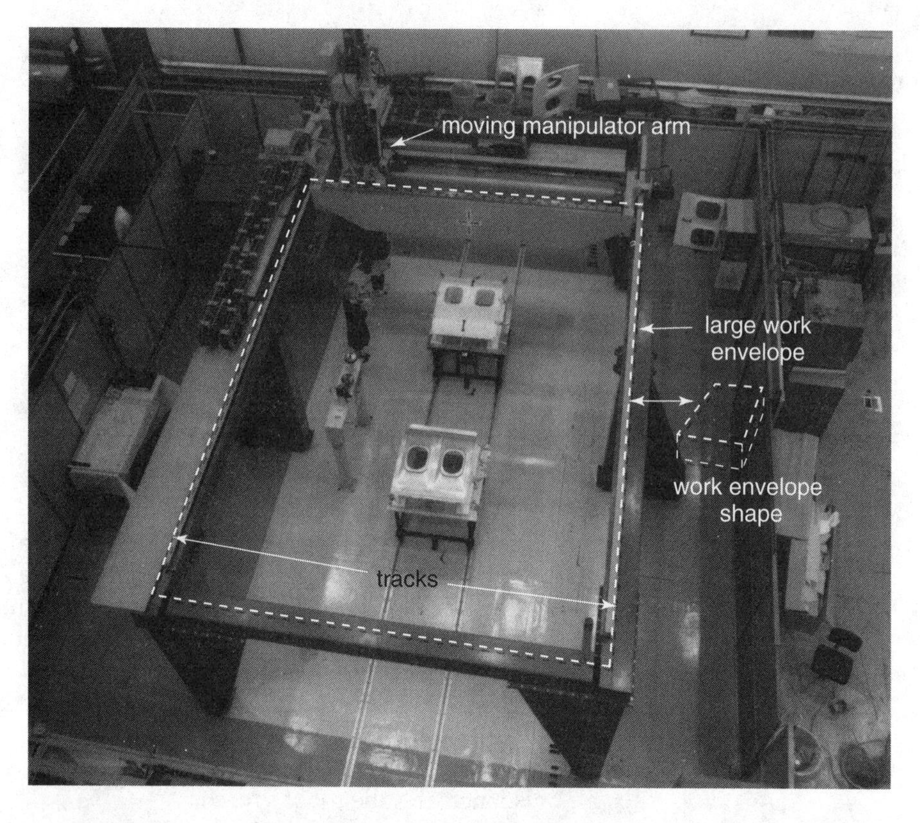

Figure 2-26
This gantry robot, viewed from above, illustrates the large work envelope obtained by mounting the manipulator arm on tracks. (Cybotech Automation Systems)

Figure 2-27
Mounting this robot on tracks gives the system seven degrees of freedom, six from the robot/gripper system and one additional degree from the track mount.

Although the robot's freedom of motion is limited in comparison with that of a human, the range of movement in each of its joints is considerably greater. For example, the human hand has a bending range of only about 165 degrees. Figure 2-28 shows the six major degrees of freedom by comparing those of a robot to a person using a spray gun.

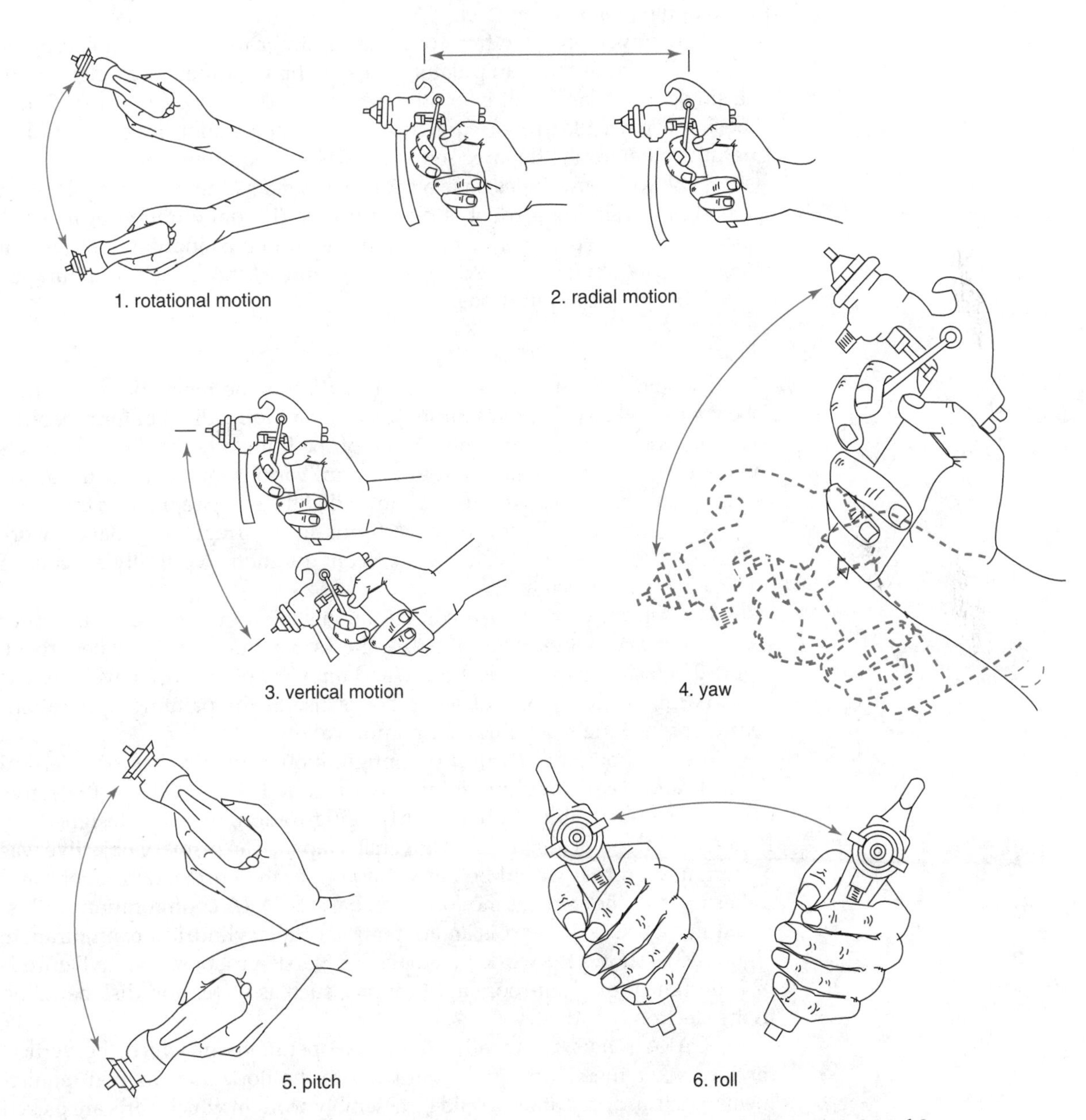

Figure 2-28 The six degrees of freedom, illustrated by a person using a spray gun. Degrees 1, 2, and 3 are arm movements; 4, 5, and 6 are wrist movements.

Robot Configurations

Robots come in many sizes and shapes. They also vary as to the type of coordinate system used by the manipulator. The type of coordinate system, the arrangement of the joints, and the length of the manipulator's segments all help determine the shape of the work envelope. To identify this maximum area, a point on the robot's wrist is used, rather than the tip of the gripper or the end of the tool bit. As a result, the work envelope is slightly larger when the tip of the tool is considered.

Work envelopes vary from one manufacturer to another, depending on the exact design of the manipulator arm. Combining different configurations in a single robot can result in another set of possible work envelopes. Before choosing a particular robot configuration, the application must be studied carefully to determine the precise work envelope requirements.

Some work envelopes have a geometric shape; others are irregular. One method of classifying a robot is by the geometric configuration of its work envelope. Some robots may employ more than one configuration. The four major ones are *revolute, Cartesian, cylindrical,* and *spherical*. Each configuration is used for specific applications.

Revolute Configuration

The jointed-arm or ***revolute configuration*** is the most common. These robots are often referred to as ***anthropomorphic*** (humanlike) in form because their movements closely resemble those of the human body. Rigid segments resemble the human forearm and upper arm. Various joints mimic the action of the wrist, elbow, and shoulder. A joint called the *sweep* represents the waist.

A revolute coordinate robot performs in an irregularly shaped work envelope. There are two basic revolute configurations: vertically articulated and horizontally articulated.

The vertically articulated configuration shown in Figure 2-29 has three revolute (rotary) joints. Figure 2-30A shows a vertically articulated robot; Figure 2-30B shows the vertical plane and top view of its work envelope. The jointed-arm, vertically articulated robot is useful for painting applications because of the long reach this configuration allows.

The horizontally articulated configuration generally has one vertical (linear) and two revolute joints. Also called the ***SCARA*** (Selective Compliance Assembly Robot Arm) configuration, it was designed by Professor Makino of Yamanashi University, Japan. The primary objective was a configuration that would be fairly yielding in the horizontal motions and rather rigid in the vertical motions. The basic SCARA configuration is illustrated in Figure 2-31A and is an adaptation of the cylindrical configuration. Figure 2-31B shows the work envelope. The SCARA robot shown in Figure 2-32 is designed for clean-room applications, such as wafer and disk handling in the electronics industry.

SCARA robots are ideally suited for operations in which the vertical motion requirements are small compared to the horizontal motion requirements. Such an application would be assembly work in which parts are picked up from a parts holder and moved along a nearly horizontal path to the unit being assembled.

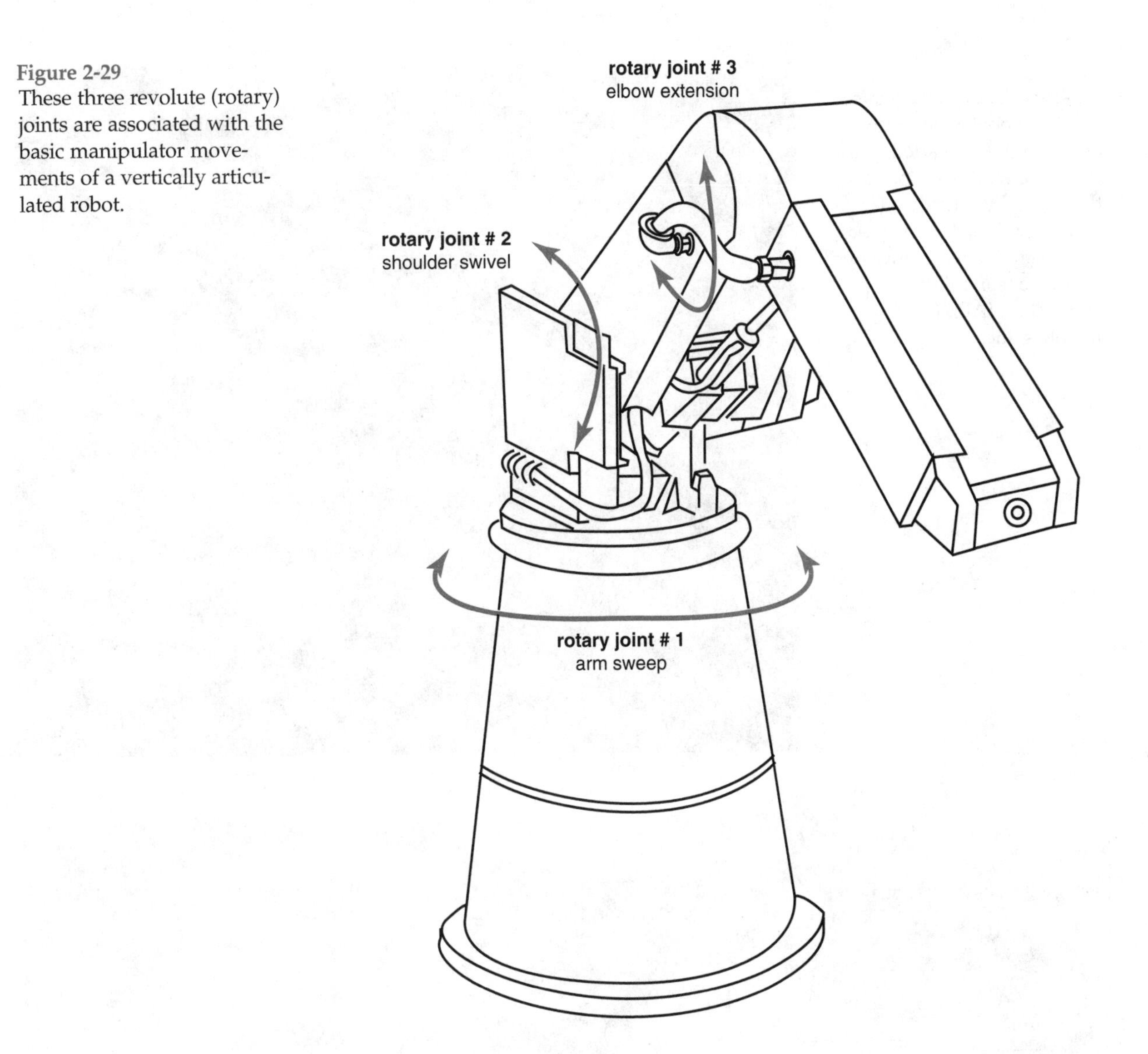

Figure 2-29
These three revolute (rotary) joints are associated with the basic manipulator movements of a vertically articulated robot.

The revolute configuration has several advantages. It is by far the most versatile configuration and provides a larger work envelope than the Cartesian, cylindrical, or spherical configurations. It also offers a more flexible reach than the other configurations, making it ideally suited to welding and spray painting operations.

However, there are also disadvantages to the revolute configuration. It requires a very sophisticated controller, and programming is more complex than for the other three configurations. Different locations in the work envelope can affect accuracy, load-carrying capacity, dynamics, and the robot's ability to repeat a movement accurately. This configuration also becomes less stable as the arm approaches its maximum reach.

Figure 2-30
A—This electrically actuated painting robot is vertically articulated. B—Shaded areas show the work envelope for the vertically articulated painting robot. The numbers shown are the work envelope dimensions in millimeters and inches. (ABB Graco Robotics, Inc.)

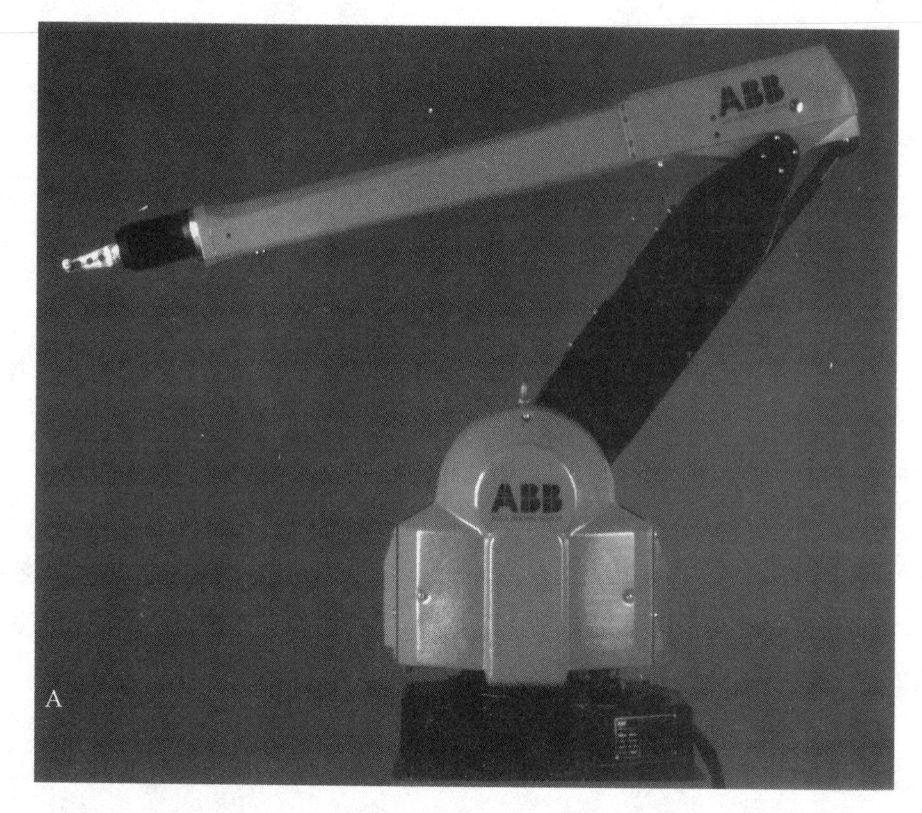

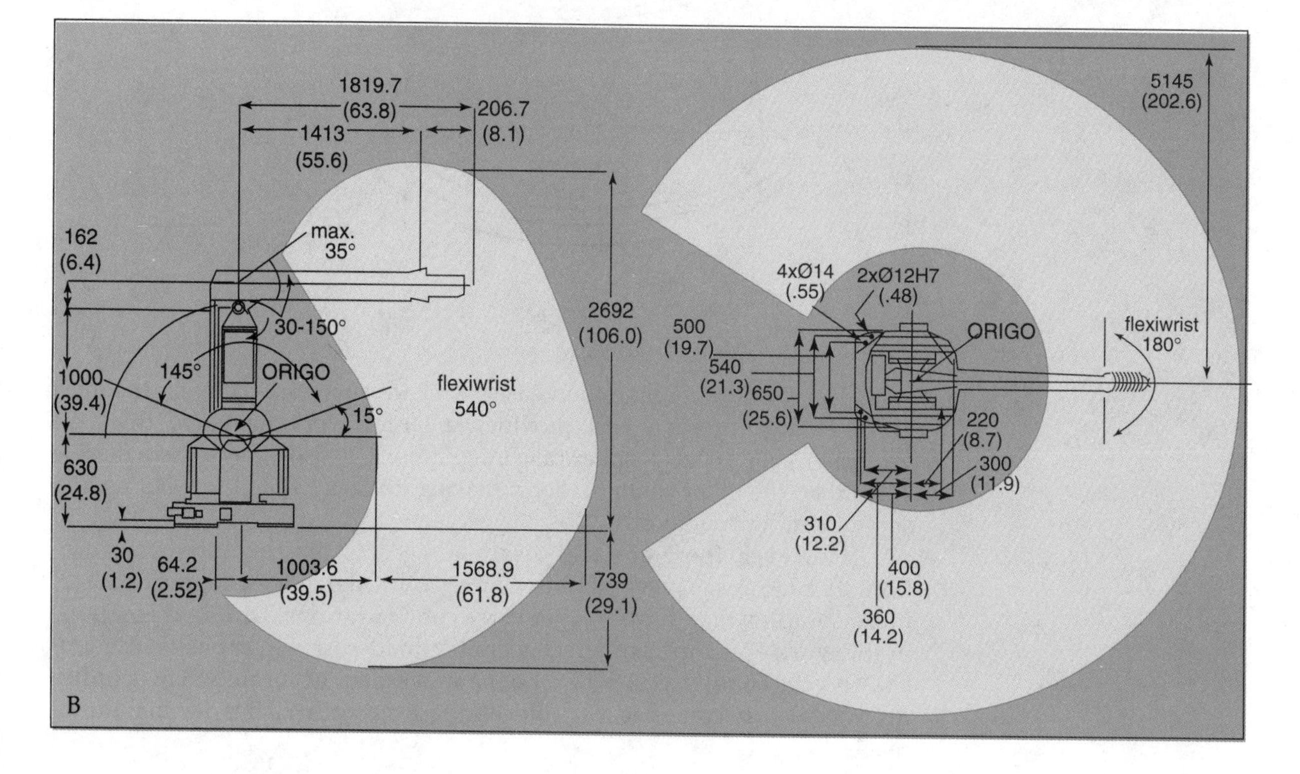

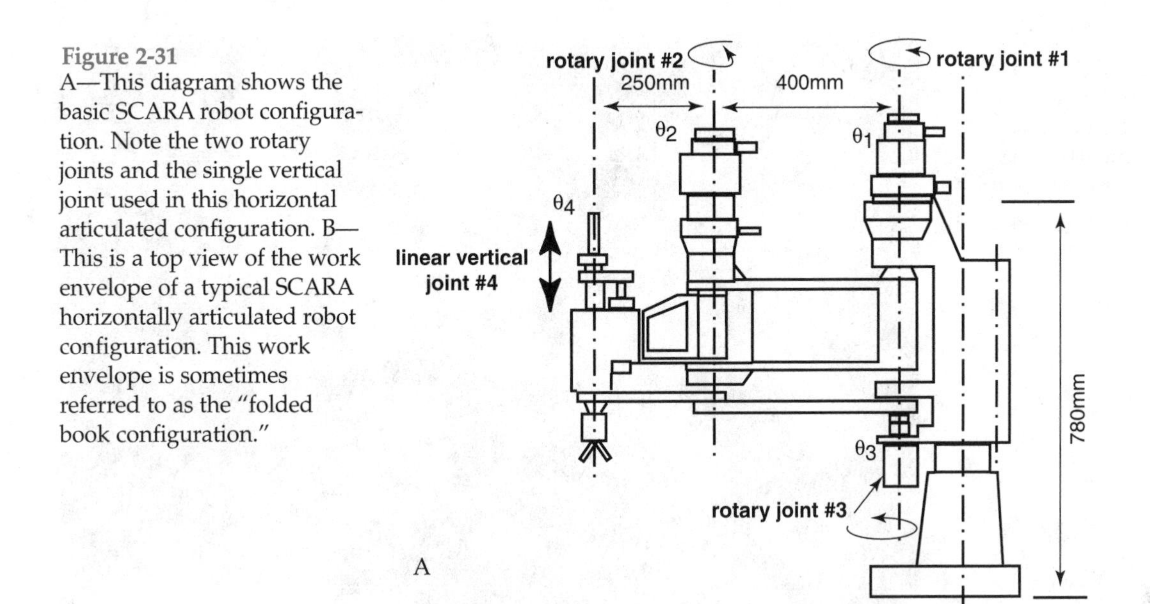

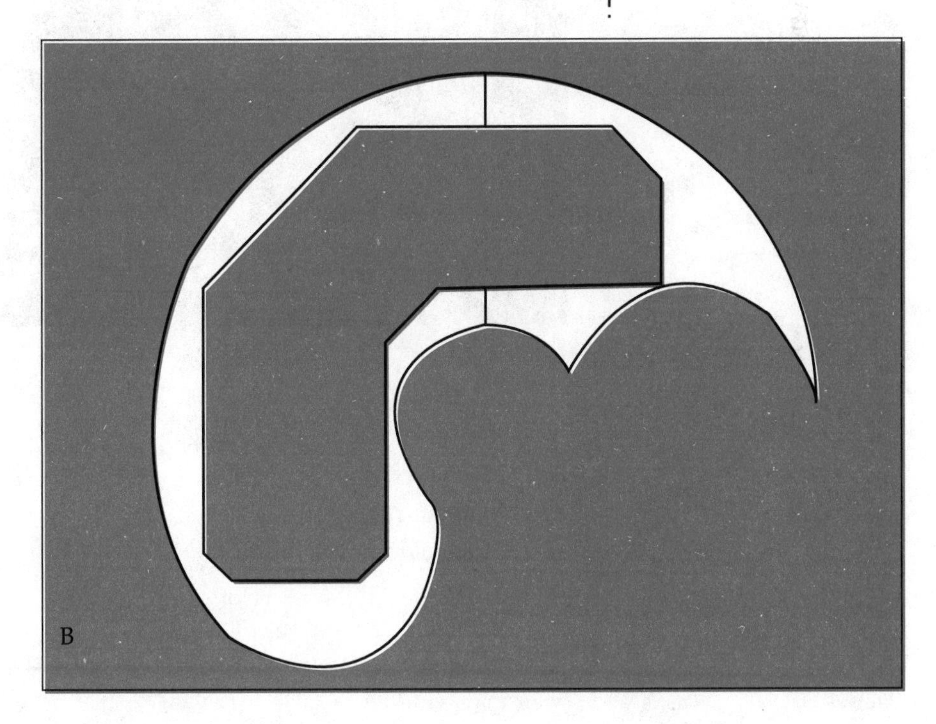

Figure 2-31
A—This diagram shows the basic SCARA robot configuration. Note the two rotary joints and the single vertical joint used in this horizontal articulated configuration. B—This is a top view of the work envelope of a typical SCARA horizontally articulated robot configuration. This work envelope is sometimes referred to as the "folded book configuration."

Typical applications include:

- Δ Automatic assembly.
- Δ Parts and material handling.
- Δ Multiple-point light machining operations.
- Δ In-process inspection.
- Δ Palletizing.

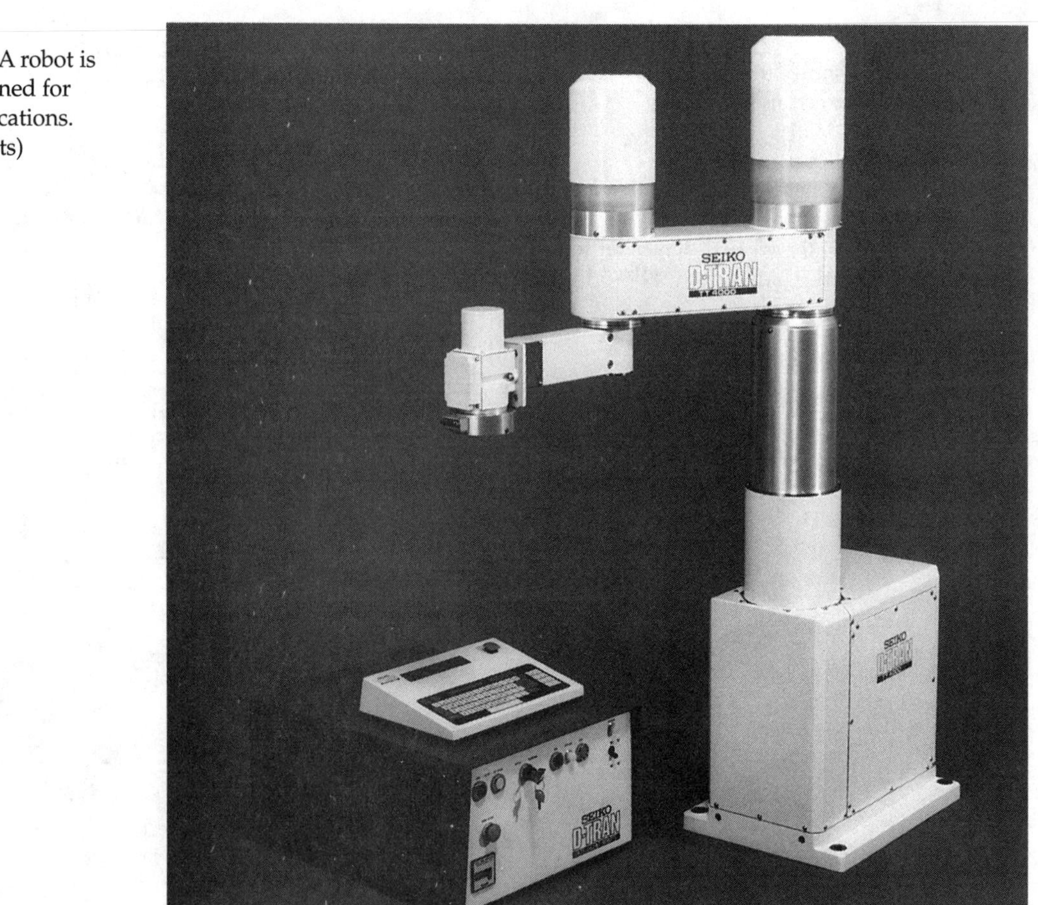

Figure 2-32
This Seiko SCARA robot is specifically designed for clean-room applications. (Seiko Instruments)

- Δ Machine loading/unloading.
- Δ Machine vision.
- Δ Material cutting.
- Δ Material removal.
- Δ Thermal coating.
- Δ Paint and adhesive application.
- Δ Welding.
- Δ Die casting.

Applications will be discussed in more detail in Chapter 4.

Cartesian Configuration

The arm movement of a robot using the ***Cartesian configuration*** can be described by three intersecting perpendicular straight lines, referred to as the X, Y, and Z axes. See Figure 2-33. Because movement can start and stop simultaneously along all three axes, motion of the tool tip is smoother. This allows the robot to move directly to its designated point, instead of following trajectories parallel to each axis. See Figures 2-34 and 2-35.

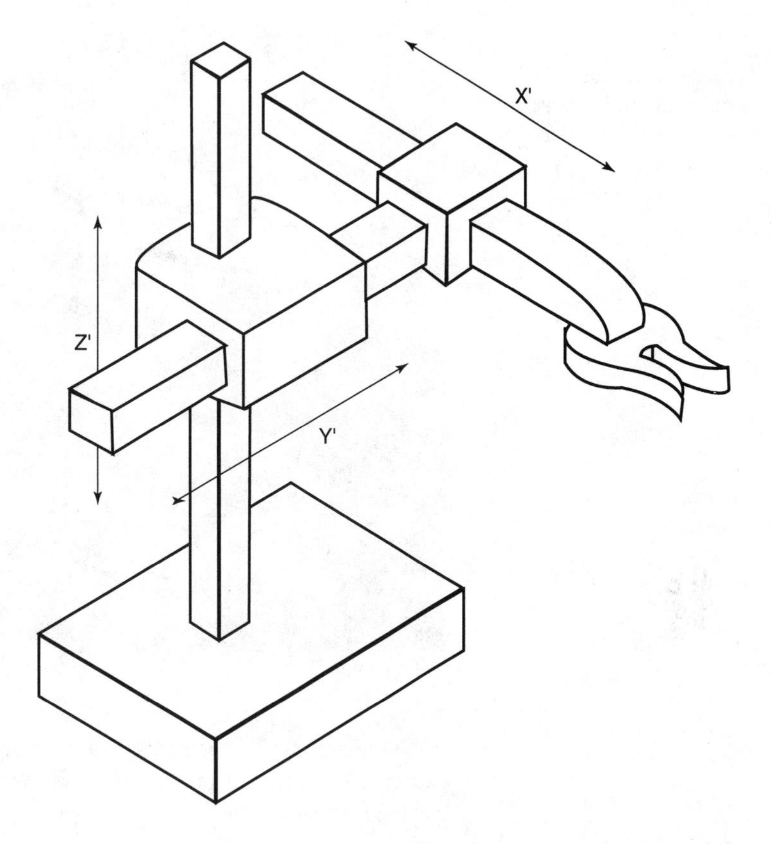

Figure 2-33
A robot with a Cartesian configuration moves along X, Y, and Z axes.

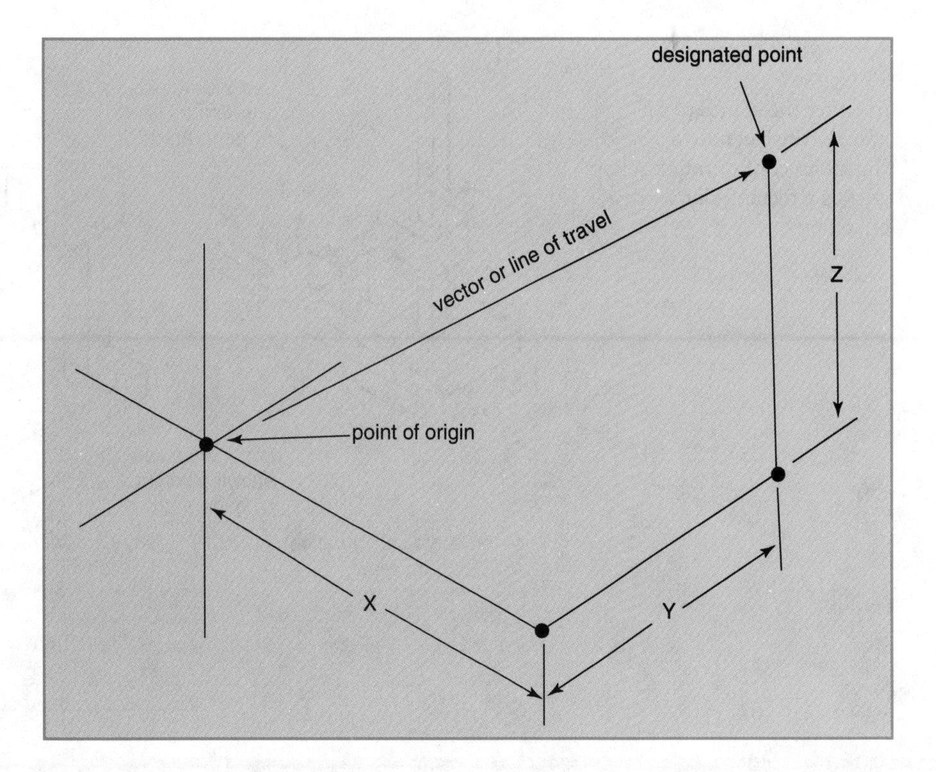

Figure 2-34
With a Cartesian configuration, the robot can move directly to a designated point, rather than moving in lines parallel to each axis. In this example, movement is along the "vector" connecting the point of origin and the designated point, rather than moving first along the X axis, then Y, then Z.

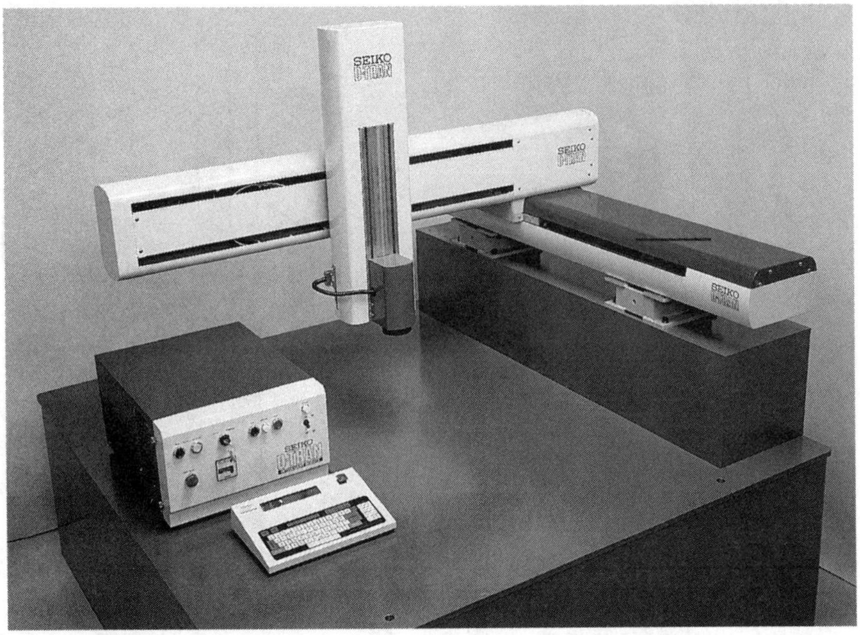

Figure 2-35
This robot has a Cartesian configuration and is used for high-precision jobs. (Seiko Instruments)

Figure 2-36 illustrates the rectangular work envelope of the typical Cartesian configuration. (Refer to Figure 2-26 for an illustration of a Cartesian gantry robot.)

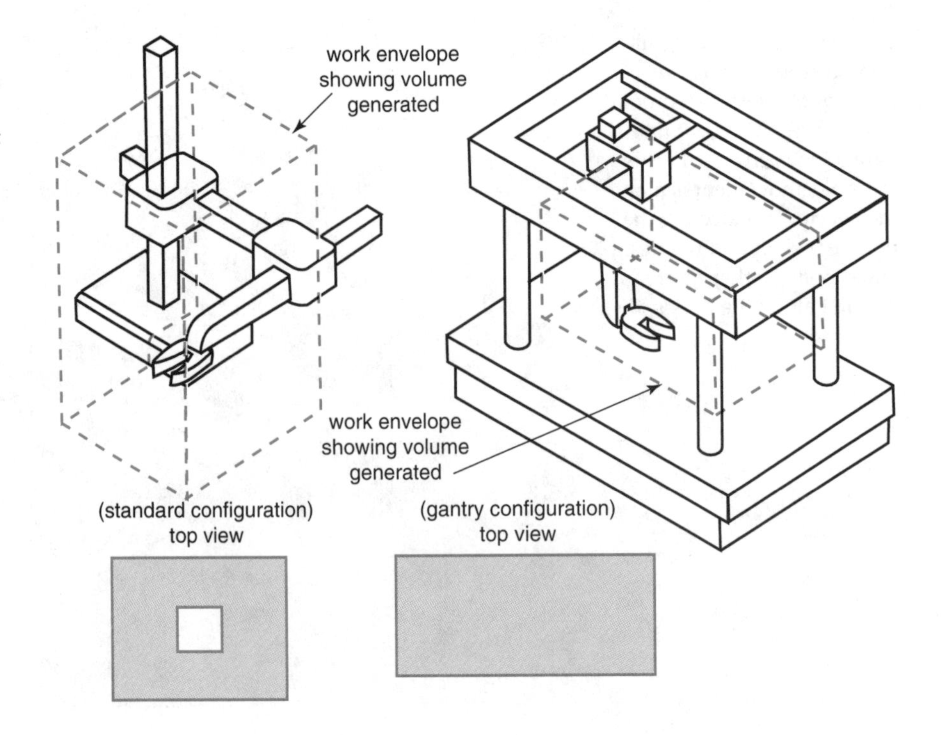

Figure 2-36
In either the standard or gantry construction, a Cartesian configuration robot creates a rectangular work envelope.

One advantage of robots with a Cartesian configuration is that their totally linear movement allows for simpler controls. They also have a high degree of mechanical rigidity, accuracy, and repeatability. They can carry heavy loads, and this weight lifting capacity does not vary at different locations within the work envelope. As to disadvantages, Cartesian robots are generally limited in their movement to a small, rectangular work space.

Typical applications for Cartesian robots include:

- Δ Assembly.
- Δ Machining operations.
- Δ Adhesive application.
- Δ Surface finishing.
- Δ Inspection.
- Δ Waterjet cutting.
- Δ Welding.
- Δ Nuclear material handling.
- Δ Robotic X-ray and neutron radiography.
- Δ Automated CNC lathe loading and operation.
- Δ Remotely operated decontamination.
- Δ Advanced munitions handling.

Cylindrical Configuration

A ***cylindrical configuration*** consists of two orthogonal (at a 90-degree angle) slides mounted on a rotary axis, Figure 2-37. Reach is accomplished as the arm of the robot moves in and out. For vertical movement, the carriage moves up and down on a stationary post, or the post can move up and down in the base of the robot. Movement along the three axes, as shown in Figure 2-38, traces points on a cylinder.

A cylindrical configuration generally results in a larger work envelope than a Cartesian configuration. These robots are ideally suited for pick-and-place operations. However, cylindrical configurations have some disadvantages. Their overall mechanical rigidity is lower because robots with a rotary axis must overcome the inertia of the object when rotating. Their repeatability and accuracy is also lower in the direction of rotary movement. The cylindrical configuration requires a more sophisticated control system than the Cartesian configuration.

Typical applications include:

- Δ Machine loading/unloading.
- Δ Investment casting.
- Δ Conveyor pallet transfers.
- Δ Foundry and forging applications.
- Δ General material handling and special payload handling and manipulation.
- Δ Meat packing.

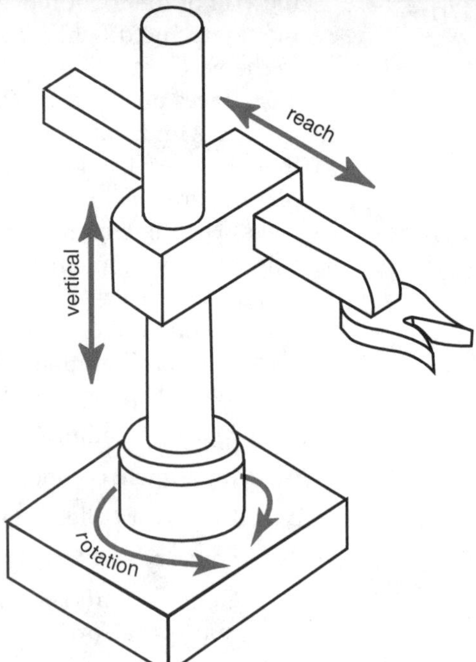

B

Figure 2-37
A—This cylindrical robot has two slides for movement up-and-down or in-and-out, and is mounted on a rotary axis. (Reis Machines, Inc.) B—The basic configuration for a cylindrical robot.

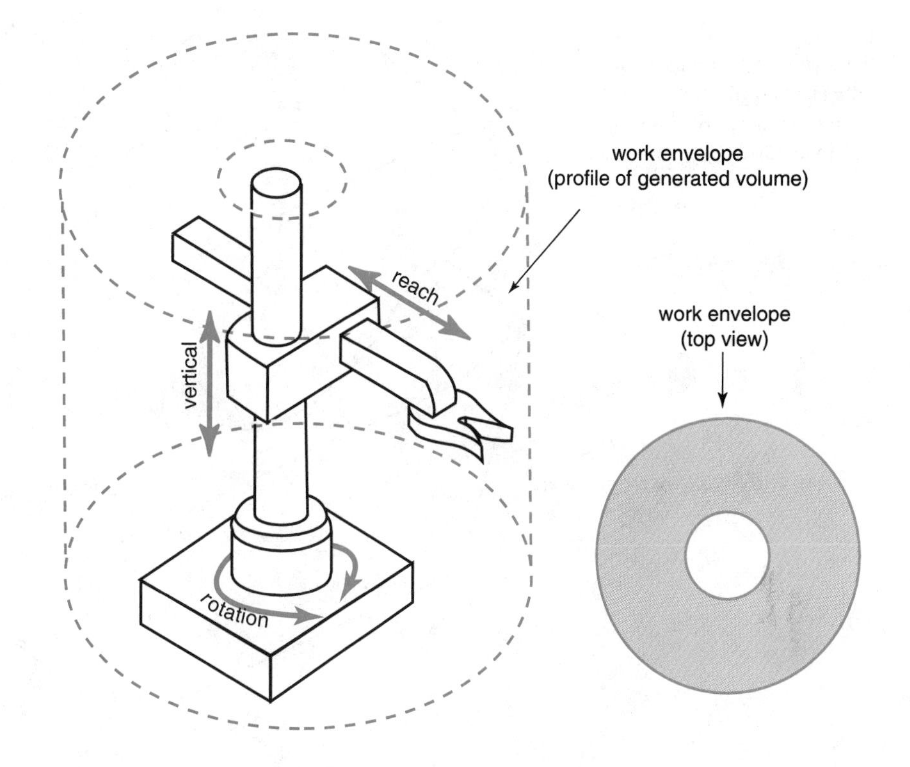

Figure 2-38
Motion along the three axes traces points on a cylinder to form the work envelope.

- Δ Coating applications.
- Δ Assembly.
- Δ Injection molding.
- Δ Die casting.

Spherical Configuration

The ***spherical configuration***, sometimes referred to as the polar configuration, resembles the action of the turret on a military tank. A pivot point gives the robot its vertical movement, Figure 2-39. Reach is accomplished, by the robot shown in Figure 2-40, through use of a telescoping boom that extends and retracts. Rotary movement occurs around an axis perpendicular to the base. Figure 2-41 illustrates the work envelope profile of a typical spherical configuration robot.

The spherical configuration generally provides a larger work envelope than the Cartesian or cylindrical configurations. The design is simple and gives good weight lifting capabilities. This configuration is suited to applications where a small amount of vertical movement is adequate, such as loading and unloading a punch press. Its disadvantages include lower mechanical rigidity and the need for a more sophisticated control system than either the Cartesian or cylindrical configurations. The same problems occur with inertia and spatial resolution in this configuration as they do in the cylindrical configuration. Vertical movement is limited, as well.

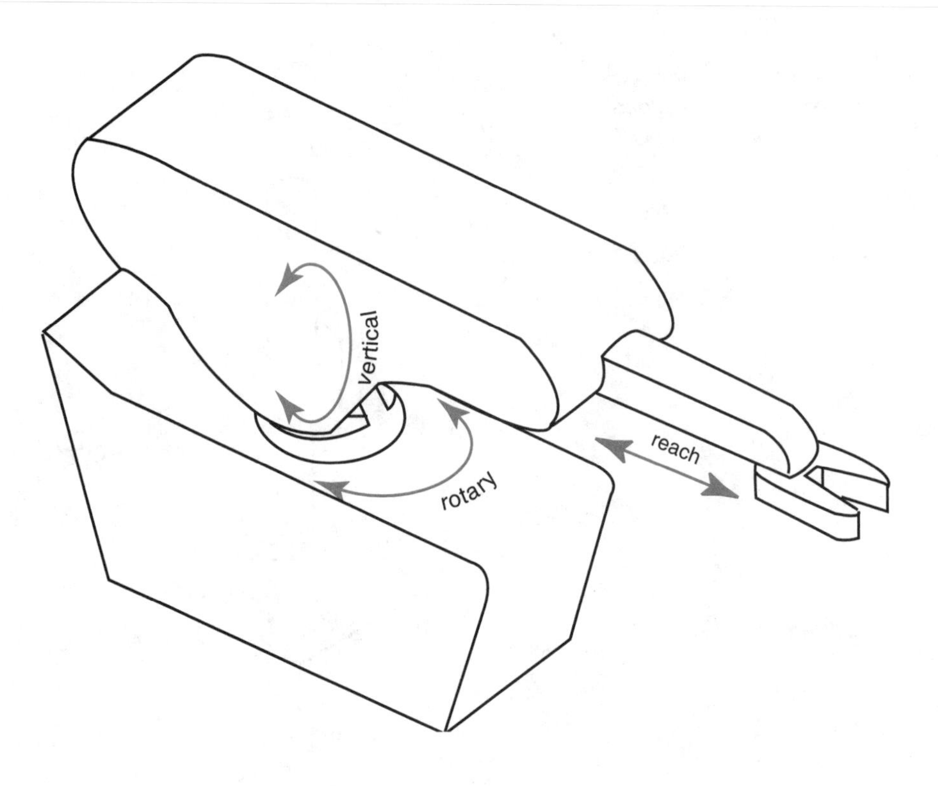

Figure 2-39
A pivot point enables the spherical configuration robot to move vertically. It also can rotate around a vertical axis.

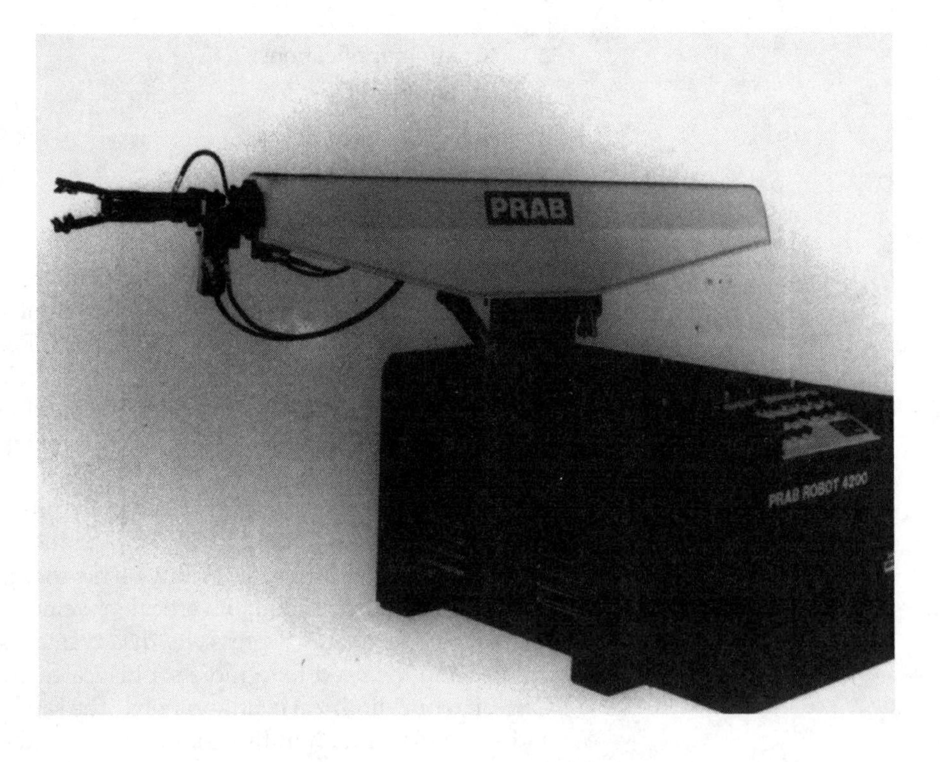

Figure 2-40
The telescoping boom of this spherical configuration robot can extend or retract. (Prab Robots)

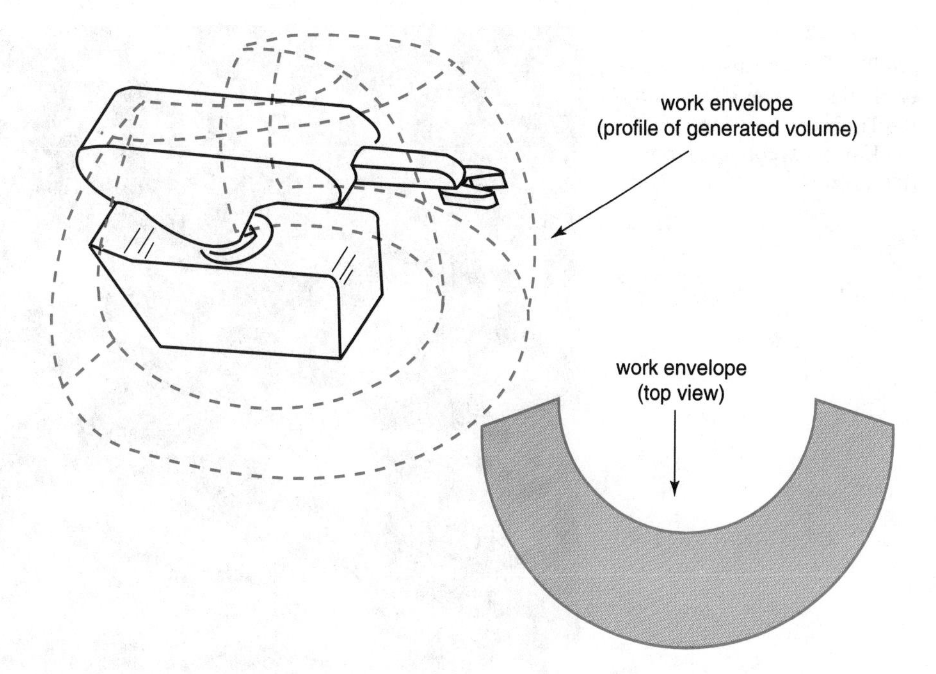

Figure 2-41
The work envelope of this robot takes the shape of a sphere.

Typical applications include:

- Δ Die casting.
- Δ Injection molding.
- Δ Forging.
- Δ Machine tool loading.
- Δ Heat treating.
- Δ Glass handling.
- Δ Parts cleaning.
- Δ Dip coating.
- Δ Press loading.
- Δ Material transfer.
- Δ Stacking/unstacking.

Special Configurations

Many industrial robots use combinations or special modifications of the four basic configurations. One example is shown in Figure 2-42A. This robot uses an articulated configuration, but its base does not rotate horizontally. It is designed to literally "bend over backwards" in order to grasp objects behind it. This feature makes it possible to install these robots very close to other equipment to minimize space requirements, while maintaining a large, effective work envelope. See Figure 2-42B. This robot has been used in such applications as spot welding and material handling.

Figure 2-42
A—This heavy-duty robot literally bends over backward.
B—The work envelope for the Kawasaki robot is large.
(Kawasaki)

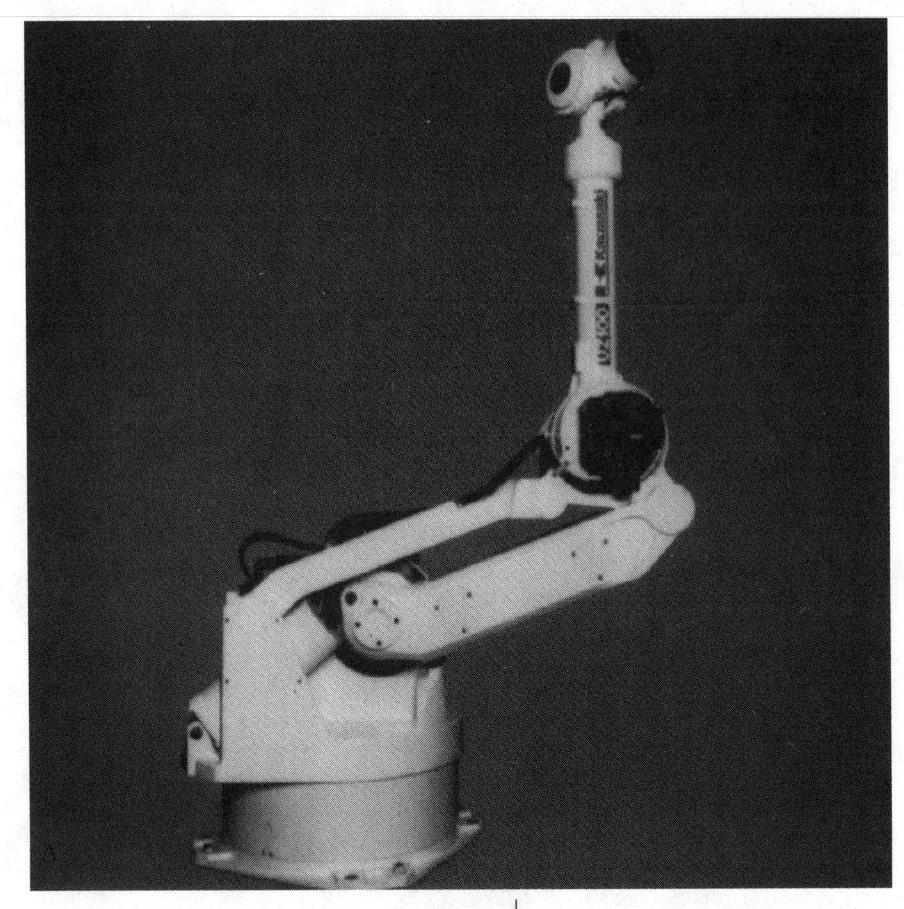

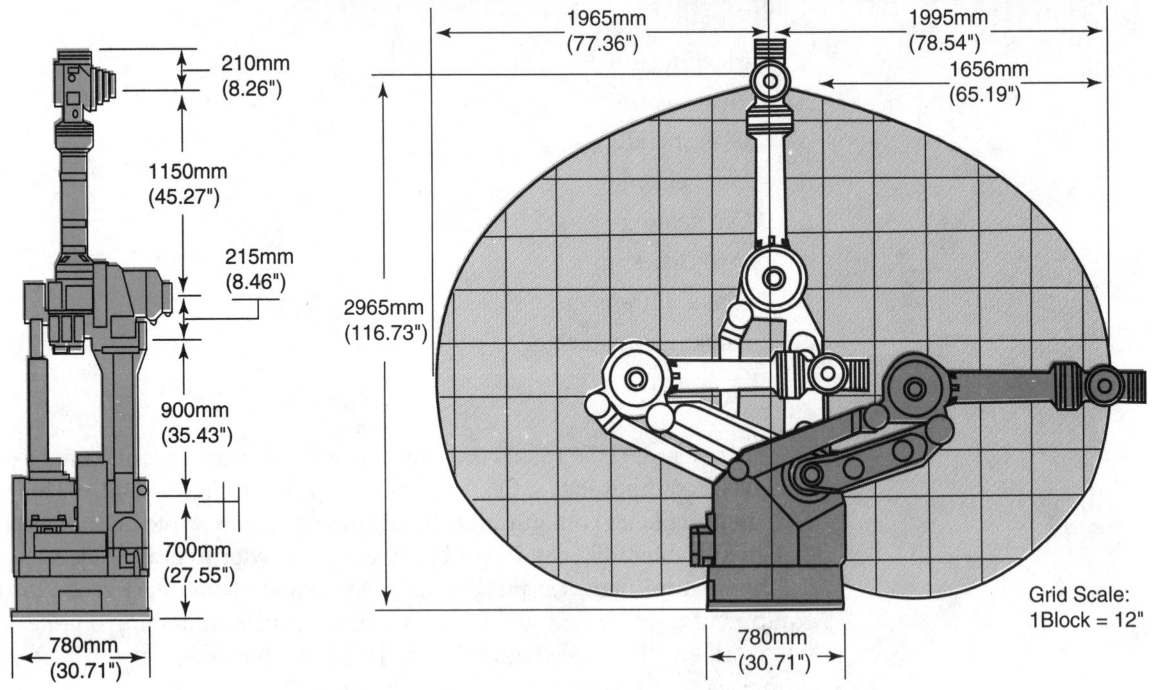

Important Terms

actuator
angular actuators
anthropomorphic
Cartesian configuration
closed-loop system
controller
cylindrical configuration
degrees of freedom
direct-drive electric motors
electric drive
end effector
error signals
hierarchical control
hydraulic drive
hydraulic system
linear actuators
manipulator
nonservo robot
open-loop system
pitch
pneumatic systems
power supply
program
radial traverse
revolute configuration
roll
rotational traverse
SCARA
servo robot
spherical configuration
tachometer
teach pendant
trajectory
vertical traverse
work envelope
yaw

Review Questions

Write your answers on a separate sheet of paper.

1. The industrial robot consists of five major components. Name these components, and explain the purpose of each.
2. What is the technical name for the robot's hand?
3. The manipulator of the robot may include such components as actuators, control valves, and internal sensors. Briefly explain the purpose of each.
4. Name the three types of power supplies used in robots today. List advantages and disadvantages of each.
5. Nonservo robots are considered open-loop. What does the term "open-loop" mean?
6. Servo robots are considered closed-loop. Sketch the diagram of a closed-loop system and explain how it works.
7. Servo robots can be classified as intelligent or highly intelligent. Explain the difference between these two classifications.
8. In terms of degrees of freedom, explain why the human hand is able to accomplish movements that are more fluid and complex than a robot's gripper.
9. List and explain the six degrees of freedom used for robots.
10. What are the common work configurations employed by robots? List some advantages and disadvantages of each.
11. What determines the shape of a robot's work envelope?
12. Why should you be concerned about the work envelope shape when installing a robot for a particular application?

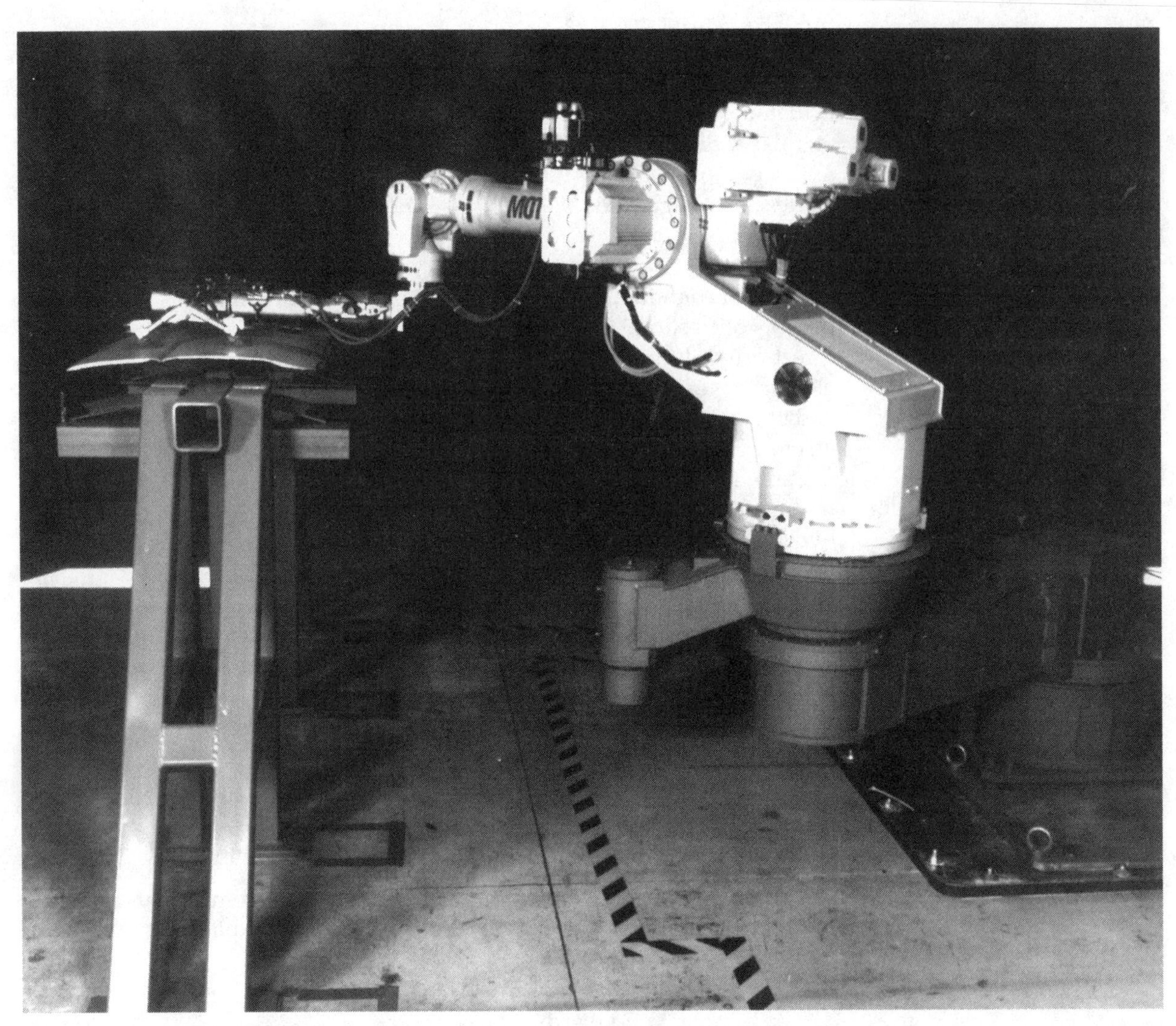

Study this press transfer robot carefully. Count the number of degrees of freedom. (Motoman)

3 Programming the Robot

Overview

Robot programming has evolved along with the robots themselves. Early robots required manual settings and adjustments. Today, state-of-the-art programming is done using a computer screen and simple menus. The easier robots are to program, the more willing manufacturers will be to use them. This chapter covers the evolution of programming, motion control, programming methods, programming languages, types of programming, and voice recognition.

The Evolution of Programming

The evolution of industrial robots can be broken down into three periods. First-generation robots were developed between the late 1950s and the mid-1970s. They performed purely repetitive tasks and did not respond to changing conditions. Most of those found outside the laboratory were open-loop, point-to-point, and pick-and-place types. These robots often had only two or three degrees of freedom. They were mainly used to transfer pneumatic or hydraulic actuators.

Programming consisted of adjusting mechanical stops and limit switches. This controlled the stroke length of each programmable axis. More complex programming was done by using a rotating drum that contained actuating switches. A series of sequential moves could be set up and executed. Pneumatic or electric relay logic circuits were also commonly used. This allowed programmed steps to be controlled by a mechanical timer. Depending on the robot's sophistication, the total number of programmable steps was generally between 10 and 100.

During the 1960s and 1970s, robotic research was being carried out in the laboratories of industry and universities like Stanford and MIT. A breakthrough in robot design occurred in 1975 with the introduction of second-generation robots that used computer-controlled manipulators. Microprocessors provided increased processing power at reasonable cost.

In the mid-1970s, robot manufacturers began to experiment with more advanced programming. In 1977, Unimation and Olivetti both introduced robots that could be controlled by means of programming languages. The earliest second-generation robots were not commercially successful. Some companies may have been intimidated by the complex programming and the higher cost of robots, compared to manual labor. It was not until the early 1980s that second-generation, language-programmable robots began to find acceptance in industry. Gradually they began to perform such tasks as welding, spray painting, assembly, and machine loading and unloading. These were all activities in which the robot was required to "think."

As second-generation robots became more complex, they were equipped with internal sensors and closed-loop control systems. This gave them limited amounts of feedback about their environment. Internal sensors were used to detect a robot's actual position. That position was compared to the program, enabling the robot to correct its position. Other internal sensors, such as strain gauges, were used to detect malfunctions. If a malfunction was detected, the problem was automatically corrected, or a warning was sent to the operator.

Third-generation robots evolved through the use of artificial intelligence (AI). Marvin Minsky, one of the fathers of AI, defines it as "the science of making machines do things that would require intelligence if done by men." Scientists and engineers have been studying AI since the 1930s. Practical use, however, was largely confined to the research laboratory until the late 1980s.

Third-generation robots are capable of sensing their environment. They can also make "intelligent" decisions about performing tasks more efficiently. Third-generation robots with vision and other sensing systems are beginning to find their way into practical applications. The next decade should see much growth in the use of AI robotic systems.

Motion Control

A robot's manipulator moves through a series of points. Robots can be classified according to their patterns of motion. The three classifications are pick-and-place, point-to-point, and continuous-path motion.

Pick-and-Place Motion

Limited-sequence robots use ***pick-and-place motion.*** The number of points the robot can move through is comparatively low. This is because programming is done by manually setting mechanical ***end-stops*** or limit switches for each designated point. It is possible to program more than two positions per axis, but the extra controls needed make it not worth the effort.

The movement of the end effector of a pick-and-place robot follows a fixed pattern. Generally, only one axis of the robot moves at a time. Figure 3-1 shows a pick-and-place motion sequence. Note that the position points are those along the various axes, not points in space.

Point-to-Point Motion

Point-to-point (PTP) motion involves the movement of the robot through a number of points in space, Figure 3-2. The programmer uses a combination of manipulator axes to position the end effector at a desired spot. The positions are recorded and stored in memory. During playback, the robot steps through the recorded points. The path of motion is a series of straight lines between the points. As in pick-and-place motion, point location is more important than controlling the path of travel.

In order to better understand point-to-point motion, look again at Figure 3-1, which shows pick-and-place motion. Suppose that the task to be taught is to take a peg out of a holder and insert it into another holder at a different location. However, instead of using end stops to control lengths of travel (as would be done for pick-and-place), the desired locations will be recorded into the robot's memory.

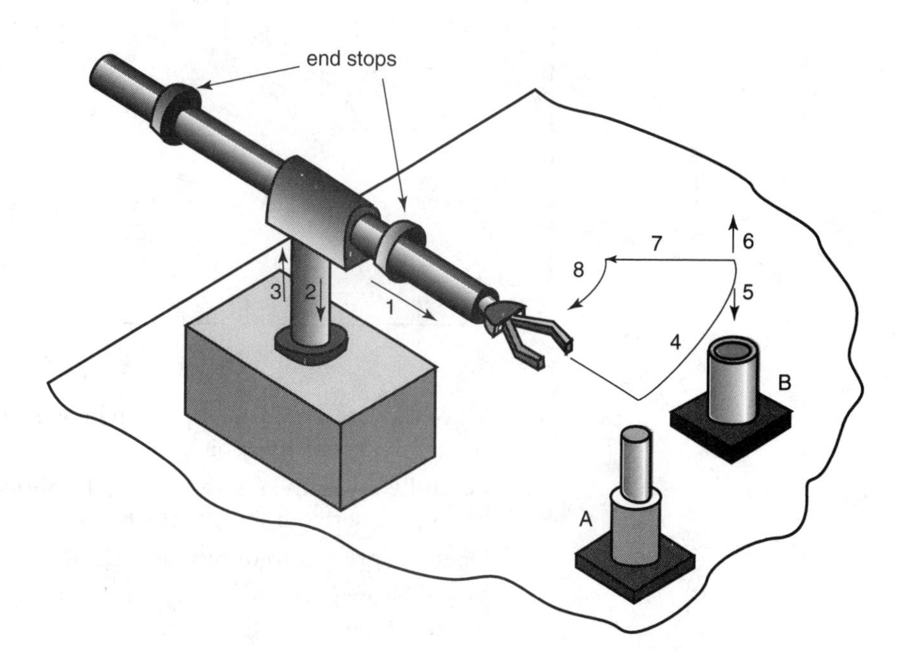

Figure 3-1
Pick-and-place motion involves only two positions *per axis.*

The peg is initially located in a holder at station A. The robot's arm is retracted and the gripper is open. The point of insertion is at station B. The steps are as follows:

1. Move the arm until the gripper is located above the peg. Record this point into memory.
2. Adjust the wrist joint of the robot until the gripper is properly aligned for grasping the peg. Record this position.
3. Move the gripper down over the peg to the point where the gripper will be able to grasp the peg when closed. Realign the gripper with the peg by adjusting the various joints. Record this point.
4. Close the gripper on the peg and record the point.
5. Carefully lift the peg from the hole vertically. After the peg is out of the hole and is at the desired elevation, record that point.

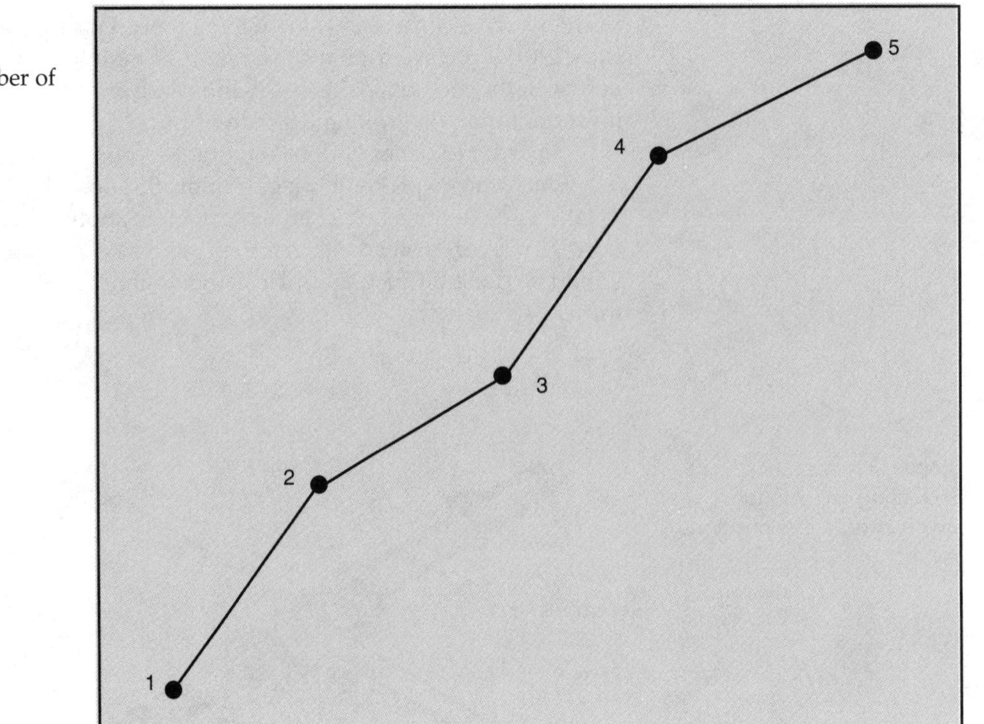

Figure 3-2
Point-to-point motion involves a discrete number of points in space.

6. Move the arm until it is approximately over the center of the hole at station B. Record this point.
7. Carefully lower the peg, adjusting the various joints until the peg is properly inserted into the hole. Record this point.
8. Open the gripper to release the peg. Record this point.
9. Move the arm until the gripper is located at some point directly above the peg, then record this point.
10. Stop the robot. Move the peg back to the hole at station A.
11. Take the robot out of the teach mode and press the play button.

The robot will return to the start position and will then step through the various points recorded into memory. The robot in the example would have to be stopped after step 9 because there are no more pegs at station A. In a real work situation, another peg would appear at A or additional steps would be added to the program.

During playback, the trajectory of the end effector is generally different from the path used by the operator when establishing the points. This is because the operator must move each axis or joint independently. However, if a joystick is used during programming, the robot will move along a straighter path. Programming time is reduced, as well.

Several stops along a given axis can be programmed. Point-to-point servo robots are capable of storing hundreds of discrete points in space, far more than the two stops in pick-and-place motion. Acceleration and deceleration between points is controlled by a device such as a tachometer.

Continuous-path Motion

Continuous-path (CP) motion is an extension of point-to-point motion. The difference is that continuous-path motion can involve several thousand points. Since more points are used, the distances between them can be extremely close, Figure 3-3. Because of the large number of points, movement is smooth and continuous. Control of the path is more important than end-point positioning. The robot generally does not come to rest at various points, as is often required in PTP.

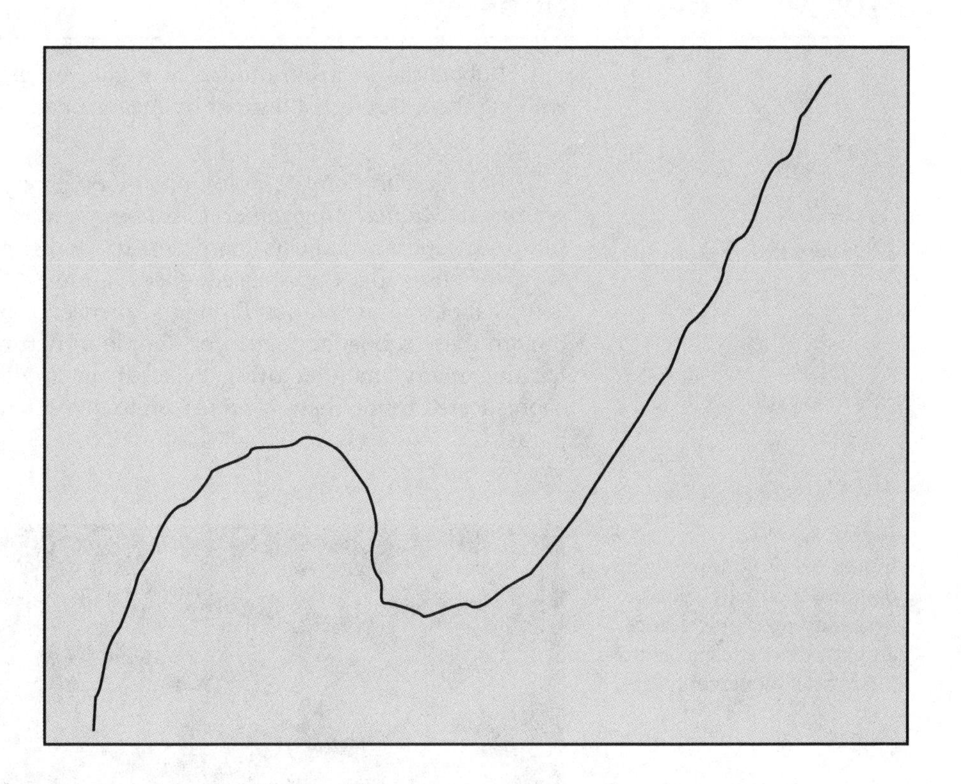

Figure 3-3
Continuous-path motion involves an infinite number of points.

Programming is done by an operator who physically moves the end effector through its motions. The positions on the various axes are recorded. Some continuous-path robots record up to 80 points per second. Programs are generally recorded on magnetic tape or a magnetic disk. Some programs may run several minutes.

Playback rate can be changed to provide the best operating speed for the task. For certain applications, such as spraying, it may be better to program the robot at a slow speed and play the program back faster. Other applications, such as arc welding, require faster programming than playback.

Continuous-path control offers several advantages. Movement is smooth and continuous. Programming is simple and no prior knowledge of programming is required. All that is necessary is that the operator understand the operation he or she is trying to teach the robot.

However, there are also disadvantages. For the controller to store all the points, a large memory is required. In the past, memory was very expensive. Today, memory chips are reasonably priced, so this is not as much of a disadvantage as it once was. Since recording during programming is constant, undesirable motions as well as intended moves are recorded into memory. The robot duplicates all these moves during playback. Also, the robot's arm must be counterbalanced and able to move freely without power or the operator will not be able to produce smooth-flowing motion.

Programming Methods

Robots can be programmed manually, by means of a teach pendant, by walking them through a task, or by means of a computer terminal.

Manual Programming

Robots with point-to-point, open-loop controllers can be manually programmed. ***Manual programming*** can best be described as a type of machine setup. An operator adjusts the necessary end-stops, switches, cams, electric wires, or hoses to set up the sequence, Figure 3-4. This type of programming is typical of first-generation limited-sequence, or pick-and-place robots. Even though these robots appear to be simple in nature, they are capable of performing many manufacturing tasks. If an application is suited for a less sophisticated robot, there is no reason to invest in a more complex and costly model.

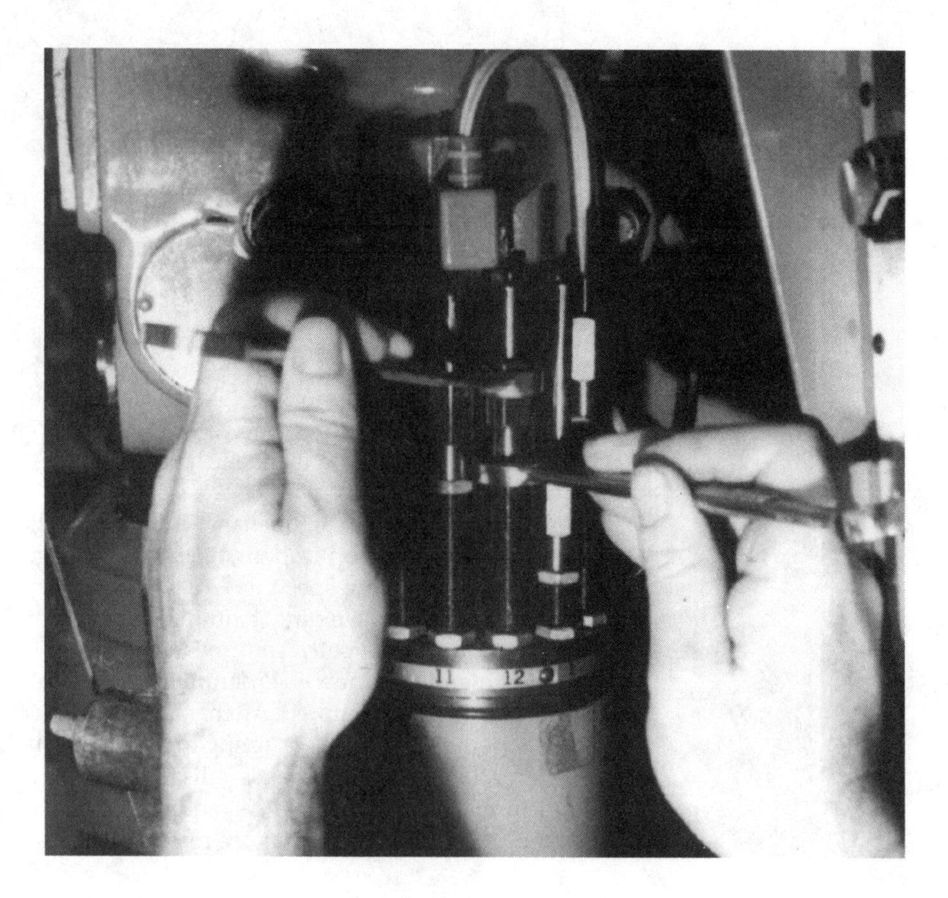

Figure 3-4
In manual programming, an operator manually adjusts necessary mechanical stops and limit switches to control the robot's movement.

Manual programming is usually easy. It does not require an operator skilled in the use of computers. The capital investment and maintenance costs for manually programmed robots are low. They are capable of high operating speeds and have good accuracy and repeatability. However, their flexibility is limited and they may have only two or three degrees of freedom. Control of intermediate points along the path is generally not available. Only two positions are programmed for each axis. Depending on the complexity of the robot, the total number of programmable steps is generally from 10 to 100.

Using a Teach Pendant

In ***teach-pendant programming,*** the operator leads the robot through the various positions. See Figure 3-5. As the end effector reaches a desired point, that point is recorded into memory by pushing buttons on the teach pendant. The recorded points are used to generate the point-to-point path the robot follows during operation.

Figure 3-5
The operator is using a teach pendant to program the robot through various moves.

The teach pendant is a popular method of programming because of its convenience and ease of use. It is simple to learn and suitable for programming many tasks found in industry. It does not require an operator skilled in the use of computers. However, complex motions and applications requiring close tolerances may require a lengthy programming time. The robot must be operating during programming. The program cannot be entered into the pendant while the robot is off-line.

Walk-through Programming

Walk-through programming is used for continuous-path robots. As shown in Figure 3-6, an experienced operator physically moves the end effector through the desired motions. While the robot moves along the desired path, as many as several thousand points are recorded into memory. The

Figure 3-6
This operator is using the walk-through method of programming. The robot is being programmed to palletize boxes.

number of points recorded can vary from one manufacturer to another. As the number sampled per second is increased, the movement of the robot becomes smoother and more fluid. However, as the sampling increases, so does the need for greater storage capacity. A floppy disk or hard drive is generally needed to store the program.

Spraying and arc-welding are the most common applications programmed using the walk-through method. Other applications include grinding, deburring, polishing, and palletizing.

Walk-through programming does not require computer experience. However, the person doing the programming must be highly skilled in the precise motion required by the task. The robot cannot be programmed off-line.

Using a Computer Terminal

Programming by means of a computer can be done ***off-line*** (away from the robot), as shown in Figure 3-7, or ***on-line*** (at the robot's console). Final testing of the program is done at the job site.

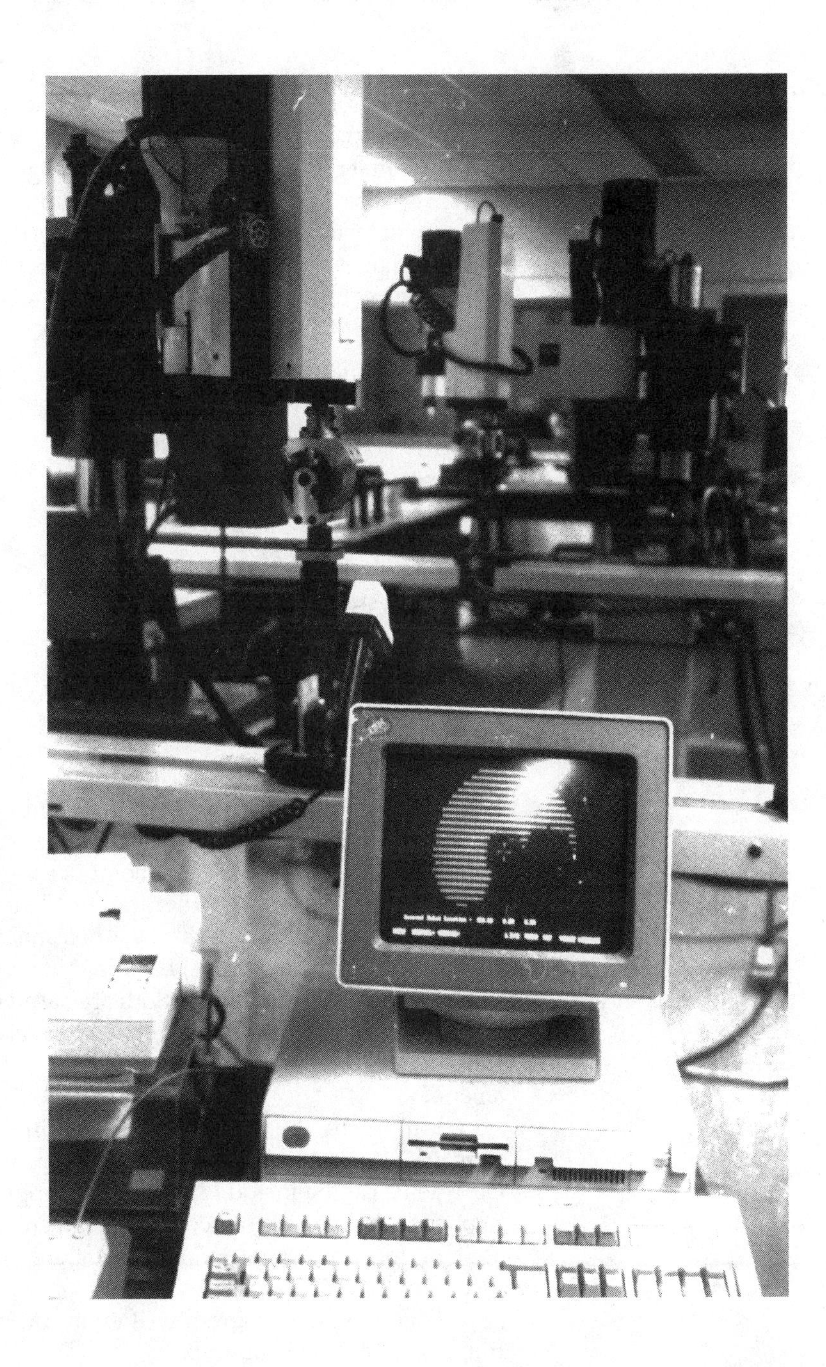

Figure 3-7
An IBM personal computer is used here to program a robot off-line.

Computer programming provides greater flexibility. The robot does not have to be taken out of operation while the program is being written and debugged. As a result, productivity is not affected. High-level computer languages allow programming of more complex operations. These languages reduce programming time, which helps to increase productivity. The primary disadvantage is that the operator must be experienced in the use of computers, high-level languages, and programming logic.

Programming Languages

The first robot programming language, known as ***WAVE,*** was developed at the Stanford Artificial Intelligence Laboratory in 1973 for research purposes. Most manufacturers that provide off-line programming have developed their own languages. No set of standards has been established in the industry.

Some common languages and their sources are:

- Δ AML-IBM.
- Δ HELP-General Electric.
- Δ VAL and VAL II-Unimation.
- Δ RAIL-Automatrix.
- Δ MCL-McDonnell Douglas.
- Δ RPL-SRI.
- Δ AL-Stanford University.
- Δ AR-BASIC-American Cimflex.
- Δ Robot-BASIC-Intelledex.
- Δ Karel-GMF Robotics.
- Δ JARS-NASA's Jet Propulsion Laboratory.
- Δ T3-Cincinnati Milacron.

High-level languages are those closer to standard English. Because computers cannot understand English, these languages are translated into machine code by means of a program called a compiler.

Even though robot programming languages have been developed by different groups, many are similar. Their main difference is in the choice of key words and commands. Also, each language has its own syntax, or structure. AML and Karel are examples. Both are based on Pascal and PLI computer programming languages. The instructions are given in the form of subroutines. A ***subroutine*** is a set of instructions within the program that has a beginning and an end.

With AML, the program name can include up to eight characters in the form of letters, numbers, and underscoring. The name starts with a letter. Comments can be included within the program by preceding them with two hyphens (--). A colon (:) is required between an identifier and a key word (such as SUBR for subroutine). Every line of the program must end with a semicolon (;), except for comments.

Following is a program written in AML to move the robot's end effector from location A to location B.

```
A:NEW PT(X,Y,R);--XYZ location to be taught
B:NEW PT(X,Y,R);--XYR location to be taught
MAIN:SUBR;
PMOVE(A);--program execution
PMOVE(B);
END;
```

With Karel, the program name can contain up to twelve characters, including letters, numbers, and underscoring. It must start with a letter. Comments can be included within the program by preceding them with two hyphens (--). An identifier follows key words. A semicolon (;) is not required at the end of each line.

Following is a program written in Karel to move the robot's end effector from location A to location B.

```
PROGRAM Exe_2
Variables
A: Position--position location to be taught
B: Position--position location to be taught
Begin--program execution
Move to A
Move to B
END Exe_2
```

Types of Programming

The majority of robots in industry today use ***hierarchical control programming.*** However, the trend is toward task-level programming, which simplifies the programming task.

Hierarchical Control Programming

As discussed in Chapter 2, hierarchical control partitions control into a number of different levels. Each level accepts commands from the next higher level and responds by generating simpler commands to the next lower level, Figure 3-8. This system uses ***sensory feedback*** (input from the environment) to close control loops. In other words, sensor inputs affect how the robot responds.

First level

Servo control functions are computed at the lowest level in the hierarchy. This is the level at which most robots in use before 1980 were programmed and controlled. Program commands are compared with feedback from position indicators. If the values are different, a drive signal is generated to move each joint until the error signal is zero.

Programming at the first level does not require the use of a computer. The arm is moved using a ***manual rate control box.*** The box consists of a knob and some switches that control the movement of each axis individually. (A rate control box should not be confused with a teach pendant. It does not have the same capabilities.) The joints are moved one at a time until the robot arm is in the desired location. The values of the position indicators are then all stored in memory. The arm is moved to another point, and the process is repeated until the desired path is stored in memory. This form of programming is time-consuming and tedious.

Second level

A computer is required to program a robot at the second and third levels. They require real-time interaction of the robot with its environment by means of sensors. Commands are issued for "primitive" operations. An example of a primitive operation would be "move along a straight line defined by the vector X, Y, Z." These commands are translated by the computer into the proper position values.

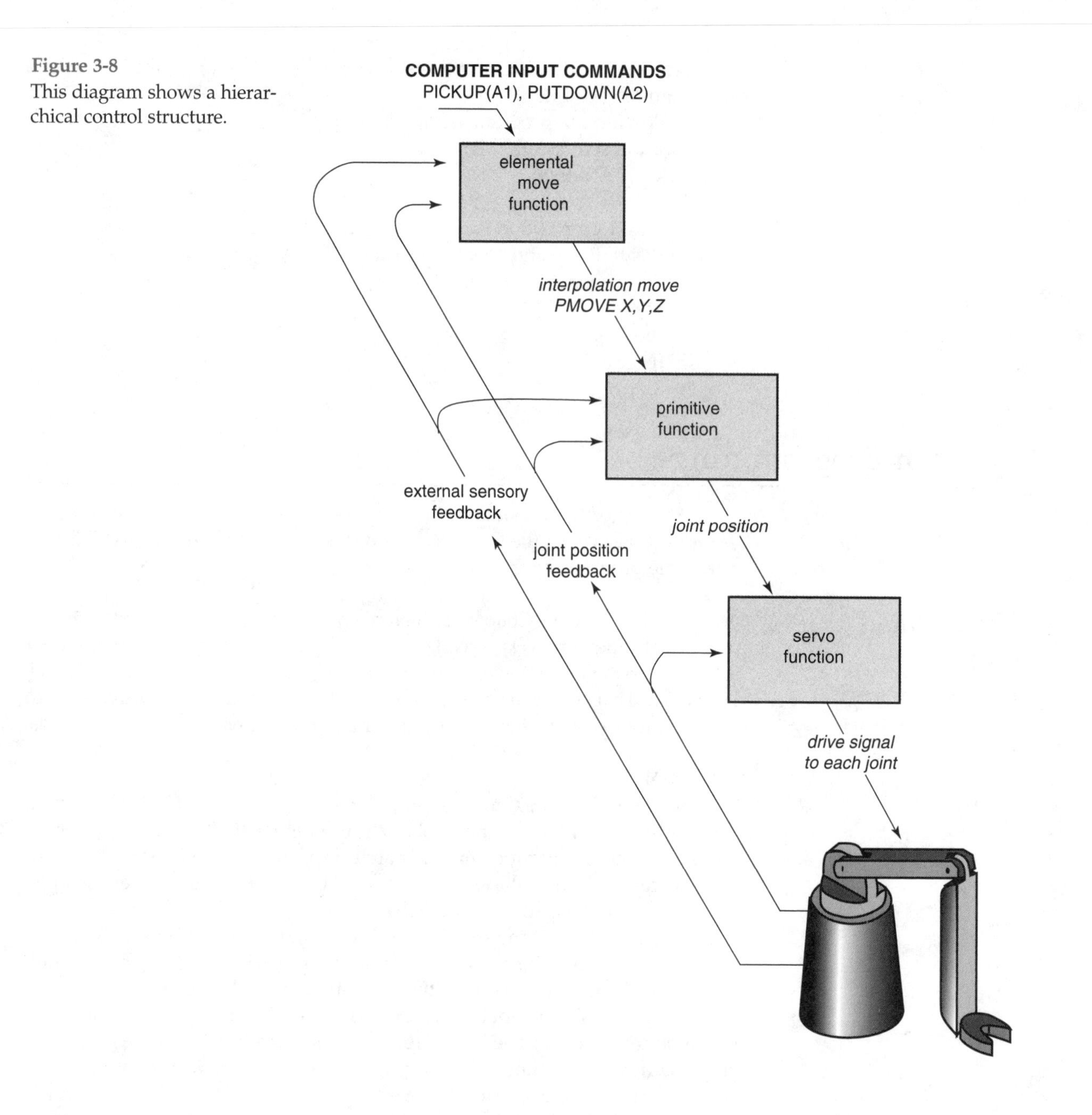

Figure 3-8
This diagram shows a hierarchical control structure.

The second level receives position feedback and generates the necessary sequence that the first control level must accomplish At this level, a joystick can be used for programming instead of the rate control box. The operator does not have to worry about moving individual joints. Programming tasks are much easier and faster.

Third level

The third level in the hierarchy receives commands for "elemental" moves. A typical command in the AML language is "PMOVE(X1, Y1, Z1)."

When these commands are issued by third-level circuitry, the control system monitors feedback signals. It then generates a sequence of commands for primitive operations. These primitive operations become the input to the next lower level.

For example, to program a simple transfer task, commands for four elemental moves are typed in the desired sequence into the computer. In the AML language this would be:

```
PMOVE(X1, Y1, Z1) GRASP
PMOVE(X2, Y2, Z2) RELEASE
PMOVE(X3, Y3, Z3) GRASP
PMOVE(X4, Y4, Z4) RELEASE
```

This set of commands causes the robot to move to location X1, Y1, Z1 and pick up an object. It then moves the object to location X2, Y2, Z2 and puts the object down. The same sequence is repeated for the next set of positions. These points are previously recorded by the use of a joystick, teach pendant, or computer keyboard. The X, Y, and Z locations can also be typed directly into memory.

During playback, the robot grasps and releases at the recorded locations. The environment must be tightly controlled. The locations of the objects to be grasped and their orientation must always be the same. A slight misalignment can cause damage to both robot and workpiece. Control of the environment often requires additional machinery and equipment that can cost several times more than the basic robot.

Programming even a simple task using hierarchical control and traditional high-level languages can take days. Additional time must be set aside for testing and making changes. Even after the system is up and running, further changes may be required.

Task-level Programming

In today's manufacturing environment, fast production time and low systems costs are critical to a company's success. Yet today's robot systems are expensive and difficult to install and maintain. A typical robot in an automated work cell might move something, assemble parts, use vision for guidance and parts inspection, control local work cell devices, respond to sensor inputs, keep statistics on processes, and communicate with a host computer or an operator.

Figure 3-9A shows the architecture for a typical custom work cell. It is more complex than it needs to be. Four separate controllers are used. A motion control system runs the robot, while the Programmable Logic Controller (PLC) provides work cell logic. The vision system is for inspection and robot guidance. The personal computer is included for user interface. What is needed is a simpler system with user-friendly hardware and software. The answer seems to lie in task-level programming.

In ***task-level programming,*** the user specifies the goals of each task rather than the motions required to achieve those goals. Instructions are input using simple English-like terms. The details for every action the robot is to perform do not need to be specified. This allows the use of instructions at a significantly higher level than those produced by languages such as AML, Karel, or Robot BASIC. Many activities are programmed automatically by the computer.

An example of a task-level programming environment is shown in Figure 3-9B. Here only one controller and one programming system are needed to work with multiple technologies. Days of programming are eliminated by straightforward menus and instructions entered in English. The operator can focus attention on the task instead of writing programming code.

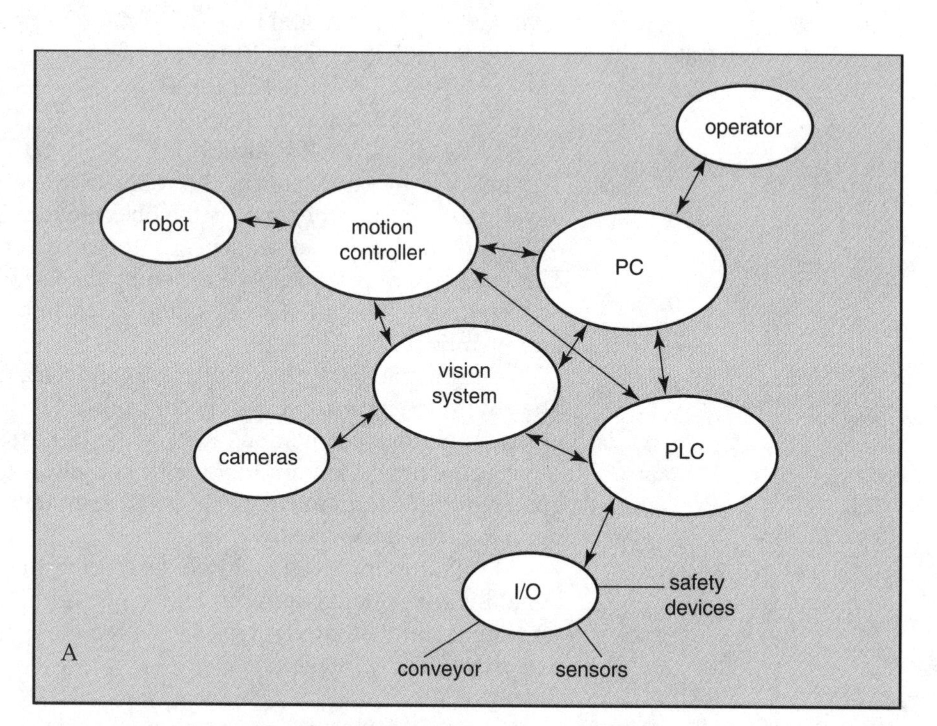

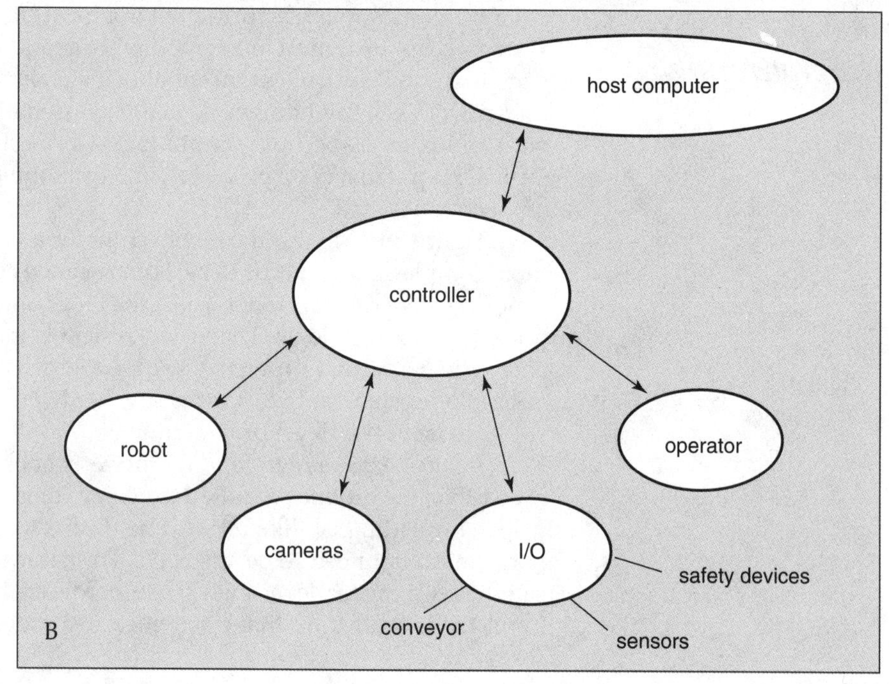

Figure 3-9
A—This is a typical custom work cell layout. The arrows indicate interactions.
B—Here, a single controller and programming system is used to work with multiple technologies. (Adept Technology, Inc.)

Figure 3-10 shows how an interactive screen is used to create, teach, debug, and run application programs. Two task statements are required. The software is so "user-friendly" that inexperienced programmers can program a robot.

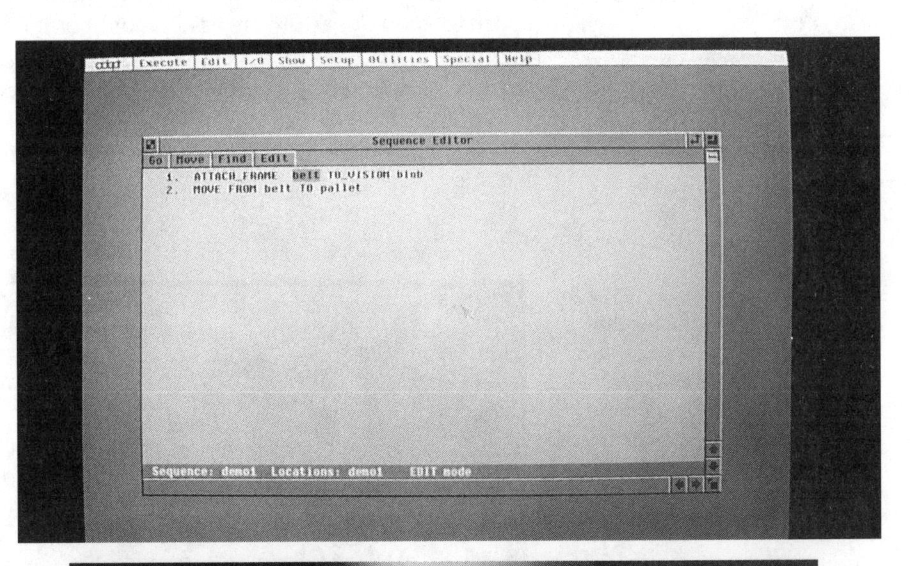

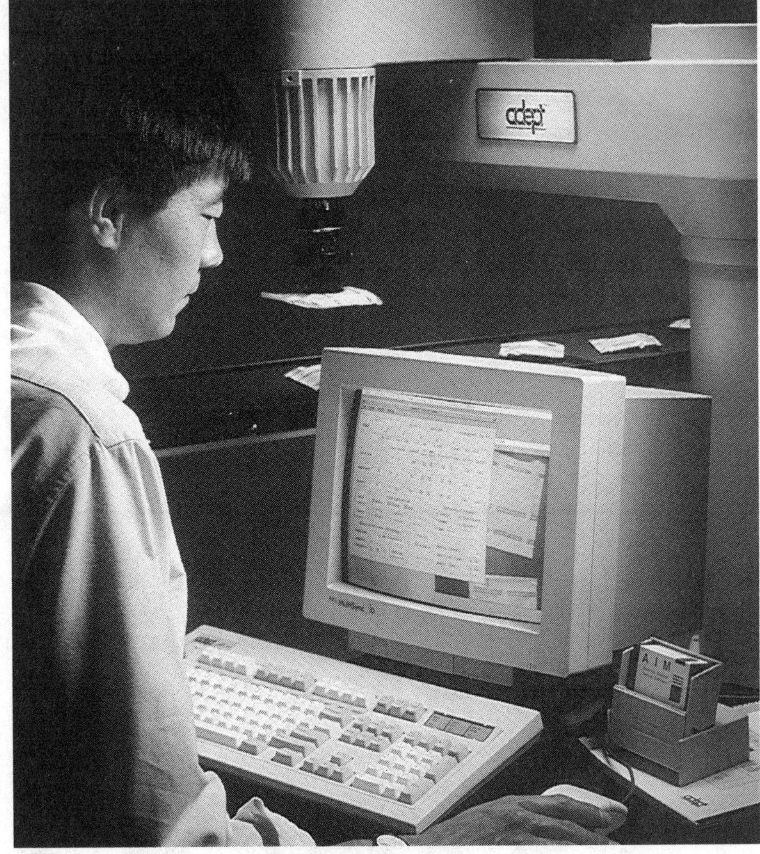

Figure 3-10
The basic program steps are defined on-screen. A MOVE statement, which allows a robot to move a part from one location to another, is shown. Below, a programmer uses on-screen sequence-editing functions. (Adept Technology, Inc.)

The first step in creating the robot program is to select the sequence of movements from the menu. An example is the MOVE statement, which tells the robot to pick up parts from one location and place them in another location. In Figure 3-10, the MOVE statement is used to move a part from a conveyor belt to a pallet.

A location database stores work cell locations. Heights of each location, approach, and departure are easily taught or modified using the screen shown in Figure 3-11. The robot's speed, type of motion, and other details are also easily selected and stored using this screen. Another commonly used statement controls automatic palletizing functions. The PALLET screen shown in Figure 3-12 allows the operator to define spacing and the number of pallet locations.

Figure 3-11
In this database, the heights of locations, approaches, and departures are stored. They can be modified using the screen shown here. (Adept Technology, Inc.)

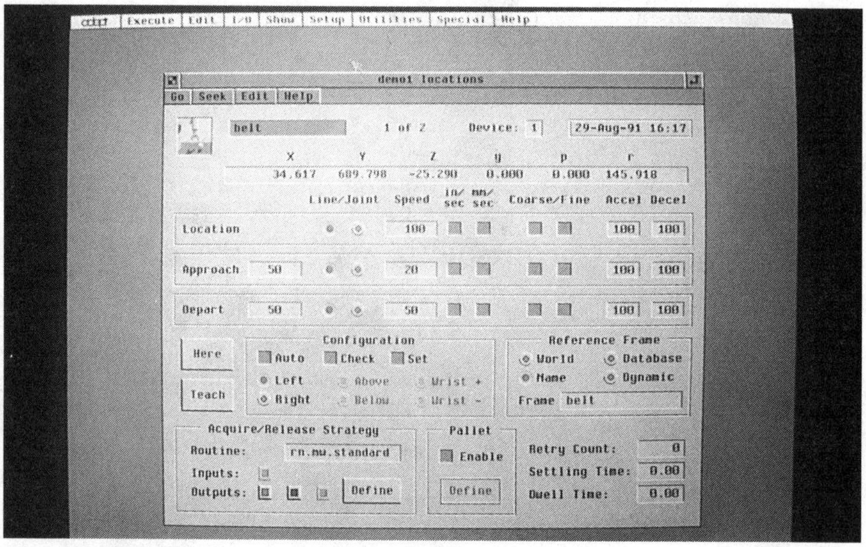

Figure 3-12
Using the PALLET screen, the operator can define spacing and the number of pallet locations. (Adept Technology, Inc.)

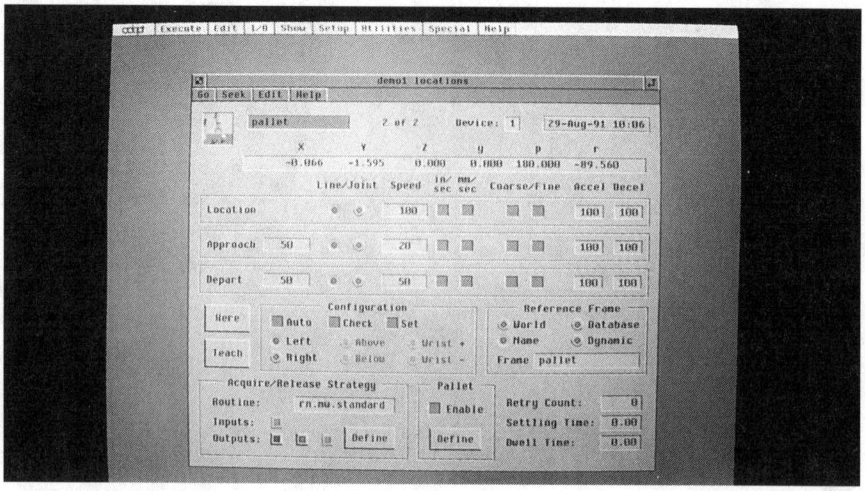

Tracking a moving conveyor is accomplished by using the ATTACH FRAME screen shown in Figure 3-13. This tells the system to use vision to determine the location of the part and update the part's position as it moves along the conveyor belt. Vision can also help locate and place parts.

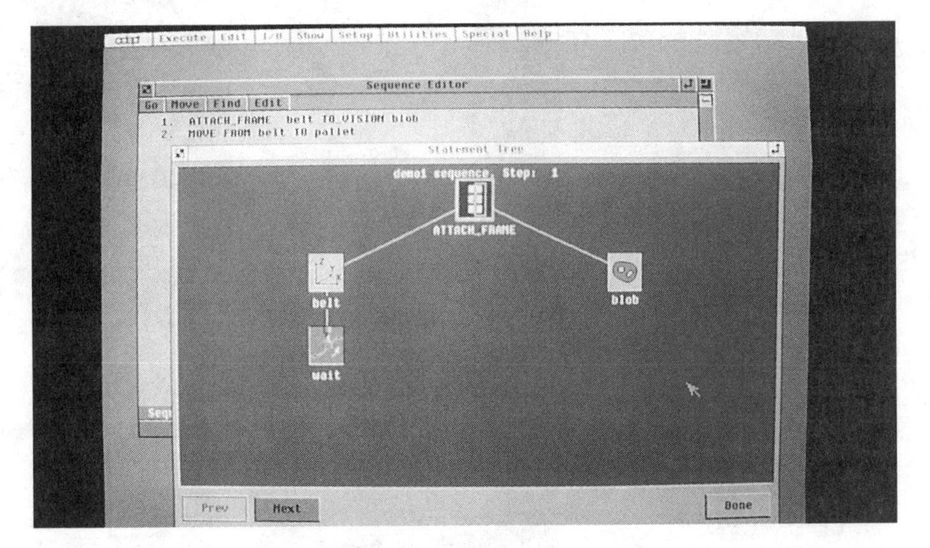

Figure 3-13
The ATTACH FRAME screen is used to direct the system to use its vision capability to determine the location of a part and update its position. (Adept Technology, Inc.)

In order to specify information needed to track the moving conveyor, the screen shown in Figure 3-14 is used. The FRAME RECORD stores the locations of the reference frame as it moves along the belt. This defines the direction the belt is moving. This same screen can be used for indexing conveyor applications. Figure 3-15 shows a screen used to specify the type of vision inspection required to locate and orient the parts. Figure 3-16 shows this vision guidance system tracking parts on a moving conveyor belt.

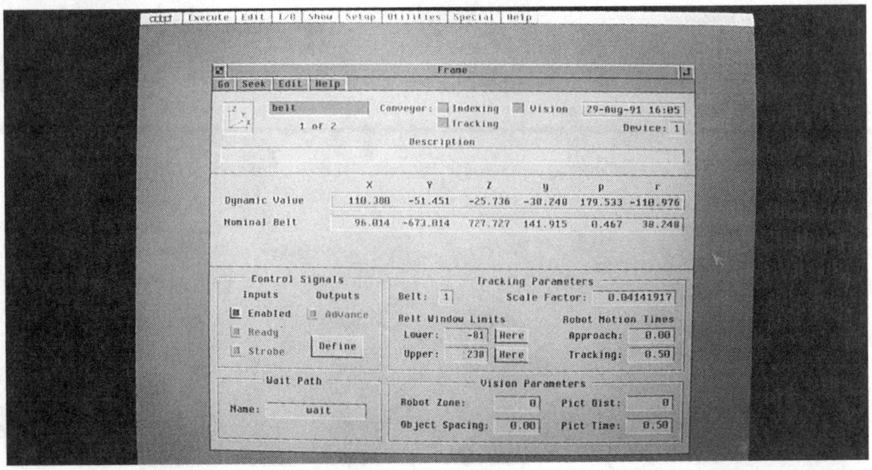

Figure 3-14
The FRAME RECORD screen stores reference information as a part moves along a conveyor belt. (Adept Technology, Inc.)

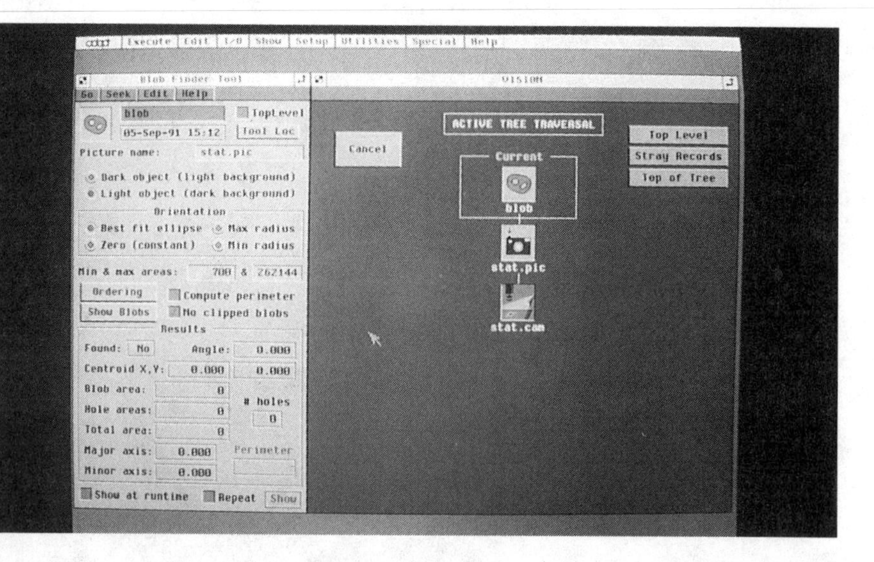

Figure 3-15
This screen is used to specify the type of vision inspection required to locate and orient parts. (Adept Technology, Inc.)

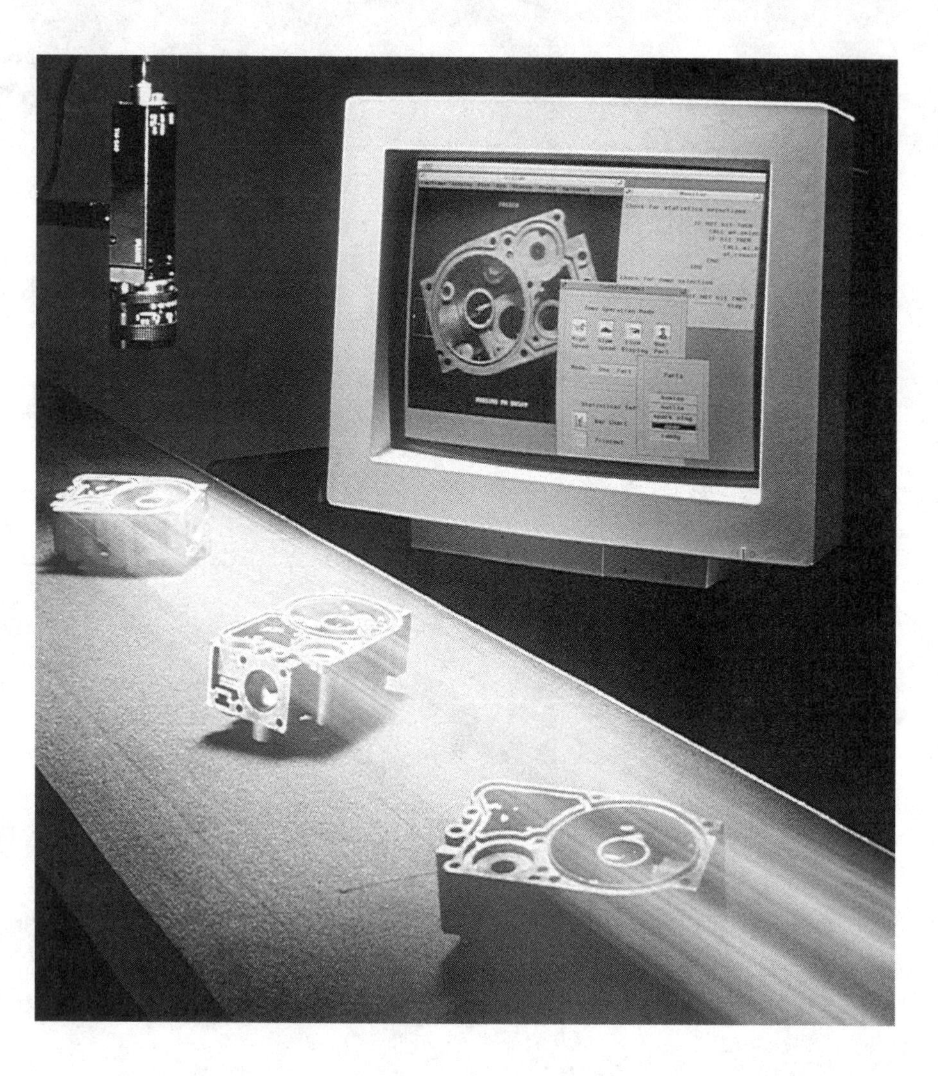

Figure 3-16
Vision guidance tracks parts on a moving conveyor belt. (Adept Technology, Inc.)

The control panel screen shown in Figure 3-17 allows the operator to start the operation, slow the speed of the robot, and step the robot through its motions to be sure the program is performing as intended. The control panel screen is also used for debugging and cell control.

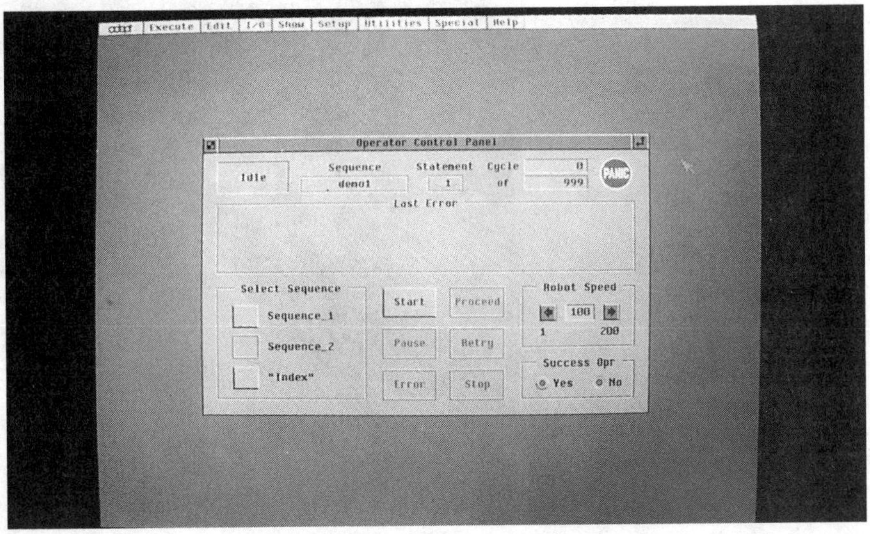

Figure 3-17
The control panel allows the operator to reduce speed, and step the robot through its motions to be sure that the program is performing properly. (Adept Technology, Inc.)

Task-level programming is at the leading edge of currently available programming technology. This highly efficient environment replaces hundreds, even thousands, of lines of programming code with a small number of menu-selected statements.

Voice Recognition

Voice recognition employs the use of a device into which the operator speaks. At this time it is still mainly experimental. As voice-recognition technology increases and costs decrease, this type of control will find more uses outside the laboratory.

The operator shown in Figure 3-18 is repeating a phrase several times into the microphone. The operator's average voice frequency is recorded in memory. Later, recognition of the operator's voice is determined by this frequency. Experts say that voice frequency is just as accurate as fingerprints for identification purposes.

After voice recognition is established, the robot responds to those commands it has been programmed to follow. If a different operator is used, the robot has to be taught to recognize that person's voice. Recognition may be affected if the operator has a cold or sinus problem.

Figure 3-18
For voice recognition, an operator repeats a phrase several times to establish voice frequency and characteristics.

Voice recognition is being considered in the manufacture of items for physically handicapped people. If a person is totally paralyzed, his or her voice may be the only practical means of communication. Probably one of the biggest hindrances has been the cost. Also, two-way communication must be possible. To prevent injury or other damage, the robot must be able to verbally repeat any commands before carrying them out.

Certain advantages make voice recognition attractive. Verbal communication is inexpensive. Voice recognition can reduce set-up time. This is important since manufacturing trends are shifting away from volume and toward variety. Operators may be intimidated by some of the other set-up methods, but with voice recognition, fear is reduced. Voice recognition can truly be described as user-friendly.

In the future, voice communication from the robot may be used to diagnose and warn of possible problems in an operation. If the robot is experiencing trouble with one of its components, the problem could be verbally communicated to the service person.

Important Terms

continuous-path (CP) motion
end-stops
hierarchical control programming
high-level languages
manual programming
manual rate control box
off-line programming
on-line programming
pick-and-place motion
point-to-point (PTP) motion
sensory feedback
subroutine
task-level programming
teach-pendant programming
voice recognition
walk-through programming
WAVE

Review Questions

Write your answers on a separate sheet of paper.

1. Name four methods used to program robots. Briefly describe each method and give an application for each.
2. There are three classifications for robotic motion control. List these classifications and explain the differences.
3. What is the advantage of using a personal computer (PC) to create robot programs?
4. The evolution of programming can be broken down into three periods. List these and discuss the differences.
5. Assume that you are teaching paint spraying to a robot that uses continuous-path motion control. What might happen if you fail to do a good job while in the teach mode?
6. What are some of the advantages of task-level programming?
7. How would programming that incorporates voice recognition be useful in an industrial robot?
8. How are pick-and-place robots programmed?

9. What method is used to convert high-level robot programming languages to machine code that the robot's microprocessor can understand?
10. Most companies use proprietary programming languages for the robots they manufacture. What is the disadvantage of this approach for the end user?

This programmer is using a teach pendant to program a robot for an assembly operation.

4 Industrial Applications

Overview

This chapter discusses applications of industrial robots, beginning with a section on integrating robots into the manufacturing process. This includes design for manufacturability and the technical factors considered when selecting robots. The next section discusses safety where robots are used. Typical uses for robots in industry also are covered.

Integrating Robots into the Manufacturing Process

Automation is a high priority for manufacturing companies worldwide. Competition, improved technology, cost, market conditions, productivity, available manpower, and undesirable work environments are forcing manufacturers to install more robots. How can robots be successfully brought into a manufacturing process? The factors include the design of the manufactured products, selection of the right robot for the job, and safety.

Design for Manufacturability

Design for manufacturability means designing products with robots in mind. In the past, companies have falsely assumed that a robot could be plugged into the production system without any changes in engineering or product design.

First, the engineering department must decide which parts of the product can be assembled by robots, and which need human assembly or ***dedicated equipment*** (a machine that performs only one function.) The next step is to design or redesign the parts of the product for ease of robotic assembly. The list of design tips below is useful to consider when designing a product for robot assembly. Following these tips will allow the robot to work efficiently by moving to a position accurately and consistently.

- Δ Minimize the number of parts to reduce complexity, Figure 4-1.
- Δ Reduce the directions of approach required to assemble the product. This allows use of a less sophisticated robot.

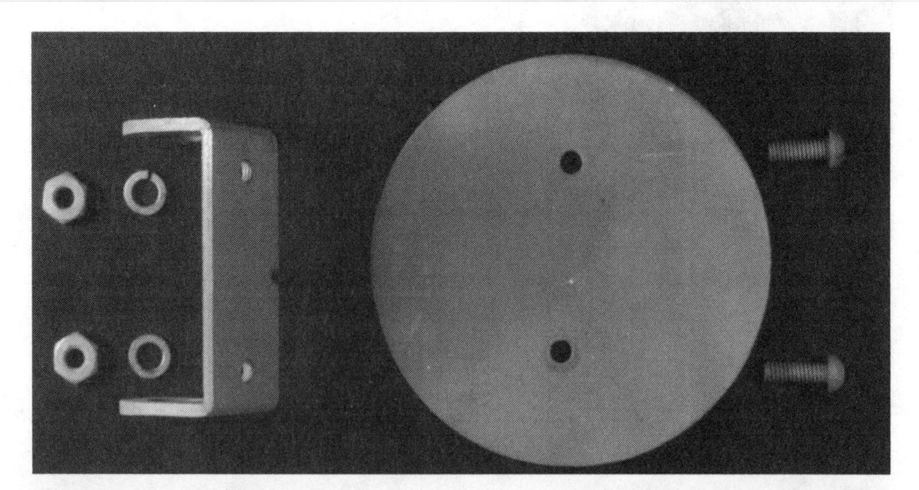

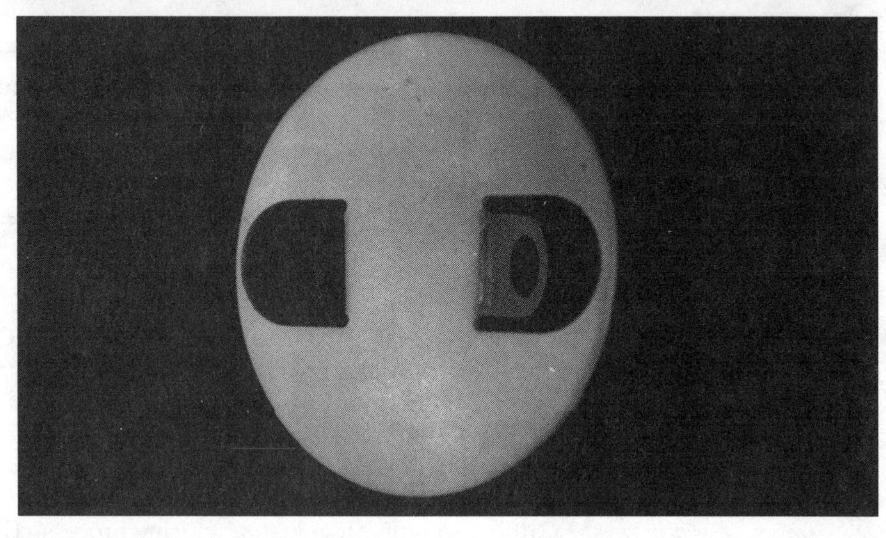

Figure 4-1
A—This subassembly contains a total of eight separate parts, making robotic assembly very difficult. B—Here, the same part has been redesigned. The brackets have been punched out of the original base plate, resulting in an 8-to-1 parts reduction.

Δ Minimize the number of obstructions so that the robot can work in a straight line.

Δ Whenever possible, use subassemblies with components that stack on top of each other or that can be assembled in sequence using downward motion, Figure 4-2.

Figure 4-2
This plastic spacer has been designed so a robot can snap it onto the steel rod using a simple downward motion. Older designs required that the rod be inserted in the hole and the spacer slid into position.

Δ Add chamfers, guide pins, ridges, and other physical characteristics that allow the robot to lock the part into its proper location, Figure 4-3. The cost of these guides is far less than the cost of robot vision systems.

Figure 4-3
The end of this pin has been rounded to allow the robot to easily insert it into the hole on the subassembly without jamming.

Δ Simplify fastening by using tabs, snaps, or other methods that allow parts to be joined in one motion, Figure 4-4.

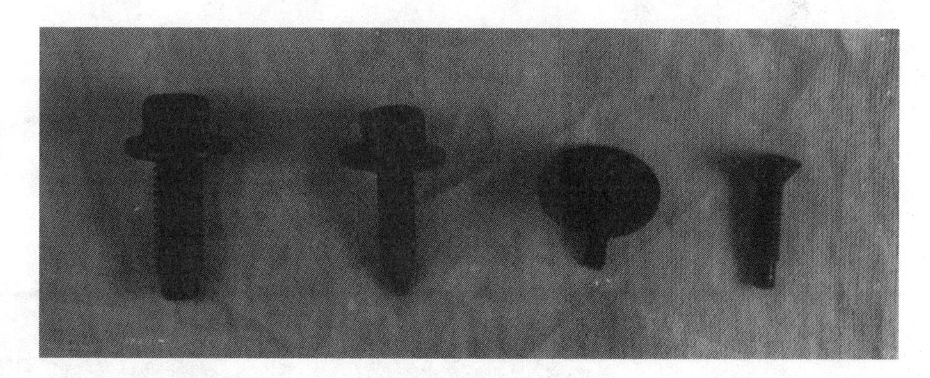

Figure 4-4
These screws have been designed specifically for ease of assembly. Note their "dog-points" (self-tapping designs) and the washer that comes attached to the screw.

Δ Eliminate screws, springs, and adjustments, Figure 4-5.

Δ Avoid compressible parts, such as wires, foils, or foams that robots do not handle well and can deform, Figure 4-6.

Δ Give parts common features. For example, no matter what their outer dimensions, all parts might have the same size hole or post in the center. That way the same end effector can pick up any of them.

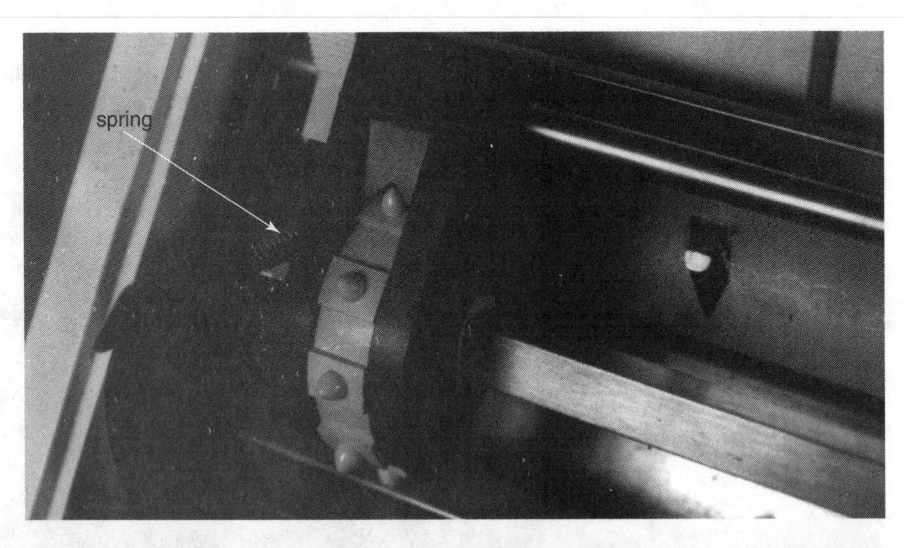

Figure 4-5
A—Note the spring used on this pin wheel assembly for a computer printer. It is very difficult and, in many cases, impossible for robots to install small springs. B—The assembly has been redesigned and the spring has been eliminated, simplifying assembly.

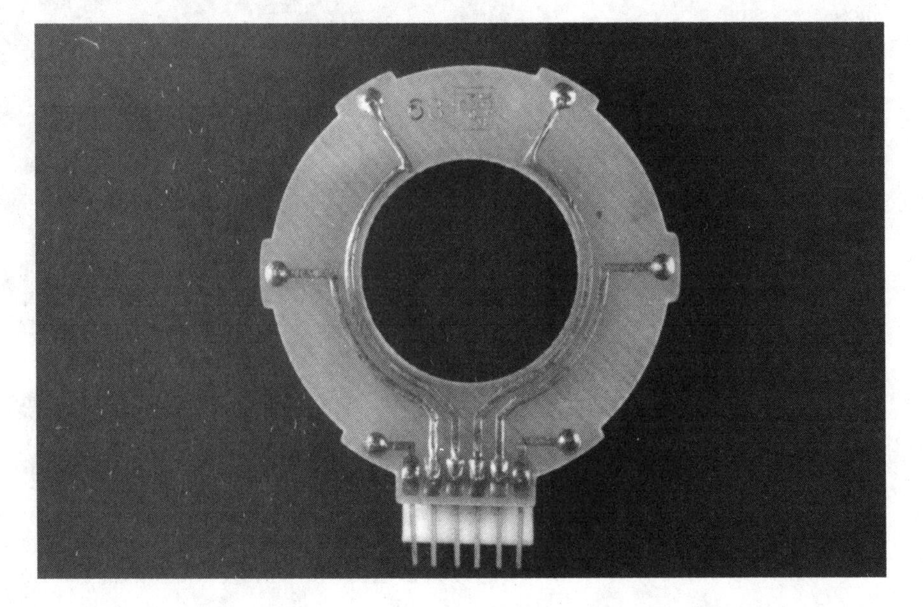

Figure 4-6
This stepper motor part has been designed for automated assembly. The use of printed circuit board techniques eliminates the use of wires that could make robotic assembly difficult.

Δ Make parts as symmetrical as possible so that they can be assembled correctly, no matter how the robot picks them up. See Figure 4-7. If a part must be asymmetrical, use the asymmetrical features to assist in orientation.

Many companies have discovered that when they design for robotic assembly, the process becomes so simple that it is faster and more productive to use human workers or dedicated machines than robots. If demand is sufficient, all three methods might be used.

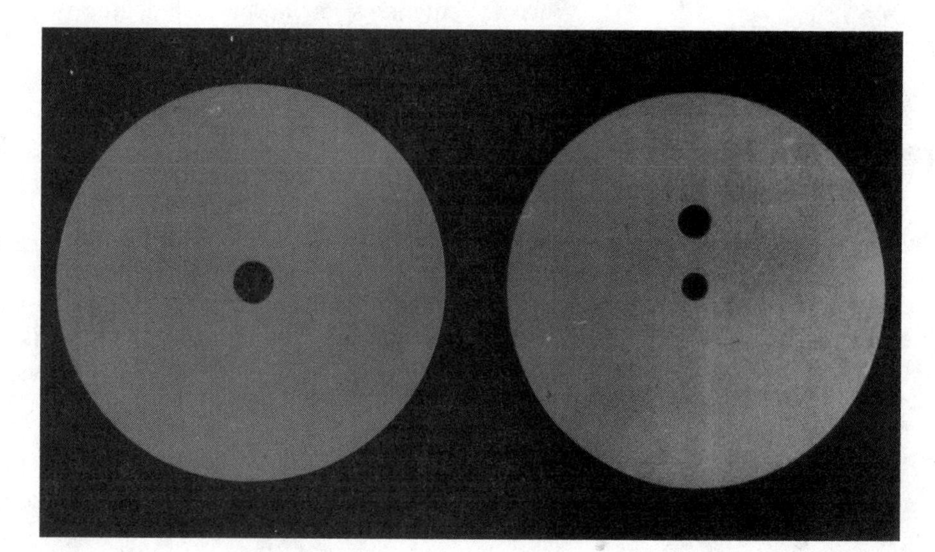

Figure 4-7
The base plate on the right contains asymmetrical holes, which means that it must be installed using one particular orientation. The redesigned base plate on the left is symmetrical and does not require any particular orientation during assembly.

Selecting the Right Robot

With hundreds of different robot models available today, selecting the right robot can be a major task. Robots occupy a position between human workers and equipment designed to perform one specialized function. Robots can perform repetitive tasks without tiring. They can operate in hostile environments. They can maintain greater precision while doing repetitive work. There are two main areas to which robots are not suited, however. They should not be used for long, high-speed runs producing many identical items. Dedicated machinery is better for that. They should not be used for short, complex tasks that require a high level of hand-eye coordination. Currently, these jobs are still best suited to human workers. Otherwise, robots are useful for many production jobs. In addition to environmental conditions and coordination with associated equipment, some of the primary things to consider when selecting a robot are the work envelope, degrees of freedom, and other measures of performance.

The Work Envelope

As discussed in Chapter 2, the work envelope is different for each robot configuration. All of the fixtures and other machinery that will work with the robot must be located within the work envelope. Many times, placement of other support equipment dictates its shape. To provide greater flexibility, some manufacturers offer optional wall and ceiling mounts in addition to the robot's standard floor mounting.

Degrees of Freedom

The number of degrees of freedom determines the robot's ability to move within the work envelope. Robots with six degrees of freedom can work anywhere within their work envelopes. However, not all applications require so versatile a robot. Each individual workstation will determine the requirements. Sometimes, an additional degree of freedom can be obtained by mounting the robot on rails. This permits it to move horizontally, increasing its overall reach.

Other Measures of Performance

In addition to degrees of freedom and configuration of the work envelope, robots can be classified on the basis of resolution, accuracy, repeatability, operational speed, and load capacity.

Resolution

Resolution is determined by the robot's control system. It deals with precision, or the smallest incremental movement the robot can make. It can also be described as the smallest segment into which the work space can be divided. ***Command resolution*** is calculated by dividing the travel distance of each joint by the number of control increments. For example, if the robot travels a distance of 36 in. (91 cm) along one axis, and the controller is programmed with a total of 8000 points, then the command resolution, or closest distance between movements, is 0.005 in. (0.013 cm), without taking into account any mechanical inaccuracies.

The manipulator uses mechanical components to position the end effector at the various points in the work envelope. Whenever mechanical members are used (gears, chains, cables, or ball lead screws), certain inaccuracies are present. Gears are prone to backlash, and chains and cables stretch. When this happens, *slippage* occurs. Also, overweight payloads can create certain inaccuracies. All these factors can affect resolution.

Positioning of the tool during programming is important. A robot must be capable of positioning the tool in order to accomplish the desired task. Tool position will vary because of the command resolution and mechanical inaccuracies. ***Spatial resolution*** describes the movement of the robot at the tool tip. Spatial resolution takes into account command resolution and mechanical inaccuracy. For example, if the command resolution of 0.005 in. (0.013 cm) has a mechanical inaccuracy of 0.003 in. (0.008 cm), then the spatial resolution would be 0.008 in. (0.020 cm). This figure is derived by adding the command resolution to the mechanical inaccuracy.

In the work envelope, spatial resolution varies with tool location. Its consistency throughout the work envelope is affected more by certain robot configurations than by others. The Cartesian configuration offers fairly constant spatial resolution. Rotary joints can affect spatial resolution and must be corrected.

Accuracy

Accuracy and spatial resolution are both by-products of command resolution and mechanical inaccuracies. ***Accuracy*** expresses how closely the robot's hand can be programmed to hit a desired point. Accuracy can be expressed as half of the spatial resolution. A robot with a spatial resolution of 0.010 in. (0.025 cm) would have an accuracy of 0.005 in. (0.013 cm).

Large robots with payloads of 100 lbs.(45.4 kg) or more have an average accuracy of ± 0.050 in. (0.13 cm). Small robots used for such tasks as assembly operations, where payloads are lighter, have average accuracies of ± 0.002 in. (0.005 cm). Improvement in accuracy will increase with improvements in technology.

Accuracy can be affected by the speed of movement and weight of the payload. As speed increases, accuracy decreases. For the manipulator to stop at a given location, speed has to be reduced to prevent overshooting that position. However, if the speed is reduced too much, valuable time is wasted. Also, the inertia overcome in moving a heavy load can affect positioning accuracy. To overcome inertia, speed may have to be reduced. As a result, there must be a trade-off between speed and accuracy.

Repeatability

The robotic novice often confuses accuracy with repeatability. Even though both are dependent on spatial resolution and mechanical inaccuracy, there is a difference. While accuracy deals with programming the robot's hand to go to a designated position, ***repeatability*** expresses how close it will actually return to that position time and time again.

Good repeatability is more desirable than accuracy. This is because inaccuracies are easier to correct. This is especially true if the inaccuracies are consistent for all moves. For example, suppose that a robot is programmed to move its gripper from point A to a target point 30 in. (76.2 cm) away. After the robot has made the move, an actual measurement is taken, which shows the robot actually moved 30.10 in. (76.5 cm). This represents an inaccuracy of 0.3 percent. If an inaccuracy of 0.3 percent is consistent for other command movements, then the programmer can compensate for this error. Adjustment for poor repeatability is more difficult. In fact, repeatability may be so poor that a more sophisticated robot may be required. However, if the error is rather small, additional tooling or alignment devices can be used to compensate.

Repeatability can change with use, especially when robots perform the same task day after day. The mechanical components are subject to wear, thus increasing mechanical inaccuracies. This increase in mechanical inaccuracies reduces the repeatability performance.

Operational speed

Operational speed, also called ***dynamic performance***, has to do with how fast the robot can accelerate, decelerate, and stop at a given point. Probably the two most important factors that influence operational speed are the desired accuracy and the payload. Other factors include the robot configuration and the location of the tool in the work envelope.

Robot manufacturers define robot speed in different ways. It may be given for each joint or various groups of joints. The range of velocity may be for no load as well as for a full load.

Since people consider robots more productive than humans, it is natural to assume that robots work at a faster pace. However, a robot's actions often are no faster than those of a human worker. In fact, many times they are slower. Increased productivity results because the robot works at a constant, steady pace and doesn't stop for coffee breaks or lunch. Therefore, basing the total cycle time on speed rates furnished by manufacturers may not yield the true cycle time. Most users want to know how fast they can move the end effector through the total cycle and still maintain the desired accuracy and repeatability. The concern is with the robot's ability to do a quality job in a reasonable time.

To arrive at cycle times that are realistic, a prototype layout may have to be built. Since several variables, such as other performance measures, machine cycle times, and conveyor speeds, must be considered, the prototype layout is probably the best overall solution.

Load capacity

The ***payload*** is the maximum weight or mass of material a robot is capable of handling on a continuous basis. The end effector weight is included in the payload in most cases. Some manufacturers' specifications indicate whether the robot's arm is extended or retracted. Other manufacturers may list load-carrying capacity for the arm as well as for the wrist joint and end effector. However, the two factors that greatly affect load-handling capabilities are the type of configuration and the placement of the end effector within the work envelope. The robot's load-handling capability is less with its boom fully extended than when the boom is retracted.

Some robots are capable of lifting as little as 1 lb. (0.45 kg). Other robots might have a lifting capacity of 2000 lbs. (907 kg). In most instances, the weight of the end effector is included in the lifting capacity. If a robot has a lifting capacity of 5 lbs. (2.27 kg) and the gripper weighs 2 lbs. (0.9 kg), then the weight of the part cannot exceed 3 lbs. (1.36 kg). Most robot manufacturers generally list both normal and maximum load-handling capacities.

The average weight carried by robots in U.S. industry is estimated to be 20 lbs. (9.07 kg). Approximately 50 percent of robots in industry today handle parts weighing less than 10 lbs. (4.54 kg). The average load-carrying capacity could be even less in the future, since smaller robots are being constructed for assembly tasks. These tasks seldom involve heavy payloads. For example, 95 percent of all parts used in an automobile weigh less than 5 lbs. (2.27 kg).

Robotic Safety Considerations

An important consideration in installing robot systems is *safety*. Robots differ from other machinery because of their degrees of freedom and often-large work envelopes. Providing a safe working environment must be considered during the design, installation, maintenance, and operation of robot systems. A robotic work station can be dangerous if workers are not properly trained or not made aware of possible hazards.

In the United States, there have been several fatal accidents involving robots. In 1984, an operator was killed when pinned between a robot used for die casting and a steel pole. Several years earlier, another worker was killed when struck from behind by an automated guided vehicle (AGV). The world's largest user of robots, Japan, has reported five worker deaths involving robots.

These accidents were similar in three ways:

Δ All the victims were experienced operators trained in the safe, correct operation of robots.

Δ All the victims entered the work envelope while the robot was in operation.

Δ All the victims were struck from behind or pinned against a stationary object, Figure 4-8.

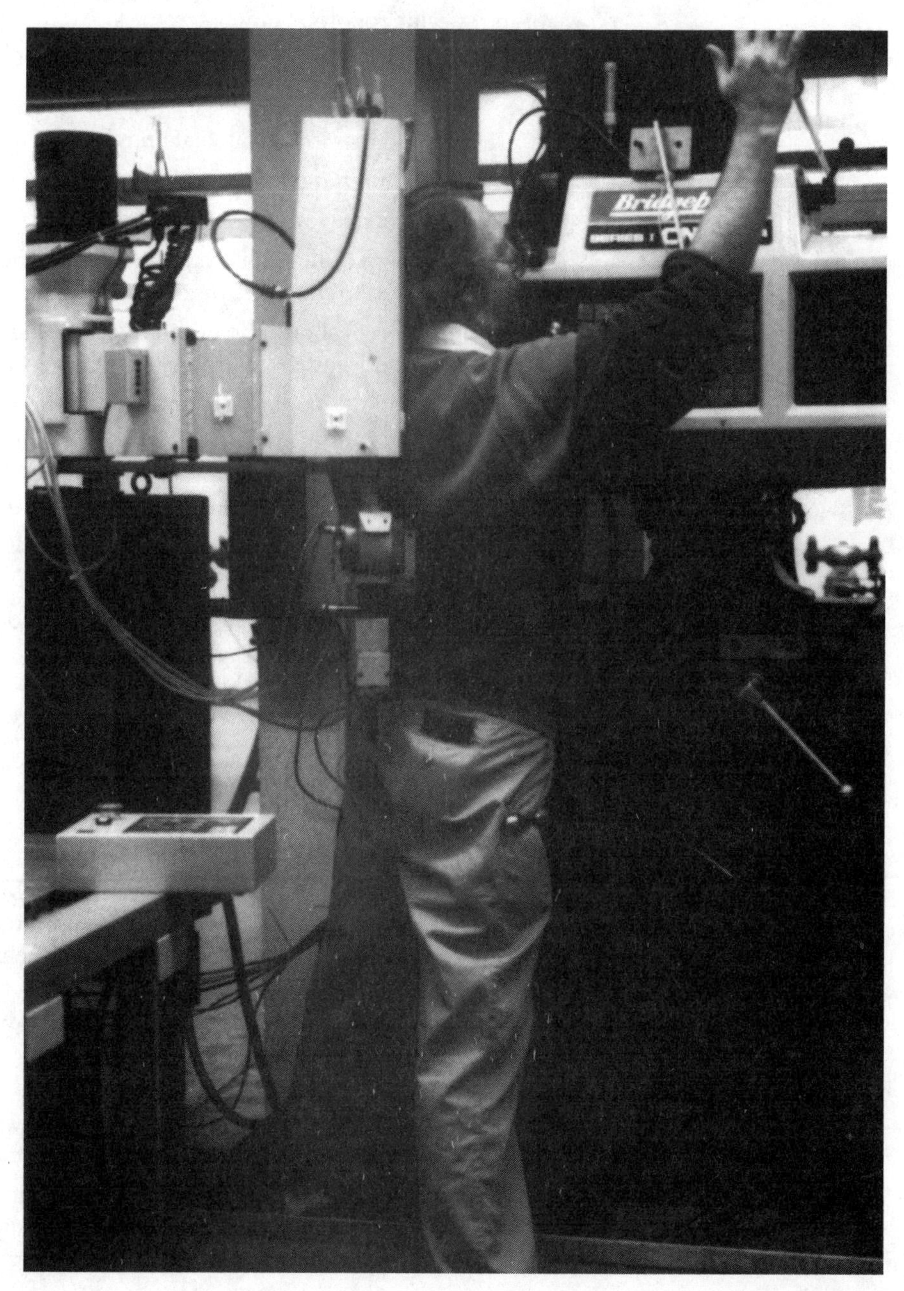

Figure 4-8
In this simulated accident, the operator is pinned between the robot's manipulator and a CNC milling machine. The result could be a serious, or even fatal injury.

Two organizations offer limited guidelines for the safe operation of robots. The National Institute for Occupational Safety and Health (NIOSH) publishes "Request for Assistance in Preventing the Injury of Workers by Robots." The Robotic Industries Association (RIA) makes available "American National Safety Standards for Industrial Robots and Industrial Robot Systems."

General safety guidelines

Safety considerations should begin with research into the characteristics of the specific robot(s) being used.

- Δ What is the size and shape of the robot's work envelope?
- Δ What methods of motion control are used?
- Δ What are the limits for a safe payload?
- Δ What is the range of operating speeds?
- Δ What other features are provided by the manufacturer?

The design of the work station should reflect safety. Particular attention should be given to the following:

- Δ Pinch points should be omitted. This means being aware of fixed objects within the work area, such as vertical poles.
- Δ Space should be planned for the end effector. The relationship of the end effector to work fixtures and other machinery or equipment must be accounted for.
- Δ Inadequate barriers that can be passed through or over should be improved.
- Δ Interlock access to the robot controller should be available.
- Δ Two-step procedures in resuming operation after a shutdown should be used.
- Δ The design should be documented. This is important to purchasing, operating, and maintenance personnel.

During implementation, changes made during installation and debugging should be recorded. Control panels should be located outside the robot's work envelope. Marks outlining the full range of the end effector should be placed on the floor using paint, colored tape, or other highly visible material.

Training of employees and a conscientious effort by management to provide safe working conditions are important ingredients in any safety program. Operating personnel, programmers, maintenance workers, and clean-up crews need extensive job- and robot-specific training from the beginning. Periodic refresher courses should be scheduled for as long as robots are in use.

Safety barriers and sensors

Safety measures for a robot are more complex than for other kinds of equipment. The two primary methods used are barriers and sensing systems.

One of the safest barriers is a fence, Figure 4-9. A safety fence prevents unauthorized entry and also safely contains flying workpieces. A safety fence is usually a minimum of 6 feet high and constructed of wire mesh, glass, or

Figure 4-9
A steel safety fence surrounds this arc welding robot. The control for the robot is mounted outside the fence. The robot cannot be started until the operator steps outside and closes the gate.

rigid plastic sheets, Figure 4-10. In toxic applications, such as paint spraying, a solid enclosure can also protect personnel from chemicals.

However, fencing makes the robot more difficult to move and work cell equipment less easy to rearrange. Fences take up valuable floor space and often obstruct the view of critical operations. Fences also make it more difficult to teach the robot or perform preventive maintenance.

Another kind of barrier is the infrared ***light curtain***, which consists of photoelectric presence-sensing devices. See Figure 4-11. If a worker enters the work envelope, power to the robot is cut off. A light curtain is programmable, allowing selective blanking of some areas. Certain equipment that is not

Figure 4-10
Glass or rigid plastic panel enclosures are also sometimes used to protect personnel.

within the reach of the end effector can be accessed. One drawback of the light curtain is that moving equipment can trip the system.

One of the simplest sensing systems is a ***switch carpet*** placed in dangerous areas. See Figure 4-12. When a worker steps on the carpet, embedded switches automatically turn off dangerous machinery.

Multiple sensing systems make use of a combination of capacitance, infrared, ultrasonic, or microwave sensors to detect a worker's presence and track that presence within the work envelope. The power is cut off whenever a worker enters the robot's work envelope.

Figure 4-11
The equipment shown here projects a curtain of light. When the light is interrupted, power to the robotic equipment is shut down.

Uses for Robots in Industry

As robot technology advances, areas where a robot might be used are probably limited only by the user's imagination and creativity. A completely automated factory is still the ultimate goal of many scientists and engineers doing research today. However, most of the robots installed in our present factories are not highly intelligent. To a large extent, they employ unsophisticated sensing systems. These factors limit their ability to sense and make decisions

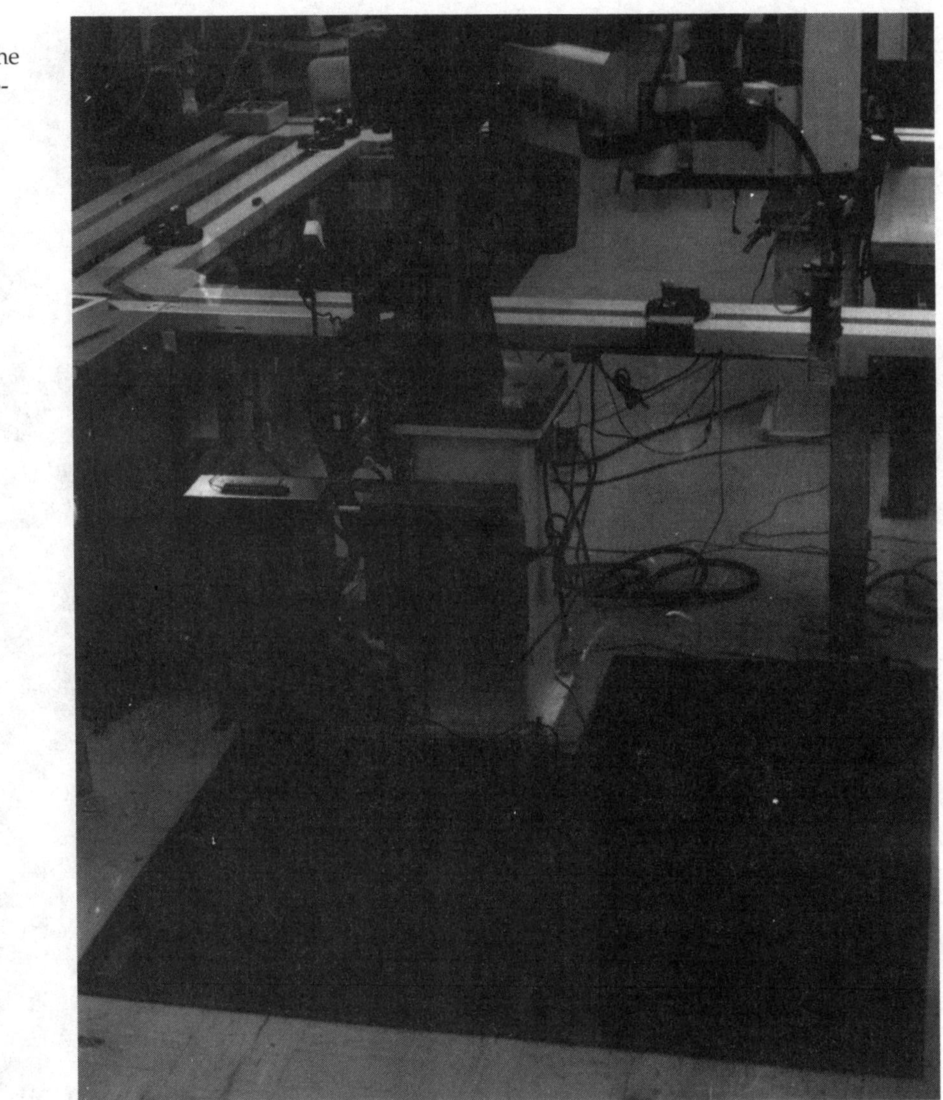

Figure 4-12
A carpet switch turns off the power to the robotic equipment whenever someone steps on it.

about objects in their environment. As newer robots equipped with vision and other complex sensing systems are brought into the workplace, the potential uses for robots will increase dramatically.

Over the years, a number of industrial applications for robots have proven to be very suitable and economically sound. For example, about 30 percent of the installed industrial robots in the world are used in the automotive industry. As a general rule, industrial robots lift heavy loads, work with dangerous materials, function in dangerous or undesirable environments, or perform work that is mundane and highly repetitive.

This section will cover a broad range of applications, from the classic uses to those that are more recent. It does not, however, include *all* the uses to which robots are being put.

Pick-and-Place

The process of picking up parts at one location and moving them to another is one of the most common robot applications. Placing parts or removing them from a uniform series of positions (***palletizing*** or de-palletizing, for example) is probably the most common form of pick-and-place. Some pick-and-place operations also are used for part orientation. Figure 4-13 shows the components used in a robot palletizing operation. Other examples can be seen in Figure 4-14.

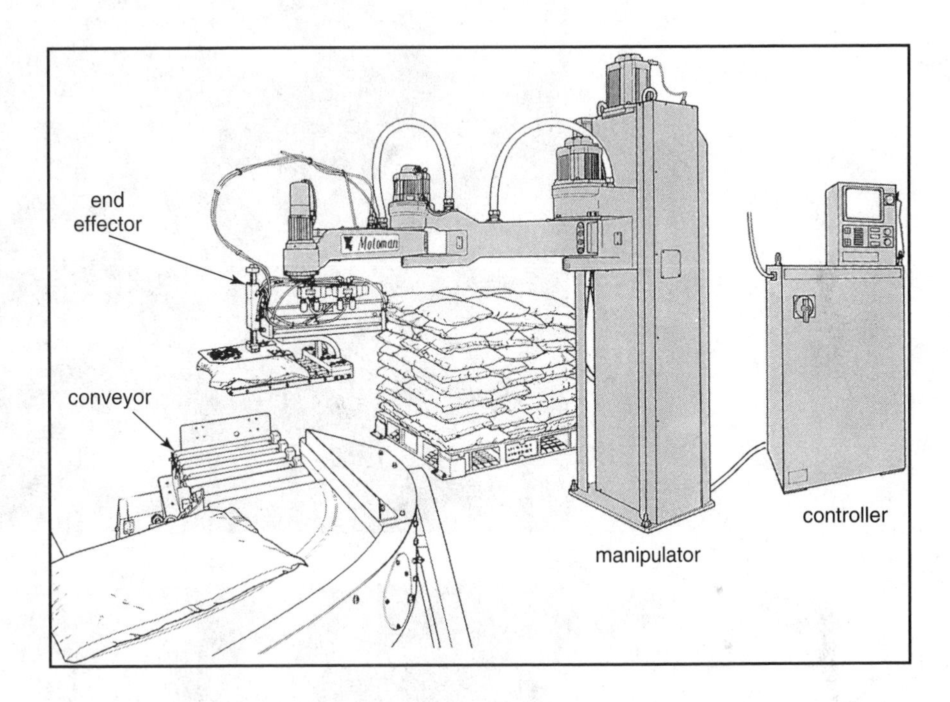

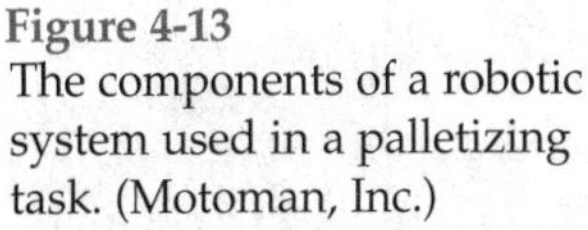

Figure 4-13
The components of a robotic system used in a palletizing task. (Motoman, Inc.)

The less-sophisticated robots used for pick-and-place offer several advantages. They have been very successful in handling fragile parts made of glass or powder metal. Those used in lightweight pick-and-place applications offer excellent speed while maintaining good accuracy and repeatability. Robots are also useful for handling heavy and very hot or very cold items.

Machine Loading/ Unloading

Possibly the second most popular use of robots in industry is machine loading and unloading. In production, robots load and unload parts associated with automated machining centers, such as CNC (computer numerically controlled) machines, Figure 4-15. A robot's machine loading and unloading skills can be applied to such operations as forging, injection molding, and stamping.

Industrial operations often expose workers to excessive heat, noise, dirt, and air pollution. Since these operations are considered unpleasant and are frequently dangerous, they are ideal choices for the use of robots. Replacing

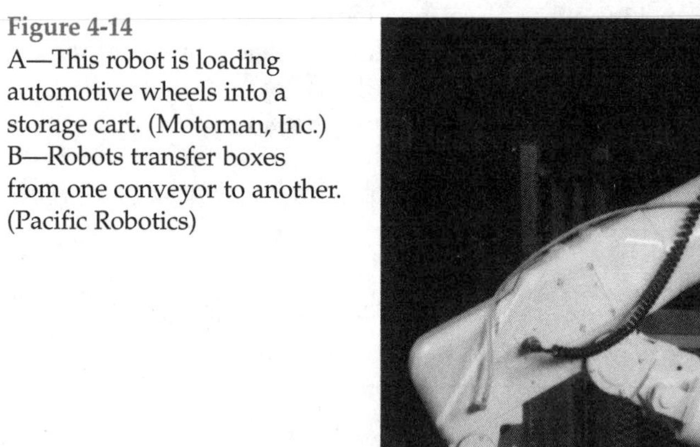

Figure 4-14
A—This robot is loading automotive wheels into a storage cart. (Motoman, Inc.)
B—Robots transfer boxes from one conveyor to another. (Pacific Robotics)

A

B

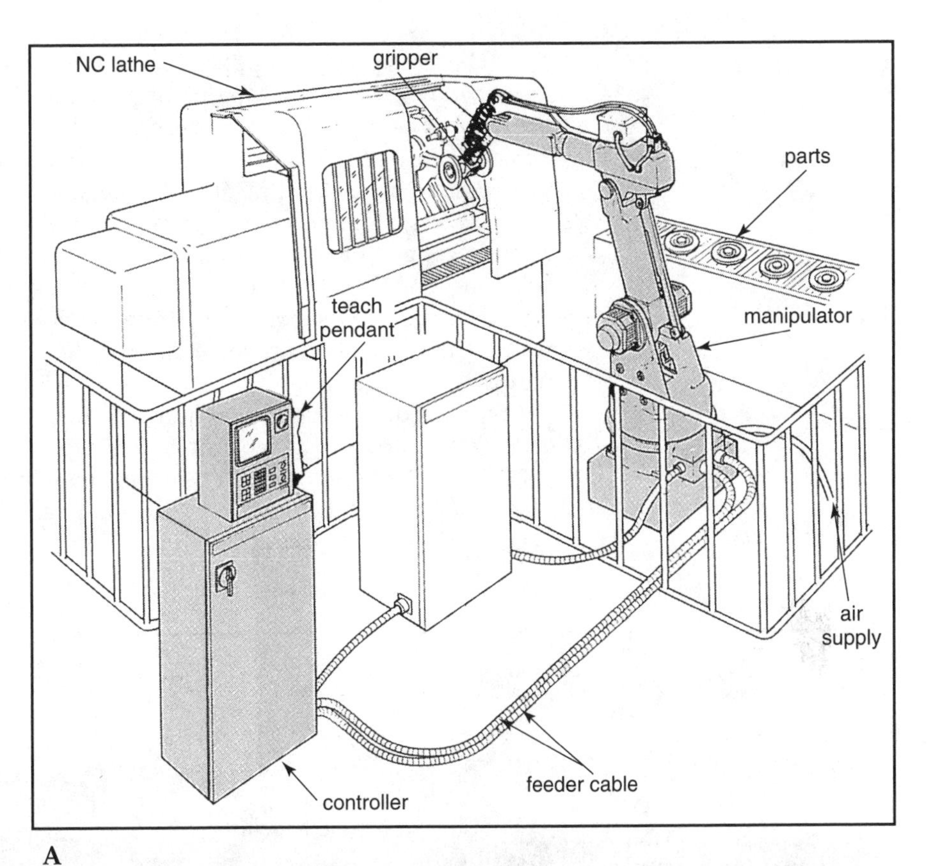

A

B

Figure 4-15
A—This robotic system is used for loading/unloading an NC lathe. (Motoman, Inc.)
B—A robot being used to load and unload parts for a machining operation.

human operators with robots reduces the need for expensive safety equipment and eliminates costs involved in cleaning up the work environment. Also, unpleasant or dangerous jobs usually have high absenteeism rates. This contributes to low productivity and poor product quality. Substantial gains in productivity and product quality generally result when robots are properly used.

Die Casting

The first practical application of industrial robotics occurred over 30 years ago when General Motors installed the Unimate robot in a die-casting operation. The use of robots in die casting has resulted in a 200 percent to 300 percent increase in productivity. After robots proved useful in die casting, they were used in investment casting, forging, and welding operations.

Die casting involves the pumping of molten metals, such as zinc, aluminum, copper, brass, or lead, into closed dies, Figure 4-16. After the metal solidifies, the die is opened and the casting is removed. The function of the robot is to remove the hot casting from the die and dip it into a liquid, usually water. This operation is called *quenching*. Next, the robot transfers the cooled part to a trimming press where excess material ("flash") is removed and the part is shaped. The robot then transfers the part to a pallet or conveyor belt, which takes it out of the work area.

Robots can work more consistently than humans with such hot castings. The increase in productivity is the result of decreasing the cycle time required

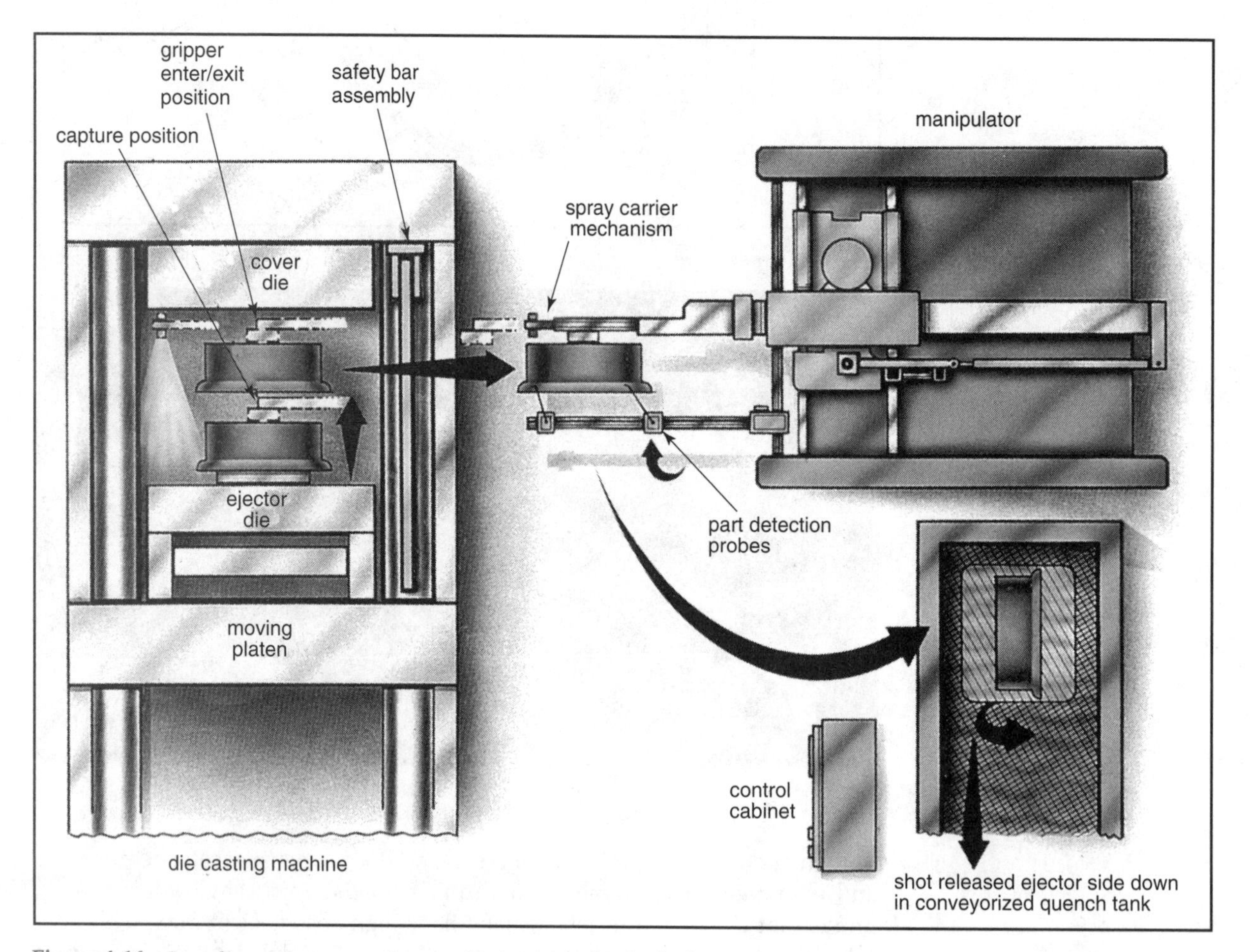

Figure 4-16 In a die-casting operation, a robotic arm unloads the hot casting from the press and moves it to the quench tank. (Sterling Detroit Company)

to produce a finished casting. Another benefit is that the reduced cycle time results in a more uniform die temperature. Less flash material becomes attached to the casting, and trimming and scraping costs are reduced.

The robots perform what is essentially a machine loading/unloading operation. It does not require critical or complex motions and is ideally suited for low-cost pick-and-place robots.

Welding

The third largest group of robots in industry are those used in resistance welding and arc welding operations. In ***resistance welding,*** electric current is passed between two metals, causing them to heat and fuse together at that point. Spot and stud welding are classified as resistance welding. See Figure 4-17. In ***arc welding,*** the weld is made along a joint rather than at one spot. The arc-welding process heats the metals until they melt at the joint and fuse into a single piece, Figure 4-18. Stick, MIG, and TIG are arc-welding applications.

Figure 4-17
A robotic spot welder at work in an automotive assembly plant. (ABB Robotics, Inc.)

Arc welding

MIG (metal-inert gas) is the most common arc-welding process. It requires an arc-welding machine, a wire (electrode) feeder, a shielding gas, a gas flow meter, and a welding gun, Figure 4-19. The wire is a consumable metal electrode. Current passing through the electrode is controlled by the electrode's speed of consumption. An inert gas is used to shield the weld from the atmosphere.

Many of the problems associated with robotic MIG welding are mainly due to poor joint fitup. Robotic arc welding requires more accurate fixturing and better-fitting joints than welding done by human operators. If joint loca-

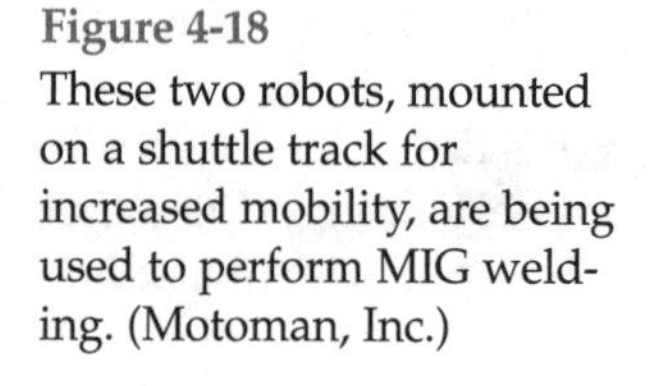

Figure 4-18
These two robots, mounted on a shuttle track for increased mobility, are being used to perform MIG welding. (Motoman, Inc.)

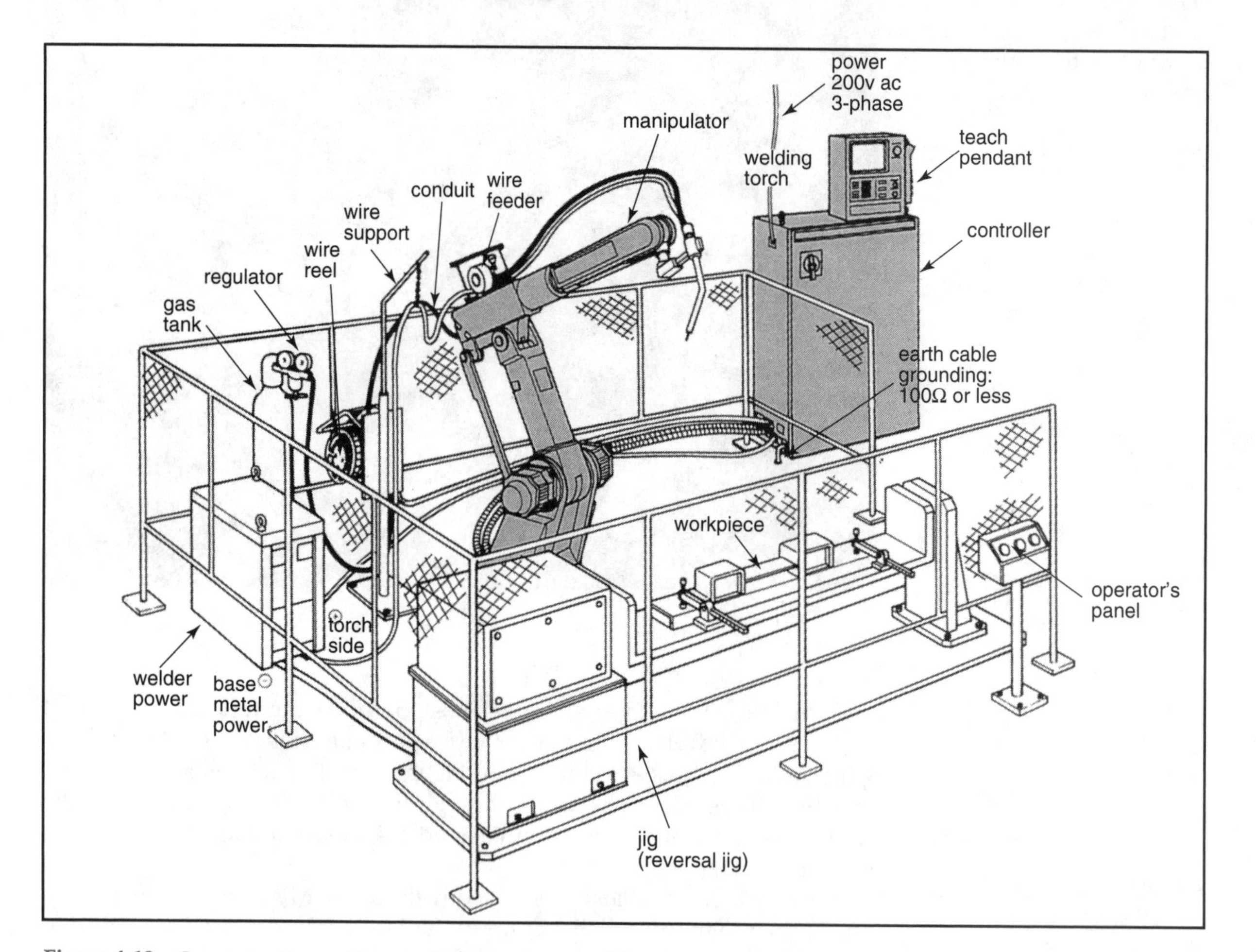

Figure 4-19 Components used in a typical robotic arc welding station. (Motoman, Inc.)

tion or the gap between parts varies, a human being can make adjustments, but most robots cannot.

Compensation for gap variation may require a change in voltage, wire feed rate, travel speed, and weave motion. To remedy the problem, robot manufacturers have developed various sensory tracking systems. One example of a robot system equipped with a visual arc sensor is shown in Figure 4-20. The visual sensor tracks the joint and compensates for deviations.

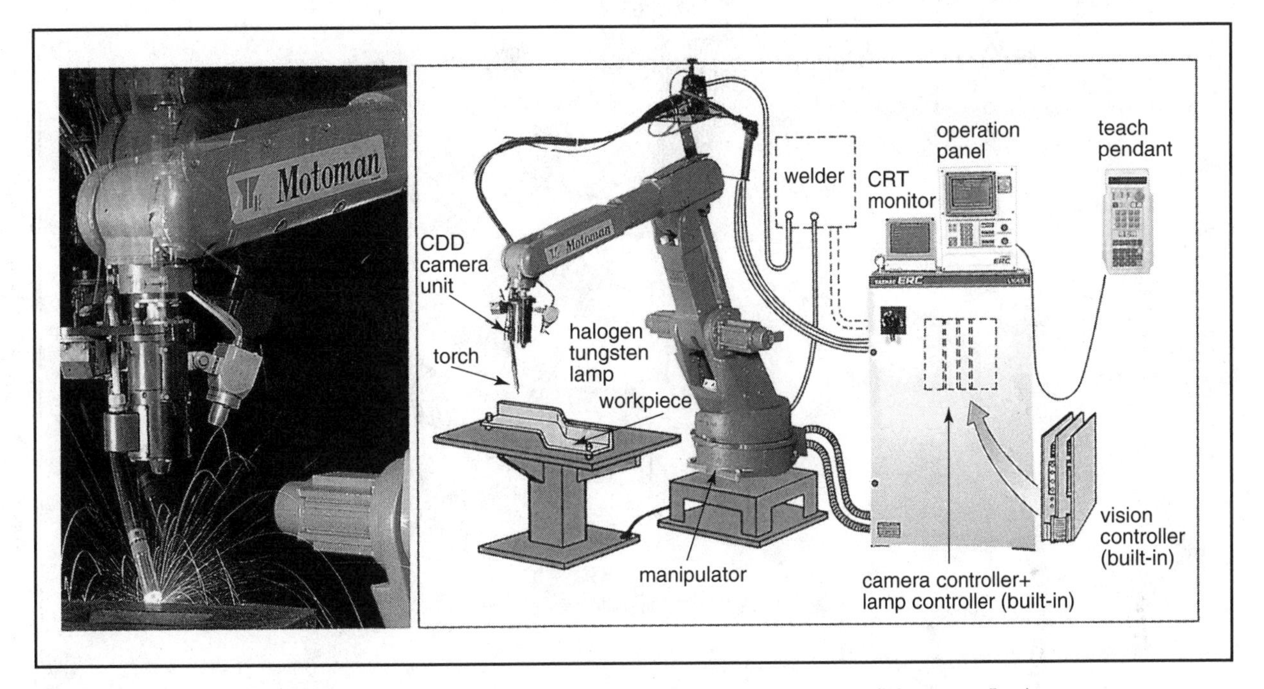

Figure 4-20 The visual sensor on this robot compensates for deviations in a joint. (Motoman, Inc.)

Resistance Welding

Of all the welding processes, spot welding is the easiest to perform and the most common. It can be done on stationary parts or on parts moving down a production line. Robots used for spot welding must return repeatedly to an exact spot. Spot welding is used extensively in the automotive industry to weld body sections and other parts together. A large work envelope is generally needed for these large and complex workpieces.

The workpieces must be pressed together at the spot where the weld is applied. This is done by the welding electrodes, which exert a force of approximately 800 to 1000 psi (pounds per square inch). Next, a low-voltage direct current is passed through the parts, causing them to fuse. The electrodes themselves do not melt. Figure 4-21 shows the components found in a typical spot-welding work station.

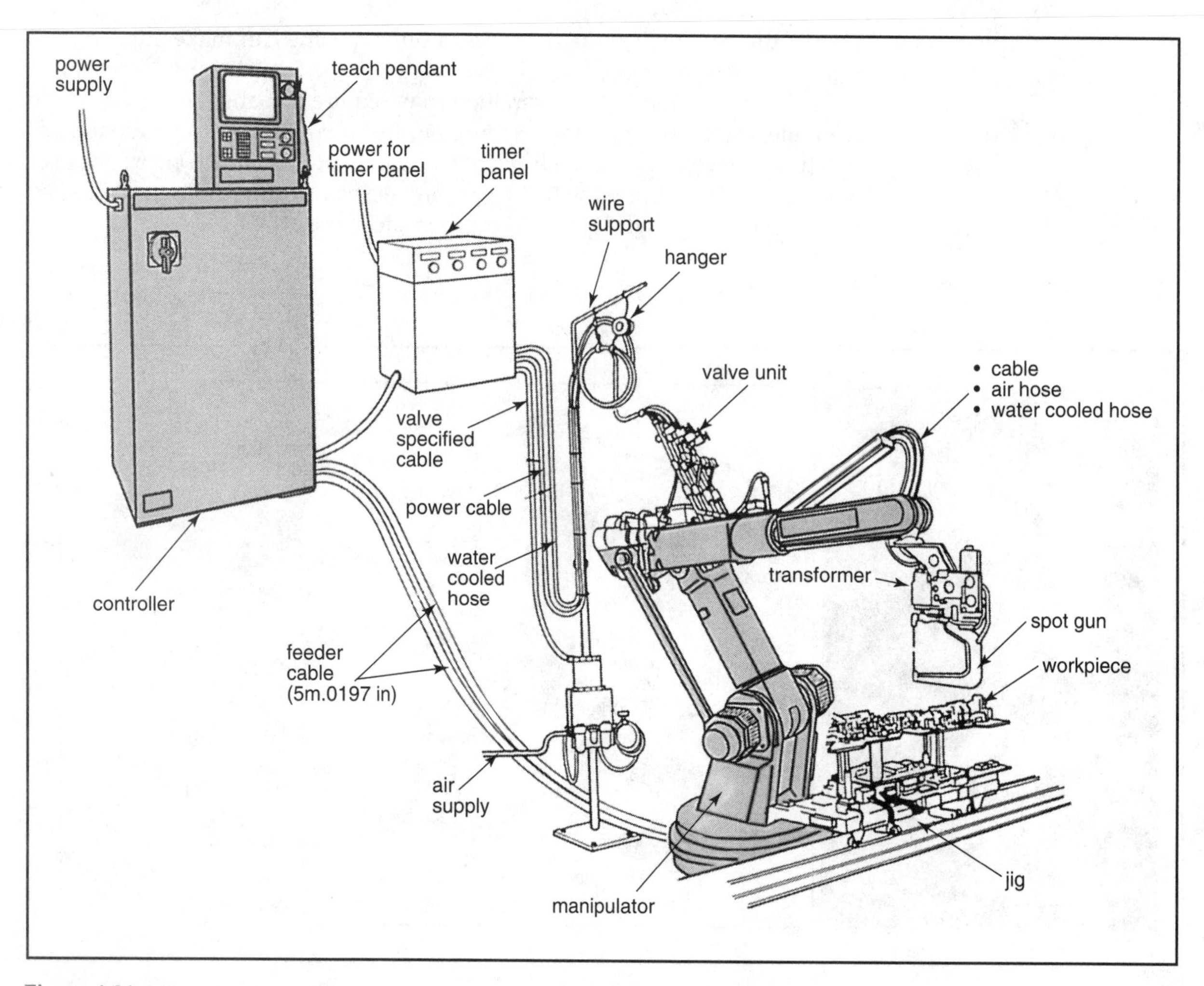

Figure 4-21 This robotic work station is set up for spot welding. (Motoman, Inc.)

Spraying Operations

Painting or ***spray finishing*** involves the application of a variety of paints, polyurethanes, or other protective coatings to the surface of a part or product, Figure 4-22. Other spraying operations are sealing and gluing.

The spray nozzle is mounted on the robot's wrist. Continuous-path programming must be done by an experienced operator because positioning is recorded on a time sample basis. Both intentional and unintentional moves will be recorded. A smooth motion is more important than the actual point location, Figure 4-23.

Several benefits can be realized by using robots in spraying applications. Spray vapors are often toxic and explosive, so expensive ventilation systems must be installed for human workers' health and safety. Since robots do not need fresh air to breathe, greater concentrations of solvent can be used. With the reduction of ventilation, energy costs are lowered. However, there is the danger of an explosion caused by a spark from the robot's electrical system. In the past, most painting robots have been

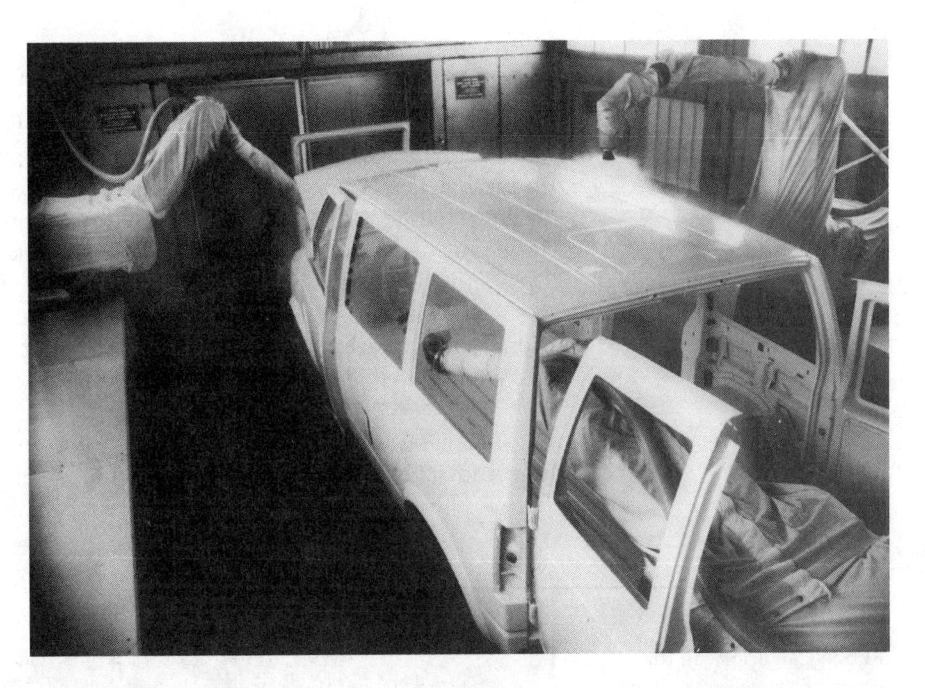

Figure 4-22
Painting robots spray a van in a typical automotive spray painting work cell. Note the third robot working inside the vehicle. (GMF Robotics)

Figure 4-23
Smooth motion of the robot arm is required for a smooth finish on this jet ski. (Kawasaki)

hydraulically powered for this reason. Today, spark-free electrical robots are available for explosive environments.

The use of robots for sealing and gluing is becoming more widespread as the quality of sealing and gluing agents and application methods has improved. See Figure 4-24. The components associated with a typical robotic sealing/gluing work station are shown in Figure 4-25. Path control is crucial in sealing and gluing work. Robot requirements are the same as for continuous-path arc welding applications. Sealing and gluing operations like the one shown in Figure 4-26 require specialized end-of-arm tooling for three-dimensional surfaces. Also, the robot's speed must not change between programmed points or glue will be dispensed unevenly.

Machining Processes

Routing, cutting, drilling, milling, grinding, polishing, deburring, riveting, and sanding are some of the machining processes performed by robots. These robots must have a high degree of repeatability. They must be combined

Types of Sealing/Gluing

bead form	purpose	material of bead
(triangle bead) 	adhesion	urethane system
(dripping) 	rustproof (mastic)	thermosetting
(round bead) 	rustproof (hemming)	thermosetting
(triangle bead) 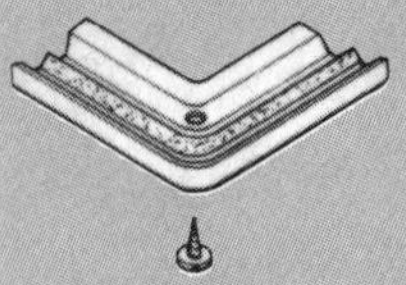	waterproof	hot butyl system
(round bead) 	waterproof	hot butyl system

Figure 4-24
Robots can apply all these types of beads for sealing and gluing tasks. (Motoman, Inc.)

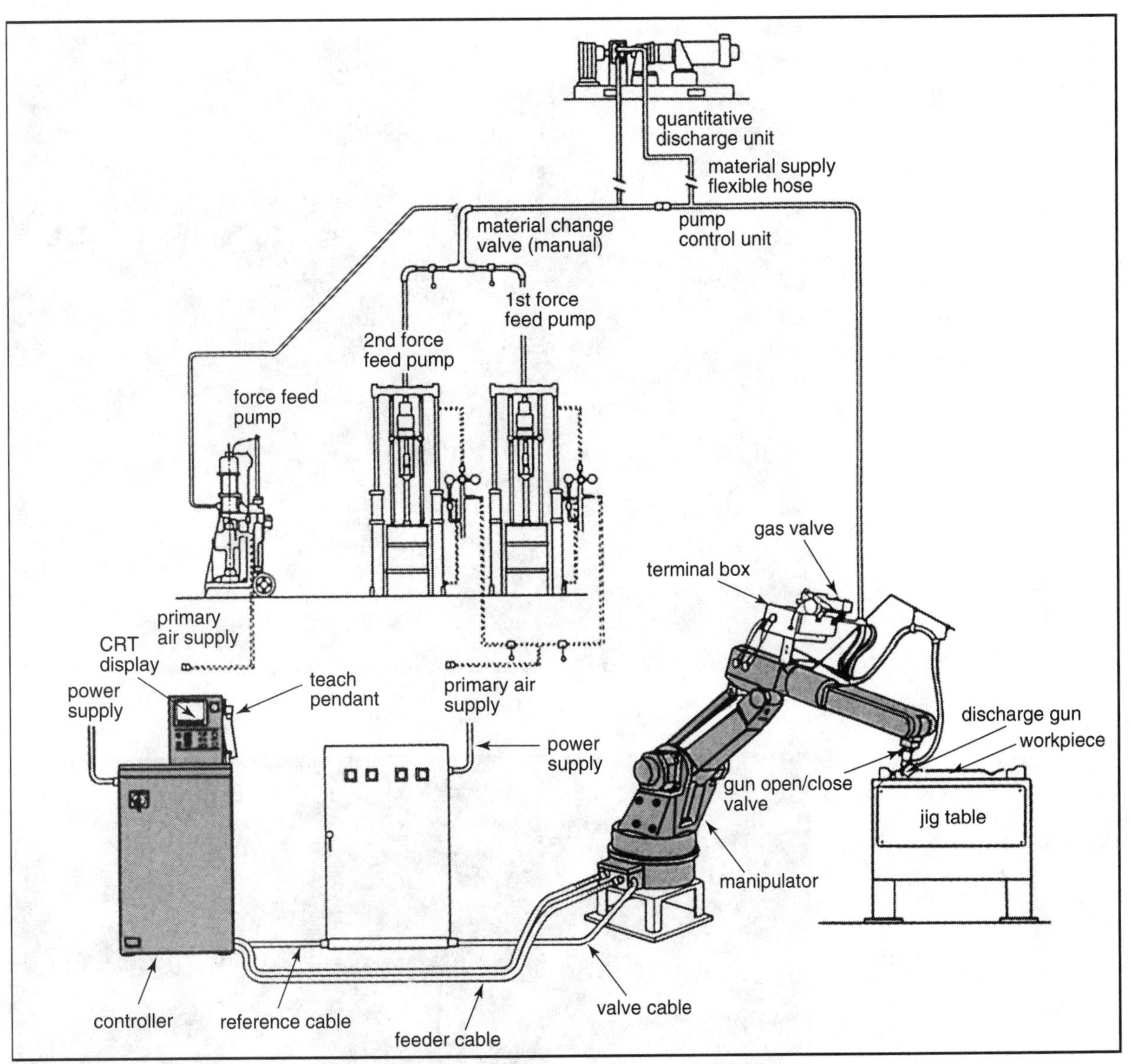

Figure 4-25 Components used in a typical sealing/gluing robotic work station. (Motoman, Inc.)

with quick-change tooling, better fixturing, and improved sensory and adaptive control. Most older generation robots do not meet these requirements.

Cutting

Cutting applications require flexible robots with tight servo control loops, directly linked drives, and dedicated cutting axes. See Figure 4-27. The chart in Figure 4-28 shows the characteristics of four common types of cutting: gas, plasma, waterjet, and laser. The components used in a typical gas cutting robotic work station are illustrated in Figure 4-29. Many cutting operations require special safety precautions, such as those shown in Figure 4-30.

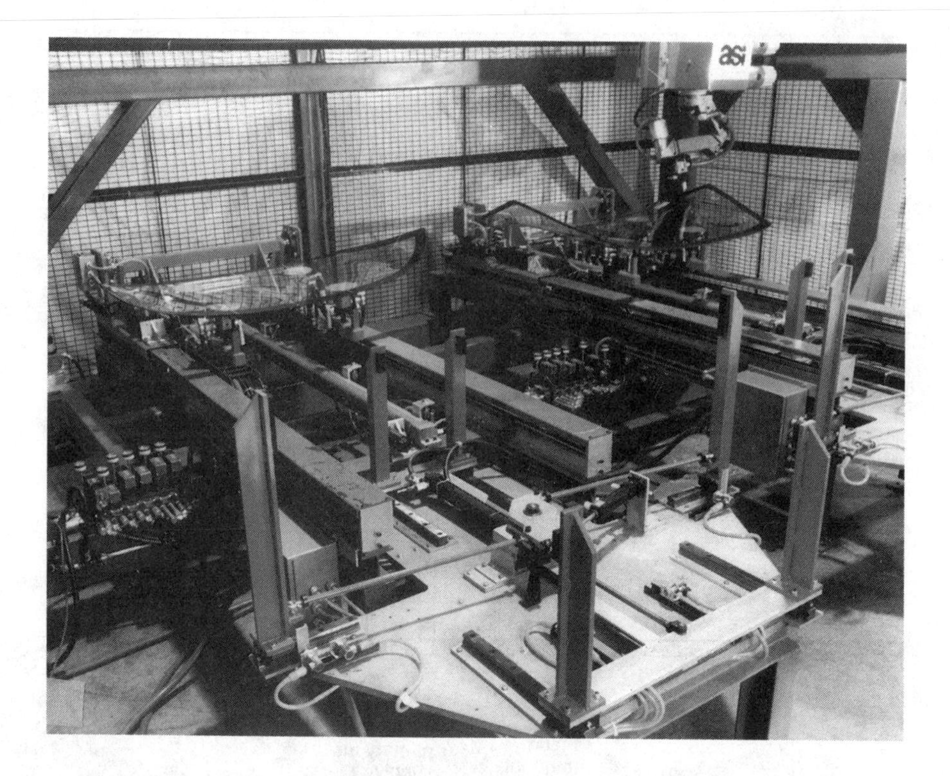

Figure 4-26
This gantry robot applies an adhesive to automotive windshields.

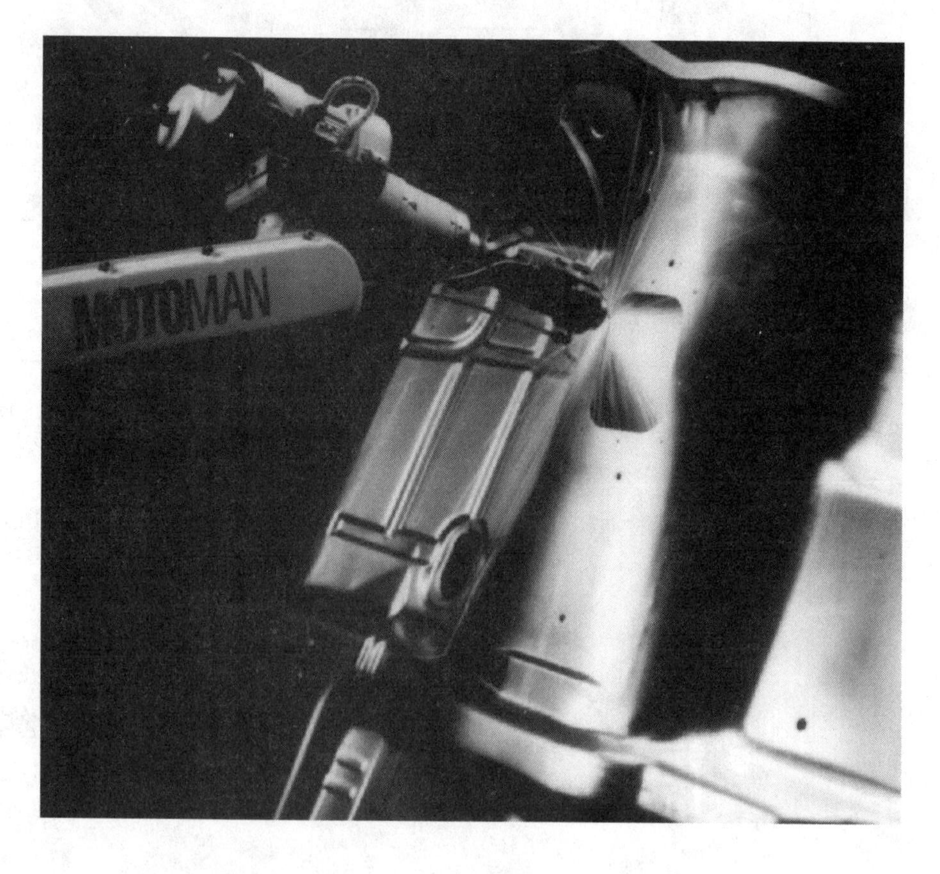

Figure 4-27
Robot using a laser to cut a floorpan for an automobile. (Motoman, Inc.)

Figure 4-28
This chart shows the features of several cutting methods used for steel and other materials. (Motoman, Inc.)

Cutting Types

Cutting	Features	Workpiece Material
gas	• the most common method for heating and cutting. • most economical method.	slab steel
plasma	• applicable for all metals. • advantageous for cutting aluminum alloys, stainless steel, etc.	slab steel non-ferrous metal resin (FRP* etc.)
water-jet	• nozzle diameter, water pressure, and cutting speed must be properly selected according to workpiece material. • excellent durability with proper plumbing and robot mounting.	steel sheet resin (FRP etc.)
laser (CO_2, YAG)	• high-energy cutting. • low heat-deformation because of cutting speeds. • applicable regardless of the hardness and rigidity of the workpiece. • by using a YAG laser, small holes in a range of 4 to 30 mm (0.157 to 1.181 in.) in diameter can be cut. (shape of cutting hole: round, square, and slotted)	steel sheet non-ferrous metal resin (FRP etc.)

Knife cutting is also applicable

* Fiberglass Reinforced Plastics

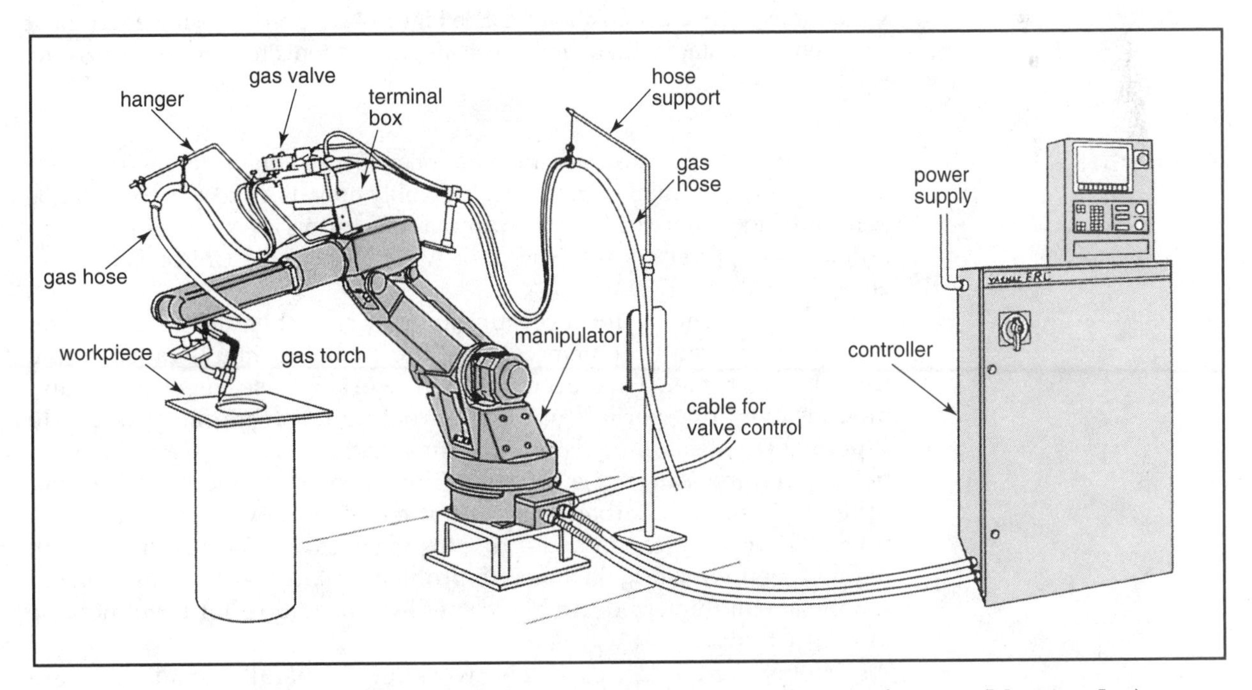

Figure 4-29 These components are associated with a typical gas-cutting robotic work station. (Motoman, Inc.)

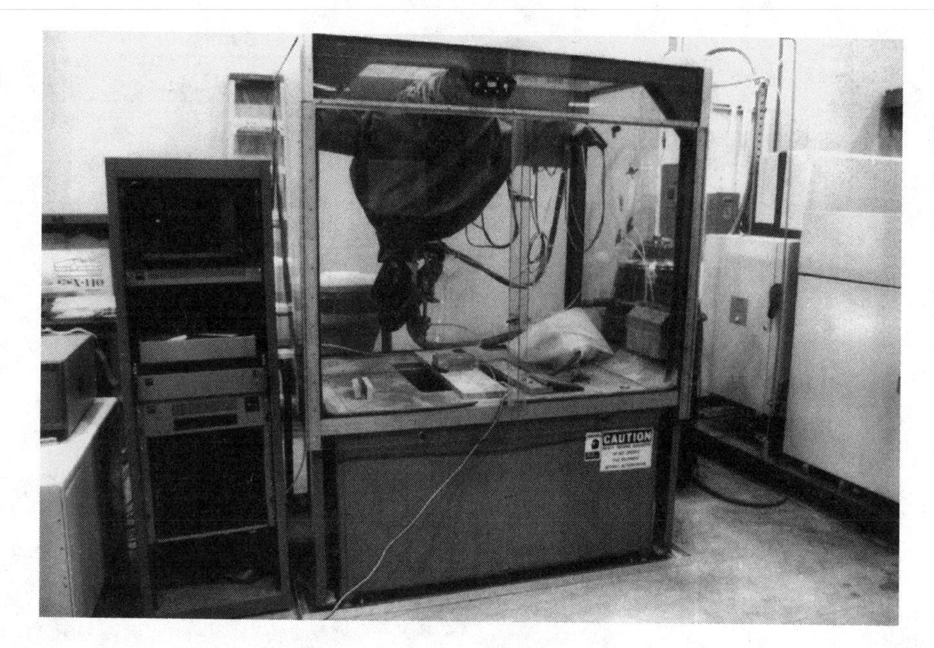

Figure 4-30
This robot is mounted on the ceiling inside a plastic safety enclosure. The enclosure is used to protect operators from the waterjet, which is under extremely high pressures.

Deburring and polishing

Deburring and polishing require robots with smooth movements and rigid wrist designs. The robot control system must be capable of automatically adjusting for such things as tool wear, burr size variation, workpiece variation, and product positioning. See Figure 4-31. Force-sensing equipment can identify contact with a workpiece and adjust tools for constant pressure. See Figure 4-32.

Assembly

In recent years, more attention has been focused on the use of robots for assembly. The potential is great since assembly operations may account for 50 percent of the labor cost involved in manufacturing a product. In 1981, an estimated 40 percent of the blue-color work force in the United States was engaged in assembly operations.

Workers who perform certain kinds of assembly tasks can develop work-related health problems. One of the most common is called carpal tunnel syndrome. This occurs when workers use repetitive hand motions over long periods of time. Carpal tunnel syndrome can lead to a permanent disability. Using vibrating hand tools over an extended period of time can cause Raynaud's syndrome (white knuckle syndrome). This is a disturbance in the normal blood flow to the hands and fingers. Psychological problems due to job stress are also associated with highly repetitive work. Such health problems make the use of automated robotic assembly very desirable. Generally, the end result is an increase in productivity.

Robots used for assembly work are generally small and are designed to move small parts accurately at high speeds, Figure 4-33.

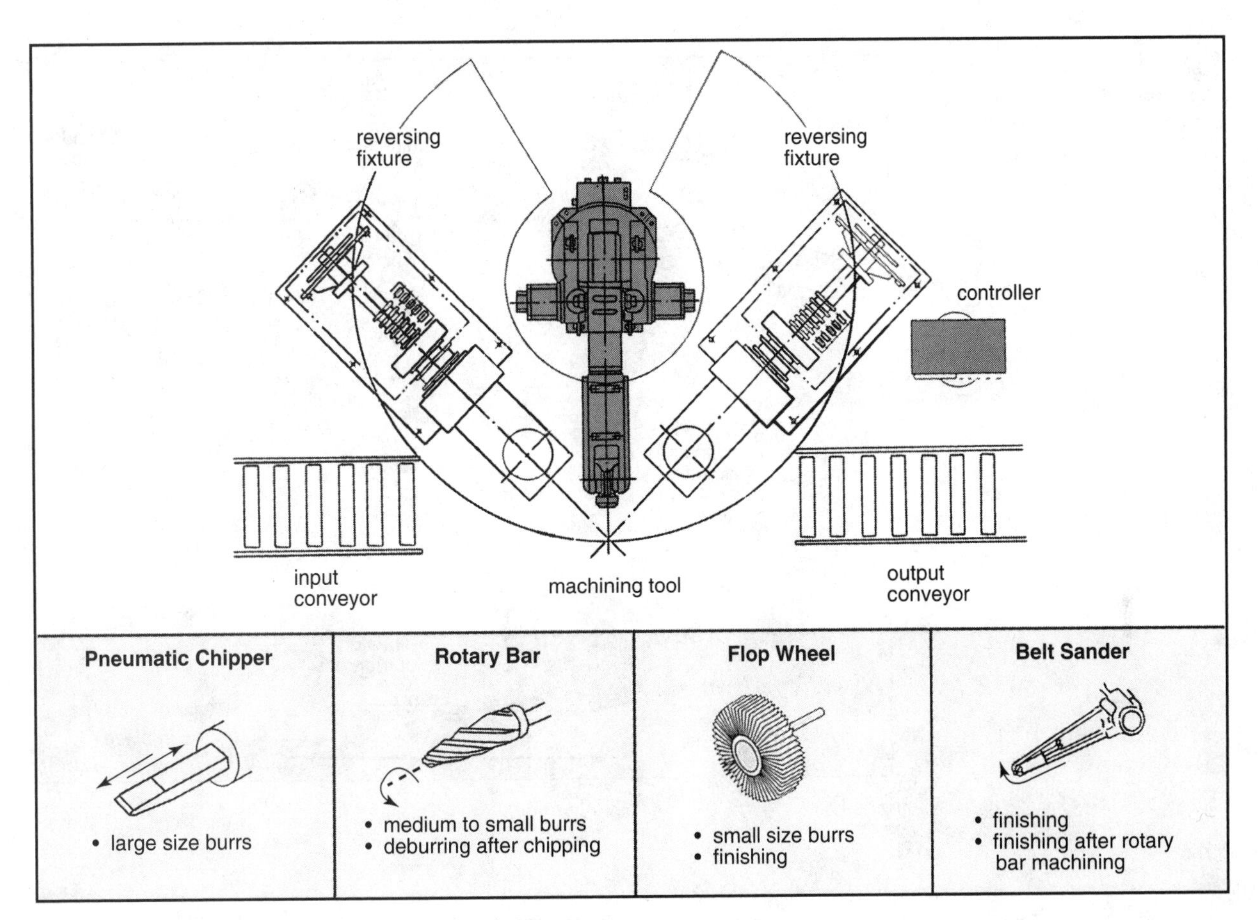

Figure 4-31 The components shown here are used for deburring and polishing. (Motoman, Inc.)

Most operations involve the fitting and holding together of parts and assemblies. This is generally done by means of nuts, bolts, screws, fasteners, or snap-fit joints.

Newer robots are being equipped with sophisticated vision systems that can detect proper orientation of parts. Tactile sensing systems attached to the end effector are being used to detect misaligned, missing, or substandard parts. The robot in Figure 4-34 is equipped with a two-camera vision system. One camera, mounted on the manipulator, guides the robot around the work cell. The robot picks up components for air motors from various parts feeders and trays and assembles them. The second camera is used for precise placement of a component within 0.1 millimeter. The robot brings the components to a turntable where assembly takes place. Two sets of eight fixtures are located around the perimeter of the turntable. All 43 models of the air motor can be assembled using either set of fixtures. Generally, each model is made in batches of eight or fewer.

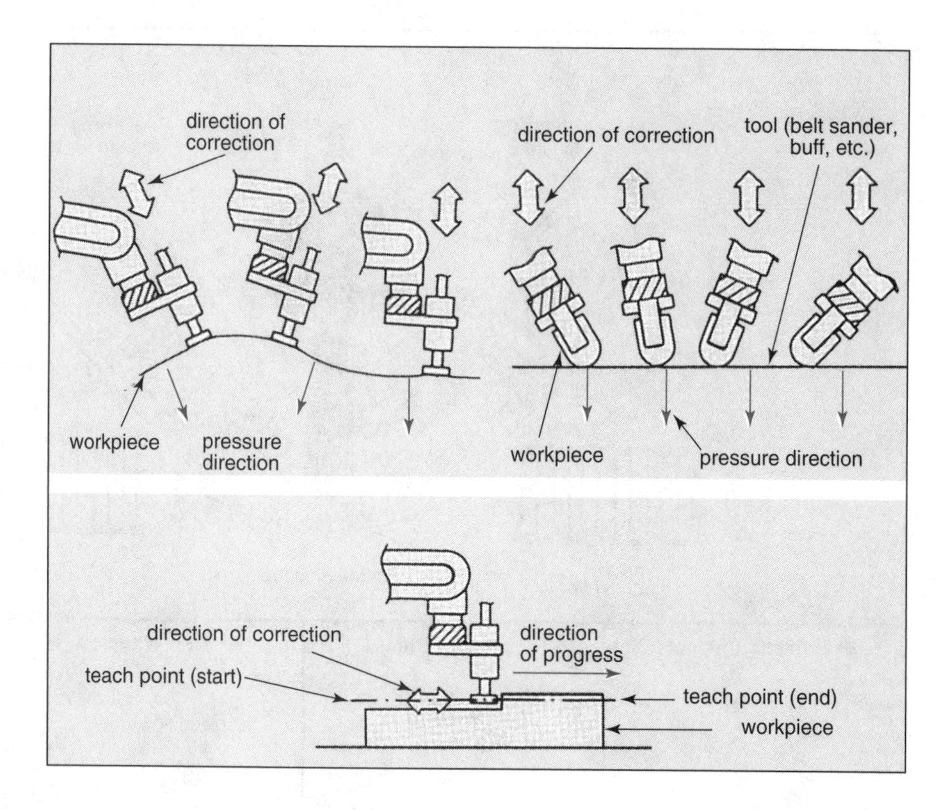

Figure 4-32
Force sensors allow for corrections in tool pressure against the workpiece. (Motoman, Inc.)

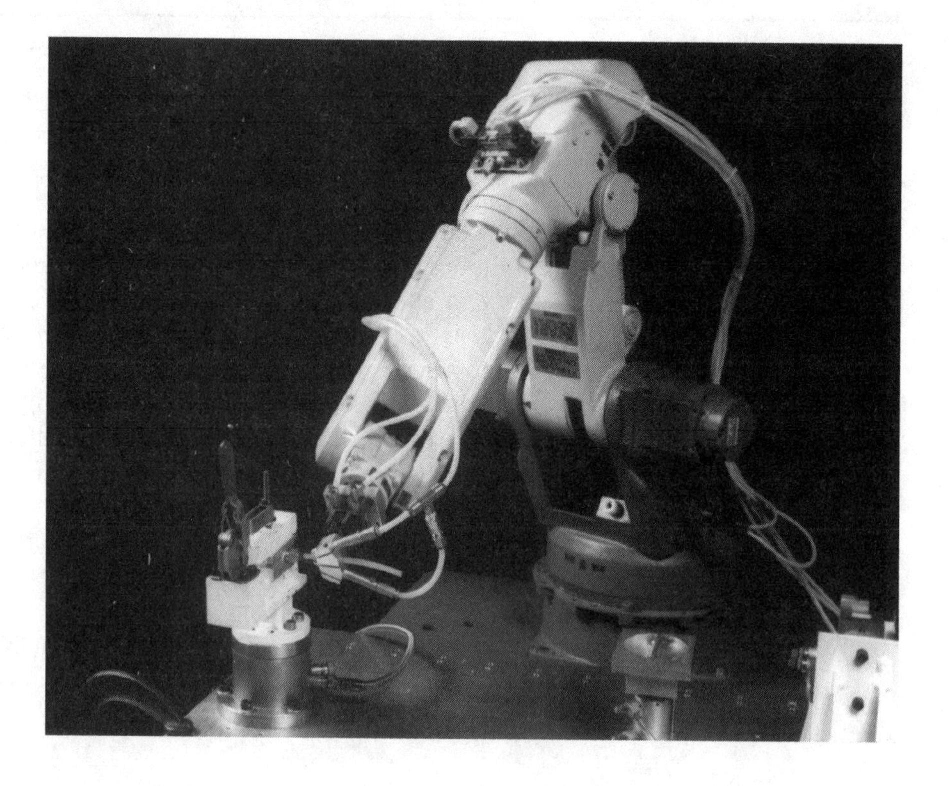

Figure 4-33
A robot inserts parts into an assembly.

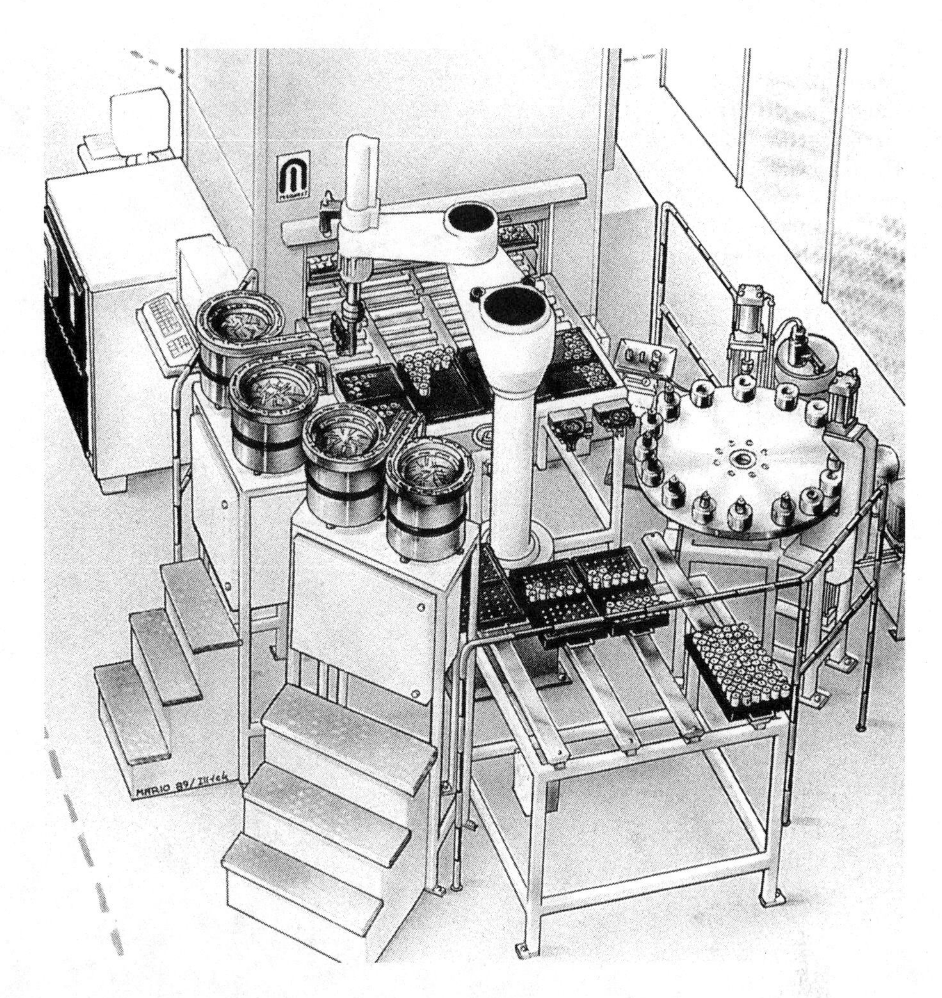

Figure 4-34
A two-camera vision system is used by this robot, which assembles air motors in a manufacturing plant in Sweden. (Adept Technology, Inc.)

Each motor is composed of nearly a dozen different components. The components are fed to the robot by four vibrating bowl feeders and a ***paternoster***, a device that holds trays of components. After assembly is complete, the robot removes the motors from the assembly fixtures and places them on an outlet tray. This allows the operator to remove the trays without interrupting production.

To obtain the necessary production mix, the operator enters the model, quantity, and due date for each order. The robot controller uses this information to calculate an assembly schedule. It signals the paternoster to deliver the appropriate trays of components before production of a new batch begins. As a result, no changeover time is required.

Inspection

Quality assurance—producing the highest quality products—remains an essential and costly part of robotic manufacturing. The demand for non-contact gauging and inspection systems has grown steadily in recent years. At the heart of these new systems is state-of-the-art electronic sensor technology, Figure 4-35. They check for tolerances, positioning, fixturing, and defects, among other things.

Figure 4-35
This profiling sensor unit is gauging a part to see if it meets specified tolerances. The sensor is guided along the edges of the part or assembly by the robot's manipulator. (Servo-Robot, Inc.)

Robotic inspection systems consist of two subsystems. One scans and/or accumulates data. The other analyzes the data and presents it in a meaningful way. The electronic sensing units used in these systems are set up in one of two ways. They may be mounted on the robot's hand and the robot is programmed to move the sensor along the part. They may also be located within the work cell and inspect the parts as the parts move along the conveyor system.

Material Handling

Automated and robotic systems are often used for material handling. Automated control and tracking of inventory and handling parts and products increases efficiency and flexibility.

The ***automated guided vehicles (AGVs)*** illustrated in Figure 4-36 are computer-controlled and battery-operated. AGVs follow a guidepath in the floor. It generally consists of a wire laid in a sealed groove beneath the floor's surface. The path does not cause obstructions and floors remain clean and uncluttered. The vehicles receive instructions and report back to the command center through this path.

Figure 4-36
An operator programs an automated guided vehicle (AGV) to carry finished parts from a machining center to a storage system.

For safety, AGVs have bumpers and automatically stop if they contact any object in their path. AGVs are bi-directional and emit a beeping sound to warn humans out of the way. Some AGVs even signal their command center when their batteries are getting low. The computer then directs the vehicle to a battery charging station. AGVs are limited in mobility by the location of the guidepath.

Service

The industrial robots described so far have been limited to six or, at most, seven degrees of freedom. Their restricted movement means that the task to be performed must be brought within the robot's work envelope. ***Service robots***, however, are mobile and have the ability to move to the work area.

One area where service robots are proving to be not only useful but economically feasible is in security, Figure 4-37. Mobile robots patrol the hallways of banks, museums, government, and other high-security buildings. The command center can be located in the same building or miles away, Figure 4-38. When the operator enters the appropriate command, the robot unplugs itself from its charging station and begins its rounds, Figure 4-39. As it patrols, the robot continuously samples the air and surveys the environment with its sensors. It also automatically responds to any emergency or intrusion. The robot can be programmed to stop and perform automatic surveys when suspicious conditions are detected. The information is then transmitted back to the command center. At the command center the operator can see the situation on a video console.

Figure 4-37
A team of mobile sentry robots is ready to go on patrol. (Denning Mobile Robots)

Mobile service robots are also being used in hazardous environments. Hazardous environments are becoming more common and there is a critical need for robots designed to assist in life-threatening situations, Figure 4-40. Mobile robots are finding applications in such diverse areas as the nuclear industry, security and law enforcement, the military, mining, firefighting, construction and excavation, and the removal of chemical or toxic hazards.

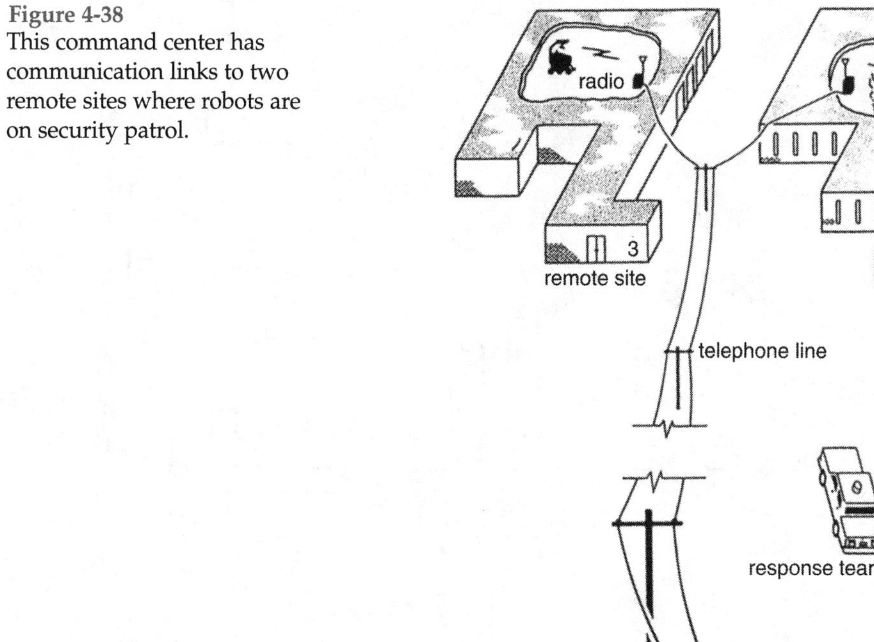

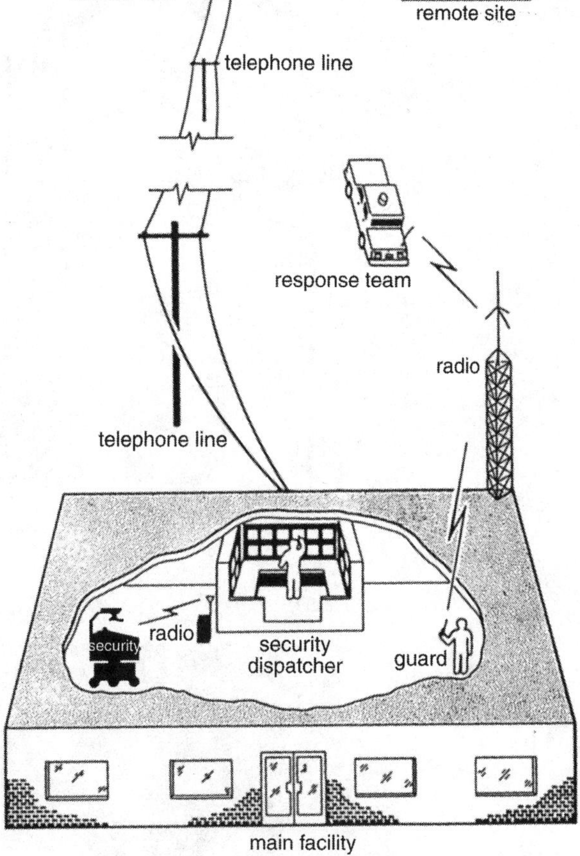

Figure 4-38
This command center has communication links to two remote sites where robots are on security patrol.

One day, mobile robots may clean your home, cook your meals, take out the garbage, provide security surveillance, and fetch your newspaper from the front yard. Joseph F. Engelberger, in his book *Robots in Service,* writes that stand-alone robots able to "pump gas, fill prescriptions, cook and serve foods, clean commercial buildings, or aid the handicapped and elderly are real prospects. They will bring the magic back to robotics." Many robotics experts agree that in the decades ahead, service will be the fastest-growing robot application, resulting in a $1-billion-a-year industry.

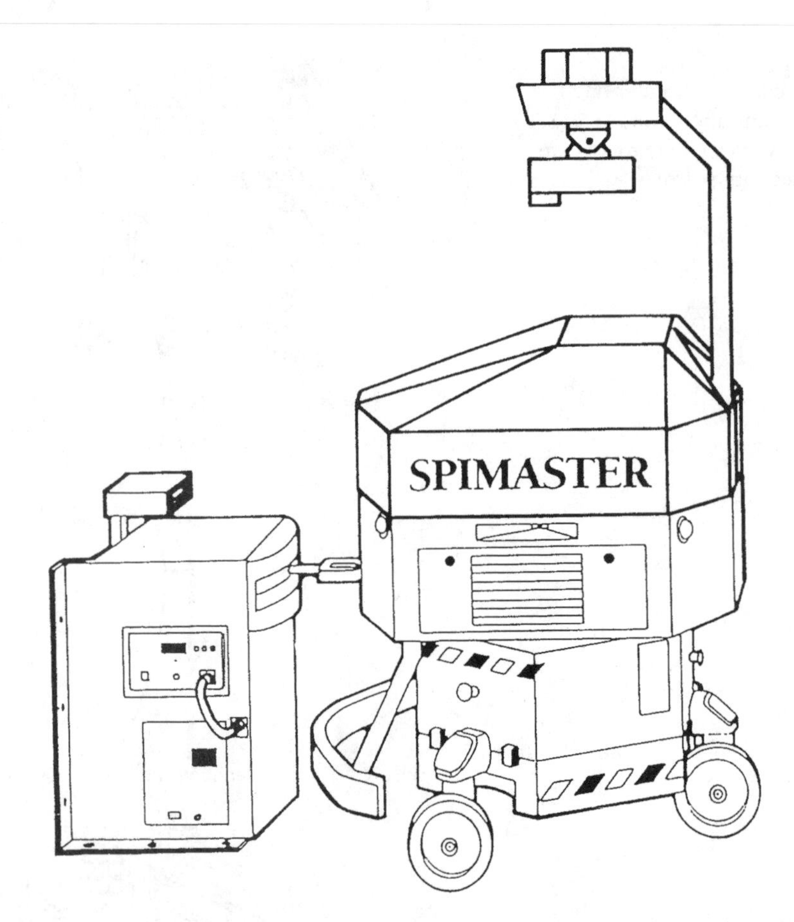

Figure 4-39
This security robot plugs itself into a battery-charging station. It can operate for 12 hours and cover distances of up to 15 miles between charges.

Figure 4-40
This small mobile robot is designed to enter hazardous environments and perform dangerous tasks.

Important Terms

accuracy
arc welding
automated guided vehicles (AGVs)
command resolution
dedicated equipment
design for manufacturability
die casting
dynamic performance
light curtain
operational speed
palletizing
paternoster
payload
quality assurance
repeatability
resistance welding
resolution
service robots
spatial resolution
spray finishing
switch carpet

Review Questions

Write your answers on a separate sheet of paper.

1. One important application of robots is for product assembly. What are five key design considerations product engineers should consider when designing parts for robotic assembly?
2. List four primary factors that must be considered when selecting a robot to perform an industrial task.
3. Define command resolution, spatial resolution, accuracy, repeatability, and operational speed.
4. What is the spatial resolution of a robot that has a command resolution of 0.006 in (0.015 cm), and a mechanical resolution of 0.004 in (0.011 cm)?
5. Spatial resolution may vary at different locations in the work envelope for certain robots. Which robot configuration offers the most consistent spatial resolution throughout its work envelope?
6. Accuracy and repeatability are often confused. Briefly explain the difference between these terms.
7. What is the maximum weight a robot with a load-carrying capacity of 50 pounds (22.68 kg) can handle when the end effector weighs 8 pounds (3.62 kg)?
8. What effect does the weight of the part handled by the manipulator have on the dynamic performance of a robot?
9. Robots can injure or kill. List four safety measures commonly used to make working around robots safer. Describe how each one functions.
10. List seven major production applications for robots. Which are the most common applications?
11. Which robot application has the greatest potential for growth in the future?

12. Can present-day robots perform every industrial task that humans now perform? If not, why not?
13. Often, companies will intentionally design parts for robotic assembly. Are there any advantages to this practice, even when robots are not actually installed on the production line?

This factory has been heavily automated. Each robot can be programmed and equipped with tools to perform separate manufacturing tasks. (Hirata)

5 The Role of Robots in Today's Manufacturing

Overview

Today, manufacturing is customer-driven. It begins by defining the customer's needs. It then determines how to meet those needs and keep the customer satisfied while maintaining satisfactory profit levels. It demands a continual striving towards quality, cost-effectiveness, shorter lead times, good customer service, and general responsiveness to changing market conditions. Where do robots fit into this picture? What is the role of the robot in today's manufacturing environment? This chapter will cover some common misconceptions about the use of robots. It also discusses their advantages and disadvantages. The chapter concludes with information on how to develop an implementation plan.

Common Misconceptions about Robots

Ten years ago, manufacturers believed that flexible automation and robotic technology were the keys to the factory of the future. Today, many major companies that invested millions in flexible automation tell stories of failure as well as success. Robots and other complex automation systems have not been a cure-all for manufacturing problems. What went wrong? We will review past experience and try to establish some guidelines for the future. This should provide a basis for making an intelligent decision on whether or not to use robotic technology.

Following are several incorrect assumptions that have caused costly mistakes:

- Δ The introduction of robots will automatically increase productivity.
- Δ Intelligent machines can do the same jobs as trained workers.
- Δ Robots will save labor costs by replacing workers.
- Δ Large investments in flexible automation will pay off over the long run.

Does robotic technology automatically increase productivity? *No.* To date, no machine has been invented that is more flexible or more efficient than a motivated human worker equipped with the right tools. Only in certain applications can robots be more productive than humans over a long period of time. This is because robots do not suffer from fatigue or require food or sleep.

However, productivity *does* increase when the product being manufactured is specifically designed for automated assembly. Manual assembly benefits from this as well, however.

Another misconception—that intelligent machines can replace trained workers—has proven disastrous in many cases. What actually happened was the replacement of production workers with computer programmers and other highly trained technicians. These people are needed to program and maintain the automated systems. As a result of this change, however, companies have been left with fewer workers who understand manufacturing processes.

Many companies also assumed that robots would replace a significant number of employees. The resulting savings was to offset the ***capital expenditure*** (initial cost) for the robotic equipment. This was a costly mistake. In general, direct labor costs rarely exceed 15 percent of the total cost of producing a product, Figure 5-1. More often, direct labor cost is within a range of 5 to 8 percent. Also, it has been estimated that a robot replaces only 1.6 workers. It is no wonder that using robots to replace humans resulted in little, if any, direct labor savings. It is hard to justify using robots on the basis of labor savings alone. Actually, experience indicates that adding robots and other kinds of high-tech equipment often will *increase,* rather than *decrease,* direct labor cost. The increase results from the need to hire additional skilled individuals to support the technology.

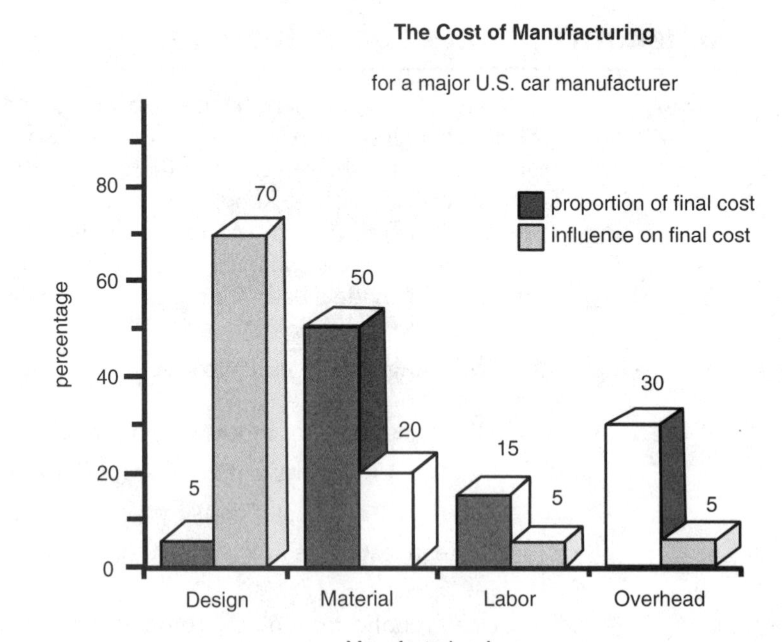

Figure 5-1
Direct labor costs related to the total cost of producing a product.

Do large investments in flexible automation pay off over the long haul? *Not necessarily.* Sometimes, after spending millions of dollars on such systems, a company locks itself into long production runs to pay for the equipment. A simple downturn in the economy can destroy such a plan overnight. Also, manufacturing methods have changed. Long production runs are a thing of the past; products are no longer made solely to build inventory. Trends such as fast product turnaround can add significant costs. This occurs because retooling flexible manufacturing systems is expensive. If robots or other systems cannot be changed over economically in response to market demands, then perhaps a large investment in them is not advisable.

Where Do Robots Work Best?

Figure 5-2 compares the use of industrial robots per 10,000 manufacturing employees in seven different countries. Japan uses more than twice as many as Sweden and five times as many as the United States. (As mentioned in Chapter 2, however, the Japanese classify more types of machines as "robots" than we do.) What role will robots and flexible automation play in the 1990s and beyond?

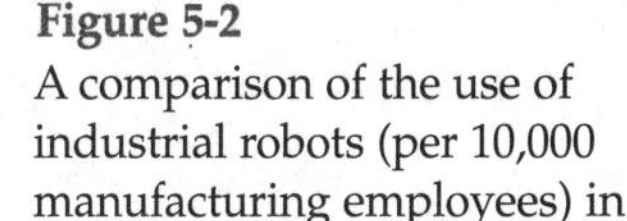

Figure 5-2
A comparison of the use of industrial robots (per 10,000 manufacturing employees) in seven different countries.

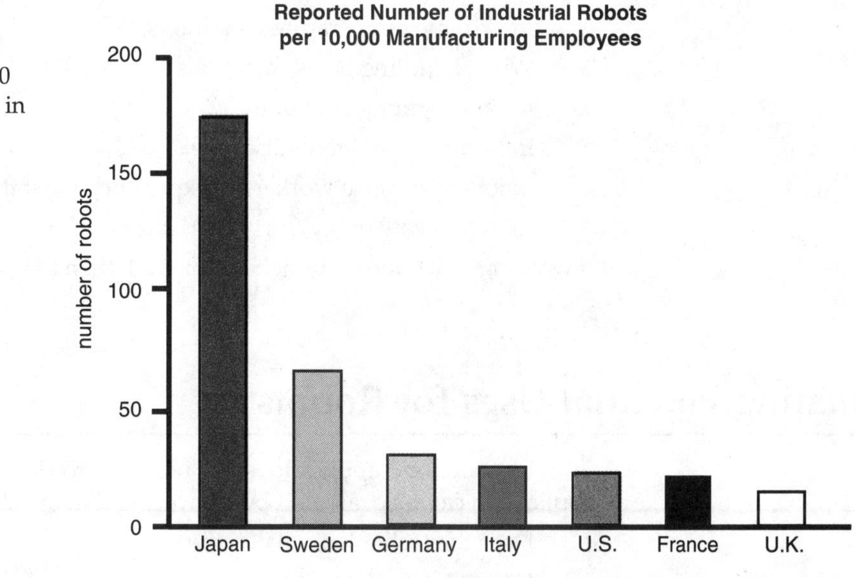

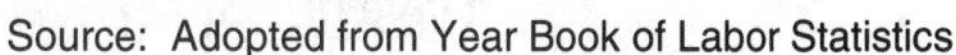

Source: Adopted from Year Book of Labor Statistics

The Automotive Industry

The automotive industry has made a large investment in robotics. The motivation for this comes from the very nature of the business. Global marketing, worldwide competition, and the attempt to meet a wide range of customer preferences has led to numerous products and models offered each year. For example, in 1970, Nissan offered 8 models and 10 body types. In 1990, it offered 27 models and 54 body types. From 3 to 5 models are changed every year.

Shorter product development time and less changeover lead time have forced the automotive industry to turn to flexible automation. In Japan, the goal has been to increase production line flexibility. Task complexity has been reduced by means of product and fixture design. Flexibility has been enhanced by the use of programmable robots. It must be pointed out, however, that a study conducted by an American university found that many Japanese plants with large investments in automation were not achieving high levels of productivity.

Undesirable Environments

The area where robotic installations have been most easily justified has been for use in undesirable environments. Whenever working conditions are unpleasant or unsafe, worker productivity is affected. The use of robots in such environments has proven to be an effective solution. Robots have been successful:

Δ In handling hazardous materials.
Δ Where there is a lack of clean air in the work area.
Δ When operating hazardous machines.
Δ In extreme temperature conditions.
Δ When handling heavy materials.
Δ In poor lighting situations.
Δ In hazardously high noise levels.
Δ In jobs that cause workers to experience mental strain or fatigue.
Δ In excessively dusty/dirty work areas.

If working conditions cannot be improved, then the operation should be automated.

Evaluating Potential Uses for Robots

Most robotic applications are not easy to define or justify. A potential application can be evaluated by answering five questions.

1. Does the technology match the need?
2. Have principal advantages and disadvantages been considered?
3. What are the noneconomic justifications?
4. What are the economic justifications?
5. Is the organization ready for robots?

Matching the Technology to the Need

A case for the use of robots or other forms of flexible automation usually exists when:

Δ Production ranges from medium- to high-volume.

- Δ Production cycles are short- to medium-length.
- Δ Heavy loads must be handled precisely.
- Δ High levels of precise positioning are required.
- Δ Lot sizes are small.
- Δ Several workpieces are produced per process.
- Δ Parts need not be oriented in a certain way.

Applications where one or more of the following conditions exist are *not* good candidates for the use of automation:

- Δ Critical aspects of the workpiece or the process cannot be measured.
- Δ Specific workpiece and process standards do not exist.
- Δ Identification of the workpiece, positioning it, or evaluating its condition is difficult.
- Δ The robot's needed range of movement cannot be readily identified; that movement cannot be contained within a limited space.

Considering the Advantages and Disadvantages

An important step in determining whether a robotic application is justified calls for very careful consideration and evaluation of the advantages and disadvantages involved. The balance of advantages and disadvantages may shift, depending upon a given situation.

Advantages

Robots can offer increased productivity, given the right conditions. As noted earlier, robots are no faster than humans. In fact, their cycle times are generally slower. Productivity gains are made through the robot's constant work pace. During an eight-hour shift, the robot will usually out-perform a human operator—especially if the task is repetitive, boring, heavy, or performed under poor working conditions. Greater gains can be realized by operating for more than one shift.

Robots also can improve product quality. This occurs because of the robot's accuracy and repeatability. An example can be seen in spot welding. A robot may be able to place a weld within 0.050 in. (0.13 cm) time after time. This is hard for a human to duplicate, since the spot-welding gun weights approximately 100 lbs. (45.4 kg). Even though the gun is counterbalanced, fatigue takes its toll after several hours. See Figure 5-3.

Product quality also is improved as a result of automated equipment requiring quality components to function efficiently. See Figure 5-4. A human operator may be able to assemble components that are a bit "out of spec." Trying to do this with automated equipment, however, is disastrous. Parts definitely must be within specifications, and must be held to closer tolerances. To take full advantage of automated production, manufacturers are starting to demand 100 percent quality parts from vendors.

Installing robots will help reduce personal injuries and increase personal safety. The passage of the ***Occupational Safety and Health Act (OSHA)*** provided an incentive for manufacturers to introduce robots in many operations that were considered dangerous. Using robots in these areas reduces the need for the complex safety procedures and equipment.

Figure 5-3
It is not difficult for a robot to lift the heavy gun when spot welding an automotive body assembly. A human operator's performance would deteriorate during a shift, due to fatigue. (NUM)

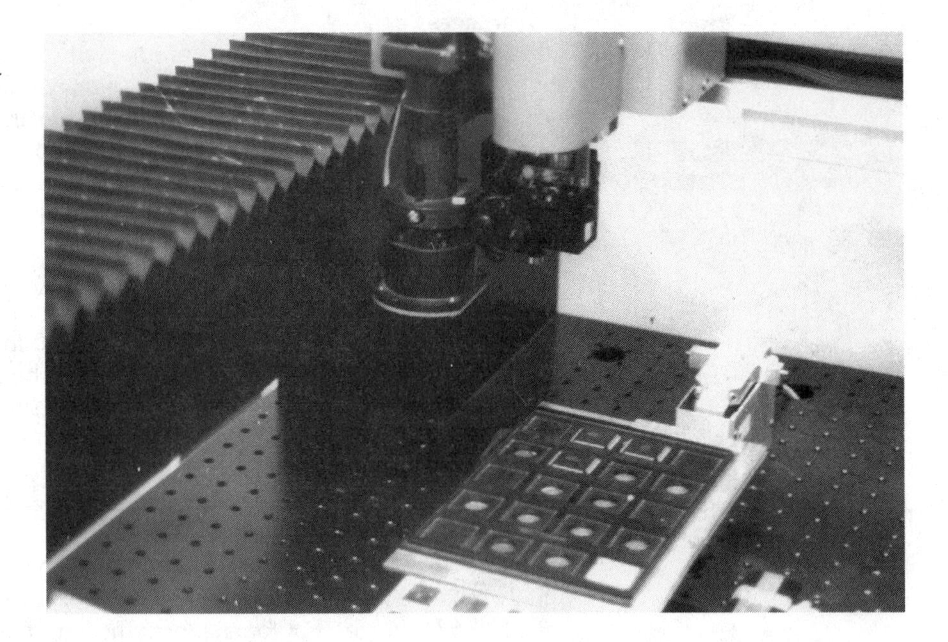

Figure 5-4
To function efficiently, assembly robots must work with components of consistently high quality. The result is a final product of higher quality. This robot is assembling electronic printed circuit boards.

Another advantage of automating unpleasant or dangerous tasks is better worker morale. If workers are moved from dirty, hostile, and hazardous environments, their attitudes improve. They can then be placed in more challenging, responsible positions.

Robots can make a contribution to holding down or reducing production costs. Some manufacturers boast of productivity gains of up to 400 percent. Others have reduced scrap and rework. (General Motors once reported that four-fifths of its rejects in spray painting came from manually-operated areas.) Savings can also be realized by reducing material use. For example, one manufacturer using robots for spray painting was able to reduce paint usage by 50 percent. Other gains can be made through a reduction of the energy required to light, heat, cool, or ventilate work areas.

Flexibility is another advantage that robots offer. Most automated equipment is *dedicated,* or designed to perform one function. Trying to adapt this equipment to new operations often proves costly or almost impossible. In contrast, robots can be reprogrammed easily to perform new tasks. An example is their ability to perform operations on different automobile models on the same assembly line. Presently, 25 to 30 percent of all manufacturing is performed on lots of 50 or fewer parts. It is predicted that this will increase to 75 percent in the future. Robots are well-suited for such a situation.

Disadvantages

One of the most obvious disadvantages of robots is that they require capital investment. Like other pieces of equipment, they have to be economically justified. If not, they should be avoided. Of course, there are exceptions. Personal safety, for instance, should come first.

Robots are certain to have an effect on the production line. Parts will have to be properly positioned and oriented. Work flow will change, and additional work space will be required. Acquiring the robot is the easy part. It is all the tooling and equipment required to interface the robot with the manufacturing line that causes the problem. This tooling and equipment often will cost more than three times the price of the robot itself.

Another disadvantage is employee opposition—some workers will feel threatened by robots, and need to be reassured. If management and workers do not develop a positive attitude regarding robots, their installation will most likely fail.

Reviewing Noneconomic Justifications

Noneconomic considerations sometimes outweigh economic ones. The use of robots in hostile environments is one example. If poor working conditions are ignored by management, productivity can decrease and product quality may suffer. Workers may become genuinely ill or may call in sick to avoid unpleasant conditions. In extreme cases, workers may cause shutdowns or even sabotage equipment. The cost of lost production can far exceed the cost of using robots.

Another justification comes from looking at the long-term impact of using robots. Over a period of years, will the use of flexible automation:

- Δ Make it possible to produce better products of more consistent quality?
- Δ Reduce new product lead time?
- Δ Increase flexibility of the production line?
- Δ Improve customer service and product satisfaction?
- Δ Improve market response time?

If the answer to several of these questions is yes, then an analysis should be done to see if the necessary capital investment can be justified. However, other alternatives should also be explored. For example, one alternative might be to modify or rebuild present equipment. The same goals might also be accomplished through retraining and motivating the workforce.

Is the market highly competitive? Can new technology help the company maintain or improve its competitive position? This, too, may justify investing in robotic equipment. Robots are often used in research and development. The expertise gained in this area can be applied later to their general use. Economic returns occur because the time needed to get robots on line is reduced.

Sometimes it takes competitive pressure to initiate change. If one competitor starts to use robots and gains a price advantage, robots can become more attractive. However, buying a robot should be based on firm economic decisions, not on blindly following the crowd.

Reviewing Economic Justifications

Even though the previous factors are considered noneconomic, each makes an economic contribution as well. However, the decision to invest in robots may be based solely on economic considerations. The subject of engineering economics requires more in-depth study than can presented here. This section is intended to merely make you aware of the various factors that may have to be considered.

An industrial robot must be subjected to the same scrutiny as any other equipment. There are two situations where economic analysis can assist in making an investment decision. The first involves investment in equipment for a new application. This is referred to as ***avoidance costs.*** The second involves the replacement of an existing method and is known as ***cost savings.***

The most efficient allocation of capital must be determined. The proposed investment has to provide a return equal to or greater than that required by the investor. Most companies have established benchmark criteria for return on investment and ***payback period*** (time required to recover, through savings in labor or material costs, the amount expended for new equipment). Investment proposals must meet these criteria. For example, a company may require a 25 percent return on investment and a payback period of less than two years.

The literature on the economic evaluation of robots focuses mainly on payback. Investment decisions based on payback can be misleading. Other factors, such as return on investment and discounted cash flows, must be considered as well. Information about depreciation, tax rates, present value, net present value, cost of capital, and production can provide insights. You should review the many books covering engineering economics.

Preparing an Implementation Plan

After it has been decided that purchasing robot equipment is a feasible means to the reaching the company's goals, the following steps can be used to develop an implementation plan.

1. **Review the available equipment.** Information about the capabilities and limitations of various models can be acquired from robot manufacturers. Another source is the *Robotic Product Database* published by

Tecspec in Orlando, Florida. This publication does an outstanding job of presenting information about the robots available today.

2. **Construct a matrix for comparing equipment.** Matching a robot to a particular situation is not an easy task. To help compare various robots, a matrix can be constructed for those that are of interest. The matrix should include such things as the model number, price, number of axes, type of controller, load capacity, work envelope, type of power supply, speed, accuracy, repeatability, and methods of programming. The matrix also can be helpful in identifying a particular robot for a given application.

3. **Identify potential applications.** One way to identify possible applications is through a plant survey. Certain job, product, or process characteristics can be used to identify possible uses for robots. For example, if the job is boring, dirty, hot, or noisy, or if it presents a possible health or safety hazard, then it is a candidate for a robot. Short product life, frequent design changes, families of parts, a minimum of tools, a minimum of parts, and multishift operations are all characteristics of jobs suited to robots. Material handling, component insertion, inspection, and testing are broad general categories with which to start the survey. Some potential applications can also be identified by reviewing personnel and safety records.

4. **Analyze potential applications.** The plant survey should yield several possible applications for robots. The next step is to select and construct profiles of the most promising ones. Typically, this will involve several trips to the production floor, as well as talking to people connected with the operations. The importance of drawing on the expertise of operators, supervisors, engineers, and other manufacturing personnel cannot be overemphasized. Familiarity with previous and future operations is also essential. Installation of a robot can have a ripple effect on the whole production line. Additional material handling and feeding devices may be required. Blueprints, specification sheets, production control records, quality records, and safety records must all be explored to gain additional information.

   Data should be obtained on the number and sequence of operations, parts flow, cycle times, product volumes, personnel requirements, product life, results of tests and quality checks, and the effects of environmental factors. This information will be useful in determining if a robot is really capable of handling a job.

5. **Match the robot to the application.** The problem now is to determine if the job can actually be handled by a robot. Consider previous operations to aid in that determination. A method for using the robot will have to be developed, as well.

   In addition to the robot itself, tooling and peripheral equipment needs must be assessed. At this point, machine cycle time should be determined. If laboratory facilities are available, a prototype can be constructed to establish cycle times and to evaluate robot performance and the process sequence. Data needed to develop the financial proposal is easier to obtain this way.

6. **Develop the proposal.** The first step is to prepare the financial analysis. It looks at investment, expense items, savings, and other economic fac-

tors to determine the most efficient allocations of capital. Return on investment, payback periods, and discounted cash flows should be considered for each alternative. Proposals without economic justification are usually doomed to rejection.

The proposal is a rather complex document. It begins with the statement of the problem, then proposes a solution supported by rationale. The rationale should include both the economic and noneconomic factors. Other areas of the proposal address personnel requirements, required resources, a time schedule, and a budget.

7. **Develop and refine the application.** After the proposal has been approved, the details must be worked out. The necessary tooling and safety devices have to be designed and ordered. The application has to be debugged and refined. Much of the project's time will be devoted to this area. This is the initial proving ground for the application.
8. **Begin implementation.** During implementation the site is prepared for production. The floor space has to be prepared and the various service drops and safety devices installed, Figure 5-5. Simply plugging a robot into a production line can wreak havoc. The object is to create as little confusion as possible.
9. **Provide training.** During implementation, everyone involved with the operation needs to be trained. Training should be provided prior to actual installation. Without adequate training, the application will be in jeopardy.

Figure 5-5
Safety fences, such as the one enclosing this robotic workstation, must be installed during the implementation phase of the project.

Many maintenance workers already possess most of the technical skills needed to work on robots. However, they still have to become familiar with the manufacturing process and with the basics of programming. Robot manufacturers usually provide training for maintenance and other personnel, either in their own facilities or at the installation site.

In-service and retraining programs also should be a vital part of any manufacturing operation. These programs are needed not only for technical personnel, but for all employees. Needs for various groups have to be identified. Most universities and technical schools often are willing to assist in providing this service.

10. **Provide maintenance programs.** Establishing a maintenance program is important to success. Maintenance personnel need to be hired. It is better to have too many than too few, since workers become sick or go on vacation, are transferred, retire, or accept positions with other companies.

    The maintenance program also should include stocking of adequate spare parts. This permits repairs to be made quickly to avoid loss of production. Robot manufacturers supply a list of spare parts to stock. As experience is gained, a more appropriate list of parts for a specific operation can be made.

Important Terms

avoidance costs
capital expenditure
cost savings
Occupational Safety and Health Act (OSHA)
payback period

Review Questions

Write your answers on a separate sheet of paper.

1. Identify and briefly discuss several common misconceptions about robots.
2. Which industry currently uses the greatest number of robots?
3. Which country makes the most widespread use of robot technology today?
4. The decision to invest in robots can be based on noneconomic as well as on economic factors. List and explain four noneconomic factors that can influence the investment decision.
5. Explain the difference between avoidance costs and cost savings.
6. List several tools used in engineering economics to assist in making appropriate investment decisions.
7. Several systematic approaches have been developed to identify, select, and implement robot applications. Summarize the approach presented by this text.
8. What must management do before the actual implementation of automation/robotics to help ensure its successful acceptance by the workers?
9. It has been established that robots are generally not faster than humans in performing many tasks. Yet in terms of productivity, the robot (when used to its full potential) is more productive than humans. Why?
10. What five questions must be answered to properly evaluate a potential robotic application?

POWER SUPPLIES AND MOVEMENT SYSTEMS

In automated applications, three kinds of motion are used: rotary, linear, and reciprocating. These motions can be produced by either electrical or fluid (hydraulic or pneumatic) power operating motors, relays, solenoids, actuators, or cylinders. The basic mechanical unit of a robot, the manipulator, has several moving joints and performs the actual work function of the machine.

ESAB North America, Inc.

6 Electromechanical Systems

Overview

A ***system*** is an combination of components or subsystems that work together to form a unit. Nearly all of the products manufactured by industry result from the use of some type of ***electromechanical system.*** Systems of this type transfer power from one point to another through mechanical motion that is used to do work. Punch presses that move up and down, rotating machinery, and robots are all examples of electromechanical systems. A robot is a unique type of system. It sometimes requires several different types of energy inputs and specialized machinery for proper operation. This chapter provides an overview of the types of electromechanical systems used with robots.

The Automated Systems Model

At one time, all manufacturing operations were manually controlled. Gas-filled tubes, magnetic contactors, and electrical switchgear served as the primary control devices. Recent developments in solid-state electronics and miniaturization have brought a number of advances in system control. Electromechanical, opto-electronic, hydraulic, and pneumatic systems are often combined in the control of a single industrial robot. These various systems become the ***subsystems*** that make up the robot as a whole. When subsystems are combined, the result is referred to as a ***synthesized system.*** Familiarity with each of the subsystems, and their locations within the overall system, is important to understanding the system itself, Figure 6-1.

A variety of subsystems is used in virtually all automated systems. An electrical power system, for example, is needed to produce and distribute electrical energy. Hydraulic and pneumatic systems are used in operations and for system control functions. Opto-electronic systems are used for inspection and in various types of sensors. Mechanical systems are needed to hold objects for machining operations and to move them on a production line.

Although each system has unique features, its main parts are basic to all automated systems. All such systems have an energy source, transmission path, control, load, and indicators.

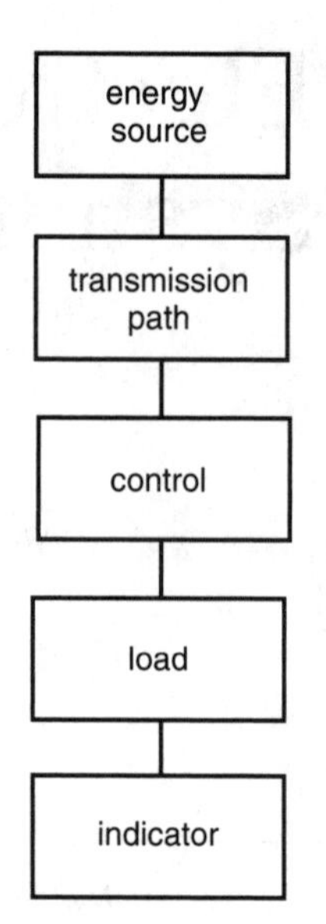

Figure 6-1
The basic parts of a system. They must work together or the system cannot function.

The energy source provides power for the system. Industry uses more than 40 percent of the electrical energy produced in the United States. The most common source of power for synthesized systems is alternating current. In ***alternating current (ac),*** electrons flow first in one direction and then in the opposite direction. Each change in direction is called a ***cycle.*** The alternating current required for an automated manufacturing system may be either three-phase or single-phase. Three-phase alternating current is ordinarily used for larger systems.

Some machines used for automated manufacturing require direct current. In ***direct current (dc),*** electrons flow in only one direction. Many robotic systems use dc servo motors and controls. Several methods are available to convert alternating current to direct current. This process is called ***rectification.*** Rectification is usually the most convenient and inexpensive method of providing dc energy to machines.

The ***transmission path*** provides a path for the transfer of energy. It begins at the energy source and continues through the system to the load device. In some cases, this path may be a single feed line, electrical conductor, light beam, or pipe. In other systems, there may be a supply line and a return line between the source and the load. There may also be a number of alternate paths. These may be connected in series to a number of small load devices or in parallel to many independent devices.

The ***control*** path is the most complex part of a system. In its simplest form, control occurs when a system is turned on or off. This type of control can take place anywhere between the source and the load device. Control alters the flow of power and causes some type of operational change in the system. Examples would be changes in electric current, hydraulic pressure, light intensity, or air flow.

The ***load*** refers to a part (or parts) designed to produce work. ***Work*** occurs when energy is transformed into mechanical motion, heat, light, chemical action, or sound. Normally, the largest portion of the energy supplied by the source is consumed by the load device.

The ***indicator*** readings display operating conditions at various points throughout the system. In some systems, the indicator is optional; in others, it is essential. When operations or adjustments are critical, system function may depend upon specific readings.

Mechanical Systems

A mechanical system produces some form of mechanical motion. As in other systems, the load is responsible for producing this action. An example is movement produced by an industrial robot. The energy source of a mechanical system is often electrical energy, but hydraulic fluid or air can also be used.

The transmission path of a mechanical system could consist of electrical conductors, belts, rotating shafts, pipes, tubes, or cables. These are used to transfer power from the energy source to the remainder of the system.

Control is accomplished by changes in pressure, direction, force, and speed. Pressure regulators, valves, gears, pulleys, couplings, brakes, and clutches are used to control such variables as force, speed, and direction.

Three kinds of motion — rotary, linear, and reciprocating — are used in automated applications. These motions can be produced by either electrical or fluid (hydraulic or pneumatic) power operating motors, relays, solenoids, actuators, or cylinders. Industrial loads are usually designed for continuous operation for long periods of time. The basic mechanical unit or a robot, the manipulator, has several moving joints and performs the actual work function of the machine.

In a mechanical power system, indicators measure physical quantities. These include variables such as pressure, flow rate, speed, direction, distance, force, torque, and electrical values. Many of these must be monitored periodically. Some indicators are used to test system conditions during maintenance. Others are designed to measure physical changes that take place.

Electrical Systems

There are many different types of electrical systems used in robotics. Electrical systems include those used for sensing, for timing, for control, and for providing rotary motion (motors). Each type of electrical system will be described in greater detail in succeeding sections of this chapter.

Sensing Systems

Sensing systems have become one of the fastest growing and most diversified areas in industrial robotics. New systems that combine optics, electromagnetics, and electronics have revolutionized automated manufacturing.

Sensing systems respond to various forms of energy, such as light. The light source, transmission path, control, and load device are essential parts of the system. The light is produced by electrical energy. Incandescent lamps, flames, glow lamps, electric arcs, solid-state light, and laser light may be used as sources.

The transmission path of a sensing system is somewhat unique. In Figure 6-2, light energy travels in a straight line in the form of an intense beam of electromagnetic waves. If the light must go around corners or be directed to unusual locations, fiber optic rods (***light pipes***) could be used as a transmission path.

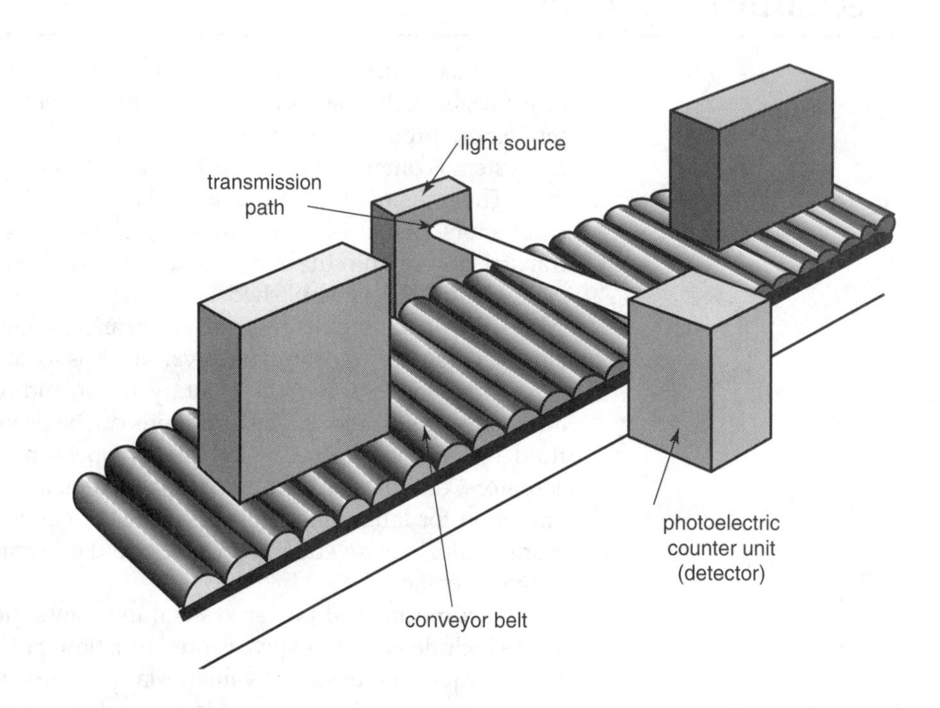

Figure 6-2
Basic parts of this sensing system respond to light energy.

The ***detector*** of a sensing system responds to energy from the source, and outputs a signal that can be used to control a load device. In some sensing systems, the load may be controlled directly by light-source energy. In others, the light energy must first be detected and amplified. There are even applications in which the detector itself serves as the load.

Sensing system control can be achieved by interrupting a light beam between the source and detector. Other control methods include altering the intensity, focus, shape, or wavelength of the light source. The detector's sensitivity can be adjusted to adapt it to specific operating conditions.

Timing Systems

Turning a device on or off at a specific time or in step with an operating sequence is done by using ***timing systems.*** Timing systems presently being used with robotics include delay timing, interval timing, and cycle timing. ***Delay timing*** is necessary when a delay is required before a load device actually becomes energized. ***Interval timing*** occurs after a load has been energized. For example, it may cause the load to remain energized only for a certain period. ***Cycle timing*** systems are typically more complex. They may include both interval and delay timing to provide energizing action in an operational sequence.

Timing systems also include thermal devices, motor-driven mechanisms, or other mechanical, electrical, or electrochemical devices. Hydraulic, pneumatic, mechanical, heat, and electrical energy sources may be used in various combinations.

Digital Control Systems

Automatic fabrication methods, packaging, and machining operations have been improved through advances in ***digital systems.*** Digital (numeric) instructions are supplied by perforated paper tape, punched cards, magnetic tape, or variations in pressure, temperature, or electric current. These instructions are then changed into a series of on/off electrical signals. The signals are processed by the logic gates of a computer and are directed to specific subsystems that then perform the necessary operations.

Electrical energy is used to energize the load device, which performs the work. The load of a system may be electrical actuators or fluid-power cylinders designed to move parts of the machine. When appropriate signals are received, these parts move some object to a specific location. Machine positions and movements, clamping operations, and material flow can all be controlled by digital systems.

Control Systems

Control of an automated manufacturing system is accomplished by human input or automatically by a physical change. During production, control systems are continually at work, making adjustments that alter machine operation. Complexity of the operation determines the number of control functions needed. In many cases, varying numbers of components are used in different parts of the system.

The control unit of an industrial robot determines its flexibility and efficiency. Some robots have only mechanical stops on each axis. Others have microprocessor (computer) control with memory to store position and sequence data. Some important factors in the selection of a control unit are speed of operation, repeatability, accuracy of positioning, and the speed and ease of reprogramming.

Nonservo, or open-loop, control systems are the most basic. They use sequencers and mechanical stops to control the end point positions of the robot arm. Pick-and-place, or "fixed-stop," robots use this type of control. Programmable servo-controlled robots are more complex. These continuous-path robots move from one point to another in a smooth, continuous motion.

Open-loop control

Open-loop systems are used almost exclusively for manual-control operations. There are two variations of the open-loop system: full control and partial control. *Full control* simply turns a system off or on. For example, in an electrical circuit, current flow stops when the circuit path is opened. Switches, circuit breakers, fuses, and relays are used for full control. *Partial control* alters system operations, rather than causing it to start or stop. Resistors, inductors, transformers, capacitors, semiconductor devices, and integrated circuits are commonly used to achieve partial control.

Closed-loop control

To achieve automatic control, interaction between the control unit and the controlled element must occur. In a closed-loop system, this interaction is called ***feedback.*** Feedback can be activated by electrical, thermal, light, chemical, or mechanical energy. Both full and partial control can be achieved through a closed-loop system. See Figure 6-3. Many of the automated systems used in industry today are of the closed-loop type.

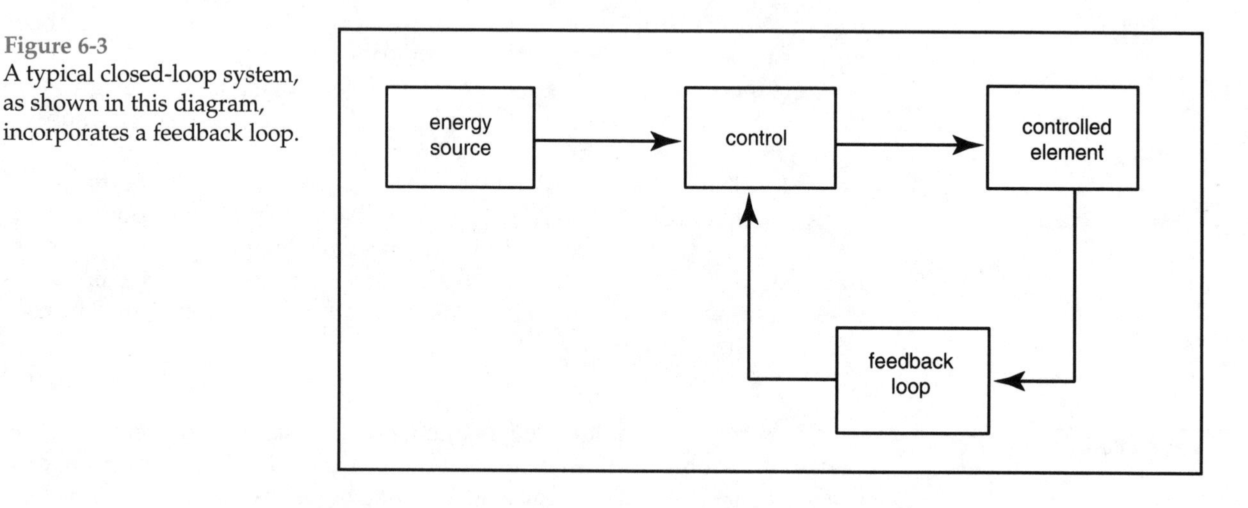

Figure 6-3
A typical closed-loop system, as shown in this diagram, incorporates a feedback loop.

Figure 6-4 shows the block diagram of a closed-loop system with automatic correction control. In this system, energy goes to the control unit and the controlled element. Feedback from the controlled element is directed to a ***comparator,*** which compares the feedback signal to a reference signal or *standard.* A correction signal is developed by the comparator and sent to the control unit. This signal alters the system so that it conforms to the data from the reference source. Systems of this type maintain a specified operating level regardless of external variations or disturbances.

Automated control has gone through many changes in recent years with the addition of control devices that are not obvious to the casual observer. These include devices that change the amplitude, frequency, waveform, time, or phase of signals passing through the system.

Electric Motors

Electric motors convert electrical energy into mechanical motion. Many types of motors are used in industry. For many industries, motors are the major power-consuming equipment. Robotic systems ordinarily use several motors.

The function of a motor is to produce mechanical energy in the form of rotary motion. Basically, an electric motor is created by placing an electromagnet, called an ***armature,*** between two permanent magnets. The north and

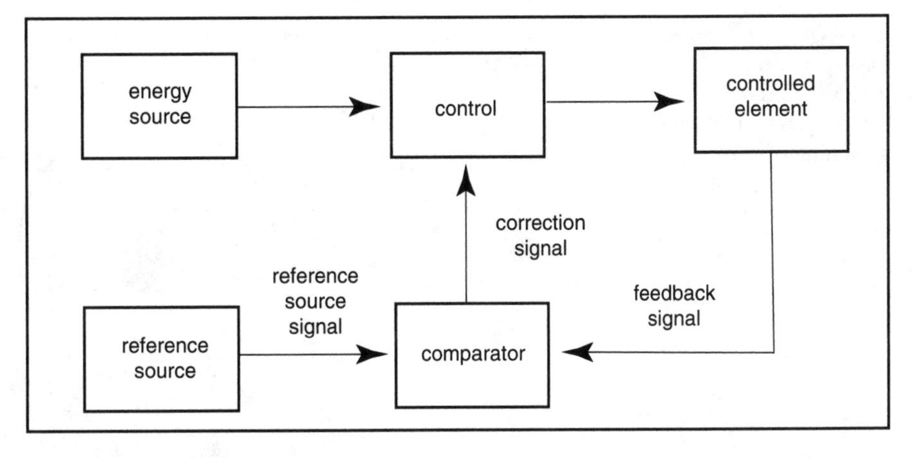

Figure 6-4
The comparator and reference source in this closed-loop control system allows automatic adjustment of a process as it is running.

south poles of the armature are aligned with the north and south poles of the permanent magnets. When a current is passed through the armature, it becomes magnetized and begins to rotate within the magnetic field of the permanent magnets. Rotation will continue until the armature's north pole is opposite the south pole of a permanent magnet, and its south pole is opposite the magnet's north pole, Figure 6-5. If the current through the armature is reversed, its poles will reverse, as well, and the armature will again rotate. The rotary motion, or *turning force,* that is produced is called ***torque.*** The amount of torque produced by a motor depends on the strength of the magnetic fields and the amount of current flowing through its conductors. As the magnetic field or the current increases, the amount of torque also increases.

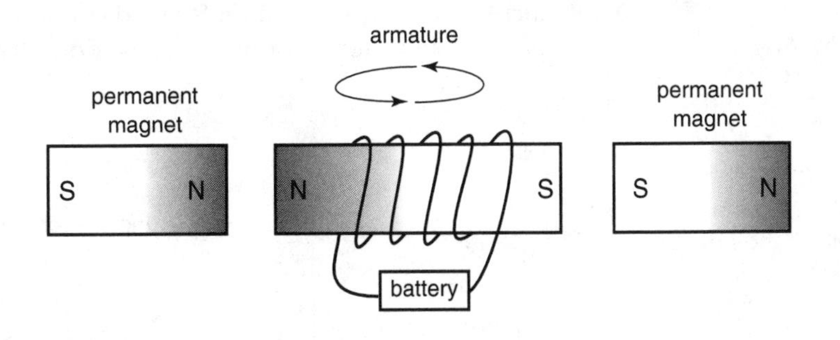

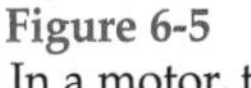
Figure 6-5
In a motor, the principle that like magnetic poles repel one another is used to cause the armature to rotate.

All motors have several characteristics in common. The ***stator*** is the permanent magnets, frame, and other stationary components. The ***rotor*** is the rotating armature, shaft, and associated parts. Auxiliary equipment includes such things as a brush-commutator assembly for dc motors and a starting circuit for single-phase ac motors.

Dc motors

Motors that operate from dc power sources are used when speed control is desirable. The basic parts of a dc motor are shown in Figure 6-6. Switching of current flow is done by the ***commutator.*** The ***brushes,*** usually

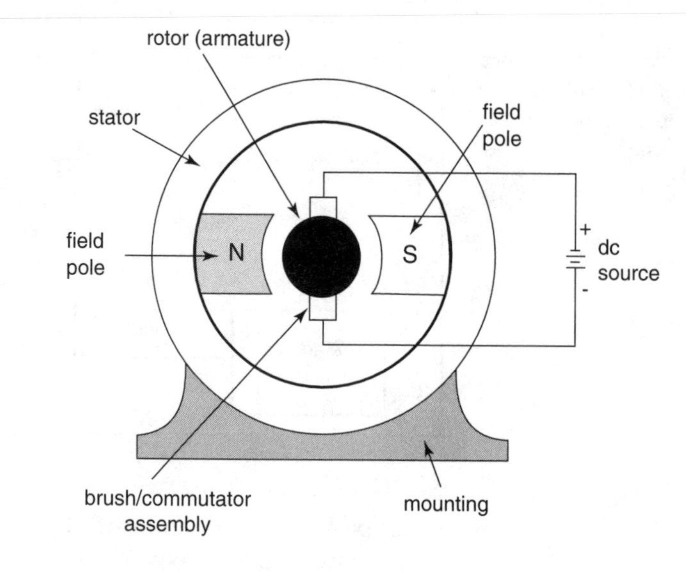

Figure 6-6
Basic parts of a dc motor. The motor includes a commutator/brush assembly for switching of current flow.

made of carbon, rub against the commutator. As current passes through the brushes and commutator on its way to the armature, a magnetic field is created. Most dc motors use electromagnets rather than permanent magnets to create the magnetic field. Coils, wrapped around iron cores, are called ***field windings.***

The operational characteristics of dc motors are shown in Figure 6-7. As the armature rotates, it generates its own voltage. This voltage is called the ***counter electromotive force (cemf).*** The cemf depends upon the number of rotating conductors and the speed of rotation. As the mechanical load increases, the speed of a motor tends to decrease and vice versa. As the speed of rotation decreases or increases, so does the cemf.

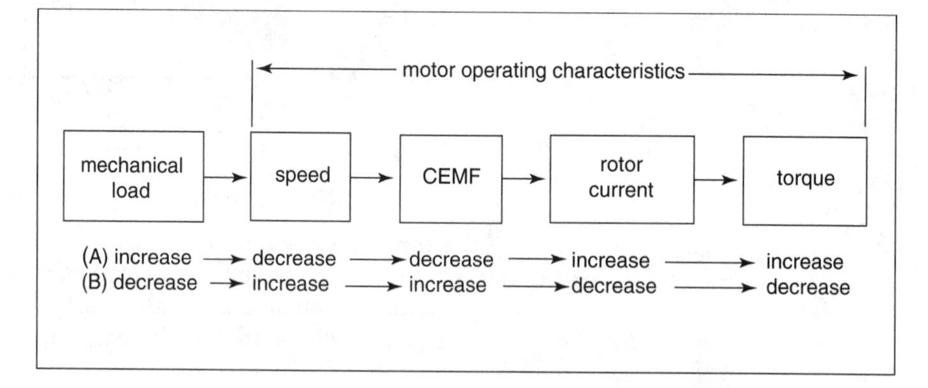

Figure 6-7
Operational characteristics of dc motors, and how they are affected by increased or decreased mechanical load.

Since the cemf is in opposition to the supply voltage, the actual *working voltage* of the motor increases as the cemf decreases. When the working voltage increases, more current flows through the rotor windings. The torque of a motor is directly proportional to current flowing through the armature. Thus, torque increases as armature current increases, and vice-versa.

Torque varies with changes in load. As the load on a motor increases, torque also increases to try to handle the greater load. As a result, the current drawn by the motor from the power source also increases.

The presence of a cemf to oppose the armature current is very important in motor operation. The lack of cemf when a motor starts explains why it draws a very large initial current, compared to its running (full speed) current. Maximum armature current flows when there is no cemf; as cemf increases, armature current decreases. To compensate for the lack of cemf and to reduce the starting current of a motor, resistances wired in series with the armature circuit are often used. Once the motor reaches full speed, these resistances are bypassed by automatic or manual switching systems. This allows the motor to produce maximum torque.

The horsepower rating of a motor is based on the amount of torque produced at the rated full-load values. This can be expressed mathematically as:

$$hp = \frac{2\pi ST}{33000}$$

$$= \frac{ST}{5252}$$

hp = horsepower rating
π = a constant
S = speed of motor, rpm
T = torque developed by motor, ft/lb

As noted earlier, an important feature of dc motors is the ability to control their speed. By changing the applied dc voltage, speed can be varied from zero to the maximum rpm. Some types of dc motors have more desirable speed characteristics than others. For this reason, we can determine the comparative speed regulation for different types.

Speed regulation is expressed as:

$$\text{Percent of speed regulation} = \frac{S_{nl} - S_{fl}}{S_{fl}} \times 100$$

S_{nl} = no-load speed, rpm
S_{fl} = rated full-load speed, rpm

The lower the percentage that results, the better the speed regulation. Thus, a motor that operates at nearly constant speeds under varying load situations will provide good speed regulation.

Commercially available dc motors fall into four basic categories: permanent-magnet, series-wound, shunt-wound, and compound-wound. Each type has different characteristics, due to its basic circuit arrangement and physical properties.

Permanent-magnet dc motor. The ***permanent-magnet dc motor*** is illustrated in Figure 6-8. It is used when a low amount of torque is needed. The dc power

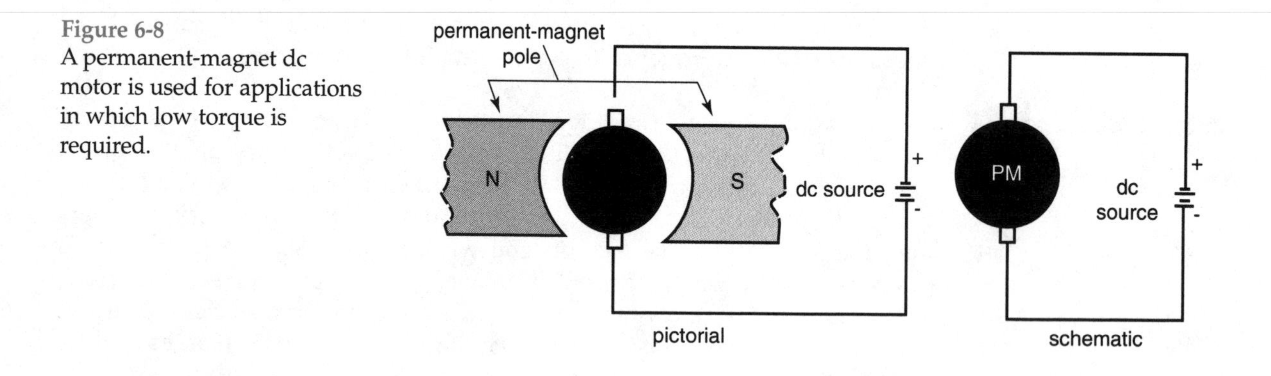

Figure 6-8
A permanent-magnet dc motor is used for applications in which low torque is required.

supply is connected directly to the conductors of the rotor through the brush-commutator assembly. The magnetic field is produced by permanent magnets mounted in the stator.

Series-wound dc motor. In a ***series-wound dc motor,*** the armature (rotor) and field circuits are connected in a series arrangement, Figure 6-9. There is only one path for current to flow from the dc voltage source, so the field has a low resistance. Changes in load applied to the motor shaft cause changes in current though the field. If the mechanical load increases, the current also increases. This creates a stronger magnetic field.

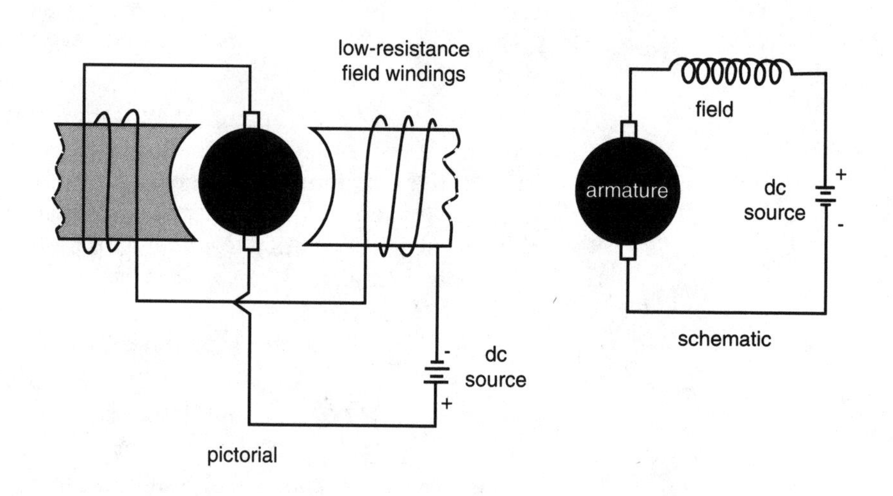

Figure 6-9
A series-wound dc motor produces high torque, but has poor speed regulation.

The speed of a series-wound motor ranges from very fast at no load to very slow under a heavy load. Since large currents may flow through the low-resistance field, the series-wound motor produces a high torque. Series motors are used when heavy loads must be moved and speed regulation is not important. They are the only type of dc motor that also can be operated using ac power. For this reason, they are sometimes called *universal motors.*

Shunt-wound dc motors. The ***shunt-wound dc motor*** is the most commonly used type of dc motor. As shown in Figure 6-10, the motor's field coils are connected in parallel with the armature (rotor). The field coils have relatively high resistance. Since the field is a high-resistance parallel path, a small amount of current flows through it.

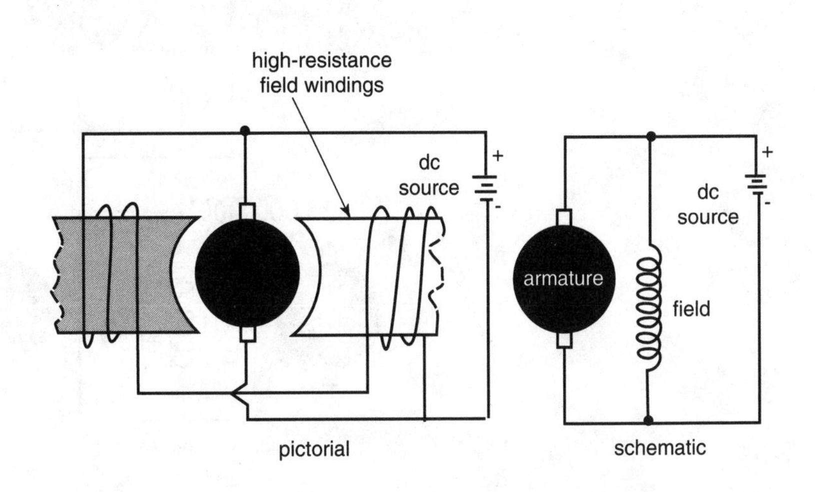

Figure 6-10
Shunt-wound dc motors are commonly used in industry because their speed is easily regulated.

Most of the current drawn by the shunt motor flows in the armature circuit. Since the armature current has little effect on the field strength, variations in load have little effect on motor speed. The field current, however, can be controlled by placing a variable resistance in series with the field windings. Since the current in the field circuit is low, a low-wattage rheostat can be used to vary the motor's speed. As the field resistance increases, the field current decreases, and vice-versa. Changes in the field current result in corresponding changes in the strength (flux) of the electromagnetic field. As field flux decreases, the armature rotates faster; as it increases, armature rotation slows.

Because its speed is easily controlled, the shunt-wound dc motor is commonly used for industrial applications. Many types of variable-speed machine tools are driven by shunt-wound dc motors.

Compound-wound dc motors. A ***compound-wound dc motor*** has two sets of field windings. One set of field windings is in series with the armature, the other is in parallel, as shown in Figure 6-11. It has the high torque of a series-wound motor and the good speed regulation of a shunt-wound motor. However, compound-wound motors are more expensive than the other types.

Single-phase ac motors

Single-phase ac motors are common in industrial as well as commercial and residential usage. They operate using a single-phase ac power source. There are three types of single-phase ac motors: universal motors, induction motors, and synchronous motors.

Universal motors. The ***universal motor*** can be powered by either an ac or a dc source. The universal motor illustrated in Figure 6-12 is built like a series-

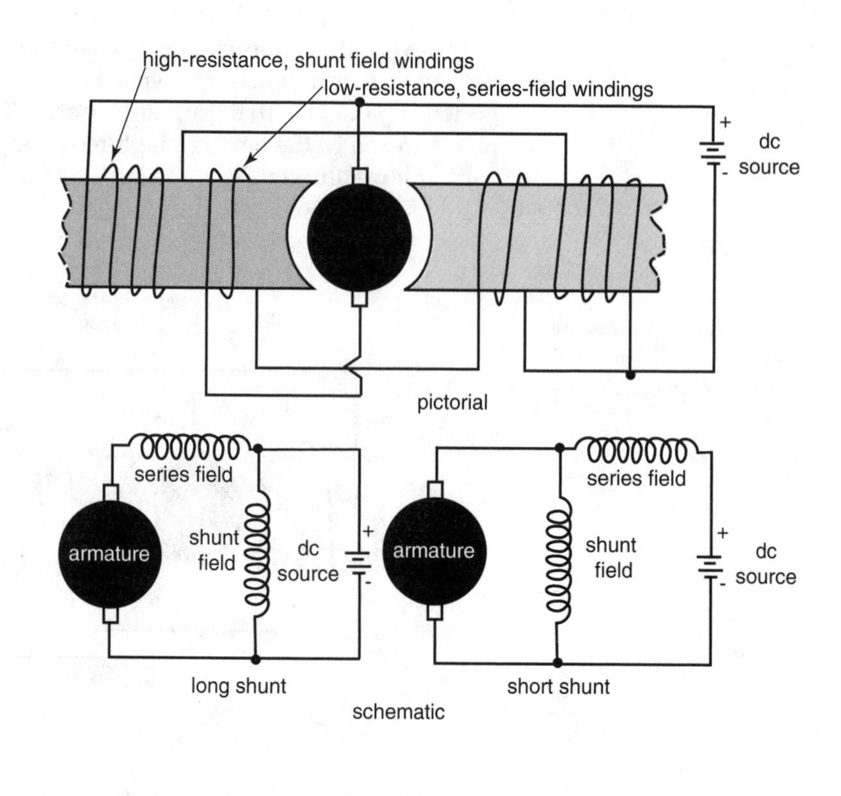

Figure 6-11
The compound-wound dc motor combines the best features of the series-wound and shunt-wound motors.

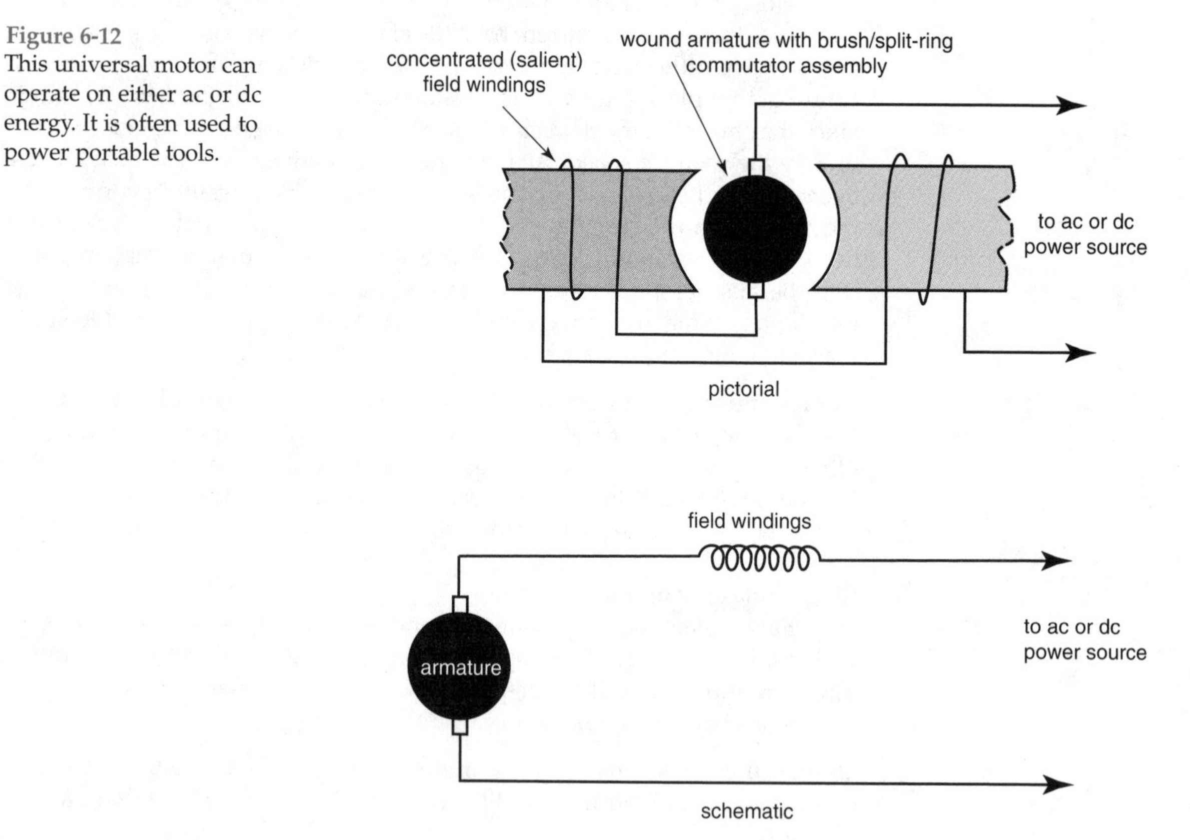

Figure 6-12
This universal motor can operate on either ac or dc energy. It is often used to power portable tools.

wound dc motor. It has concentrated field windings and speed and torque characteristics similar to those of dc series-wound motors. Universal motors are used mainly for portable tools and small equipment.

Induction motors. The ***single-phase induction motor*** has a solid rotor, referred to as a ***squirrel cage rotor,*** Figure 6-13. Large-diameter copper conductors are soldered at each end to connecting plates. When current flows in the stator windings, a current is induced in the rotor. The stator polarity changes in step with the applied ac frequency. This develops a rotating or revolving magnetic field around the stator. The rotor becomes polarized and will rotate in step with the stator's magnetic field. However, due to inertia, the rotor must be set into motion by some auxiliary starting method.

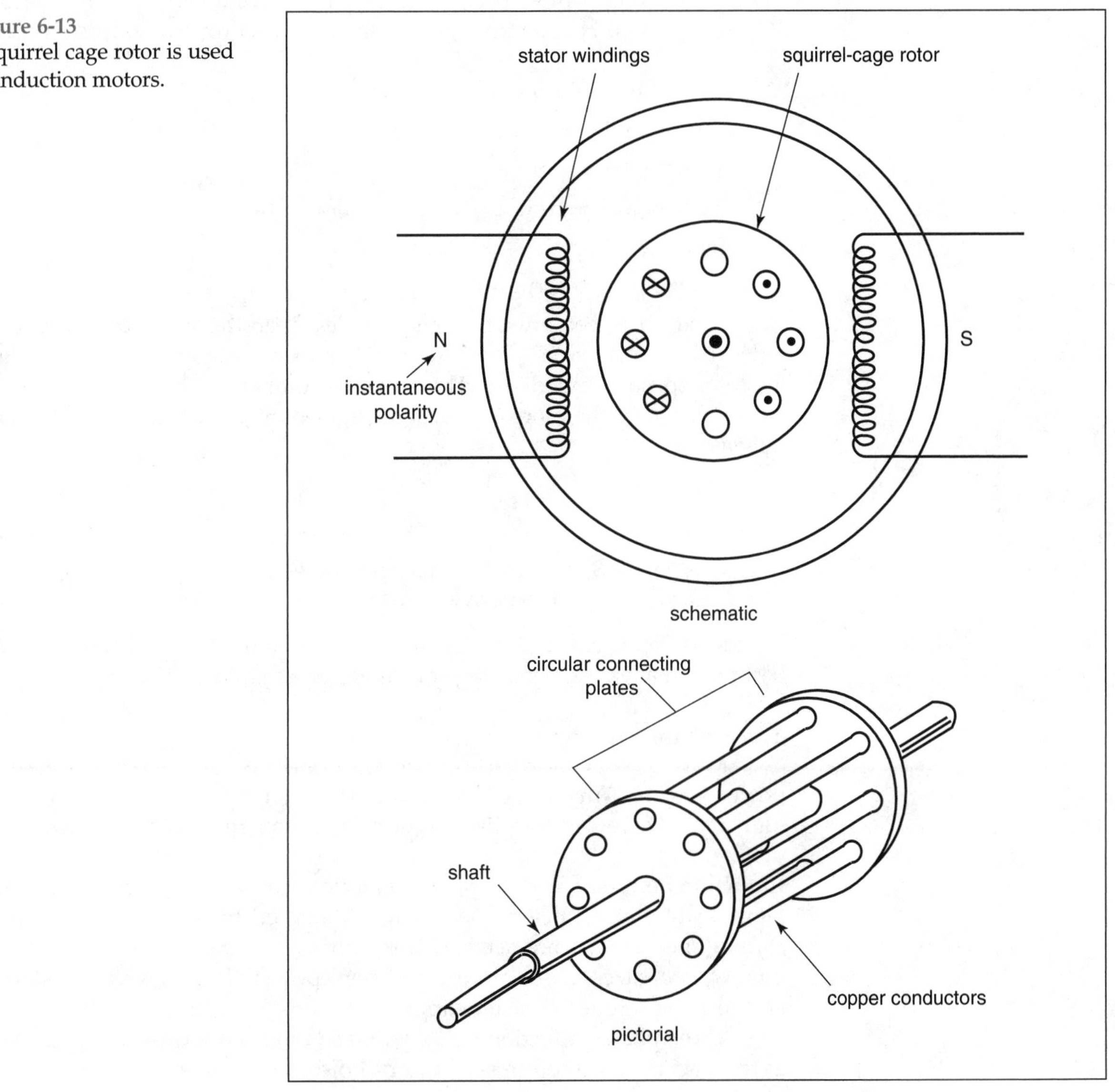

Figure 6-13
A squirrel cage rotor is used on induction motors.

The speed of an ac induction motor is based on the speed of the rotating magnetic field and the number of stator poles. The speed of the rotating field can be expressed as:

$$S = \frac{f \times 120}{n}$$

S = speed of rotating stator field, rpm
f = frequency of applied ac voltage, Hz
n = number of poles in stator windings
120 = conversion constant

The stator speed is also referred to as the ***synchronous speed.*** A two-pole motor operating from a 60 Hz source would have a synchronous speed of 3600 rpm. For 60 Hz operation, the following synchronous speeds would be obtained:

- Δ Two-pole, 3600 rpm
- Δ Four-pole, 1800 rpm
- Δ Six-pole, 1200 rpm
- Δ Eight-pole, 900 rpm
- Δ Ten-pole, 720 rpm
- Δ Twelve-pole, 600 rpm

The rotor speed must be somewhat less than the synchronous speed in order to develop torque. The difference between the synchronous speed and the rotor speed is called ***slip.*** The more the rotor speed lags behind the synchronous speed, the more torque is developed. Slip is expressed mathematically as:

$$\text{Percent slip} = \frac{S_s - S_r}{S_s} \times 100$$

S_s = synchronous (stator) speed, rpm
S_r = rotor speed, rpm

As the difference between the rotor speed and the synchronous speed becomes smaller, the percentage of slip becomes smaller, as well.

Three-phase ac motors

Most motors used in industry are operated from a three-phase ac power source. ***Three-phase ac motors*** are often called the "workhorses of industry." There are two basic types: induction motors and synchronous motors.

Induction motors. The ***three-phase induction motor*** has a squirrel-cage rotor, Figure 6-14. Since three-phase voltage is applied to the stator, no external starting mechanisms are needed. Three-phase induction motors are made in a number of sizes, based on integral horsepower. They have good starting and running torque characteristics.

Three-phase induction motors are used in such industrial applications as for machine tools, pumps, elevators, hoists, and conveyors.

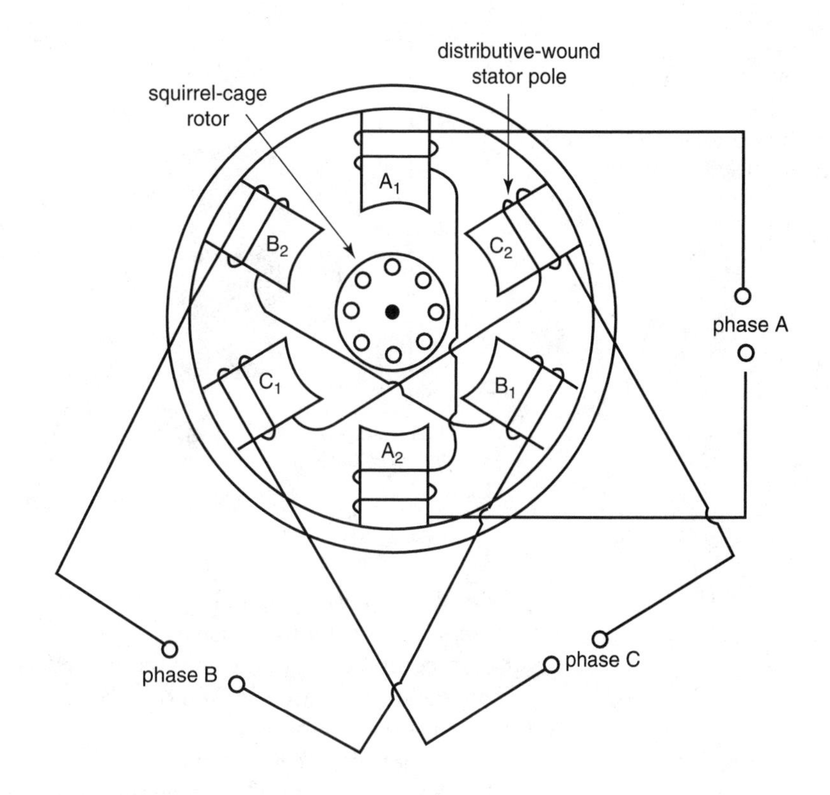

Figure 6-14
A three-phase ac induction motor, widely used for industrial applications, does not require an external starting mechanism.

Synchronous motors. The ***three-phase synchronous motor*** is unique and very specialized. It delivers constant speed and can be used to correct power factors of three-phase systems. Direct current is applied to the wound rotor to produce an electromagnetic field. Three-phase ac power is applied to the stator, which also has windings.

The three-phase synchronous motor, in its pure form, has no starting torque. Some external means must be used to initially start it. A synchronous motor will rotate at the same speed as the revolving stator field. At synchronous speed, rotor speed equals stator speed and the motor has zero slip.

Rotary Electric Actuators

Robots require ***rotary electric actuators*** to produce a type of rotary motion different from that produced by an electric motor. This type of rotary motion controls the angular position of a shaft. Through these devices, rotary motion is transmitted between locations without direct mechanical linkage.

Robotic systems use either electrical, hydraulic, or pneumatic actuators. Electrically operated systems are usually driven by dc stepping motors. Such systems are not as powerful or as fast as hydraulic units. However, they have better accuracy and repeatability and require less floor space.

Synchro systems

Synchro systems are motor-generator units that are connected together. Angular shaft positions are transmitted by electromagnetic signals, Figure 6-15. When an operator turns the generator shaft to a given position, it automatically rotates the motor shaft at a remote location to the same position.

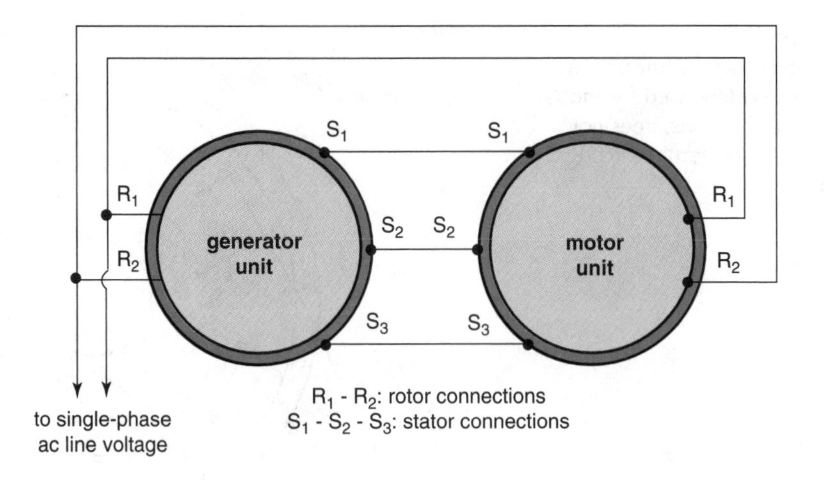

Figure 6-15
This circuit diagram shows a basic synchro system, in which a generator and motor are electrically connected.

With this type of system, it is possible to achieve accurate control over great distances. A ***servomechanism*** is a special type of ac or dc motor that drives precision equipment in angular increments. They are used in synchro systems that require increased torque or precise control. Systems that include servomechanisms generally require amplifiers and error-detecting devices to control angular displacement of a shaft.

As a general rule, the generator and motor units in a synchro system are electrically identical. Physically, the motor unit has a metal flywheel attached to its shaft to prevent shaft oscillations or vibrations. On a circuit diagram, the letters *G* or *M* are used inside the electrical symbol to denote the generator or motor functions.

Single-phase ac voltage is used to power the system. The voltage is applied to the rotors of both the generator and motor. The stator windings are connected together. When power is applied, the motor positions itself according to the location of the generator shaft. (If calibrated dials were attached to the two shafts, they would show the same angular displacement.) Systems of this type are used in automatic process control applications.

Servo systems

Servo systems are machines that change the position or speed of a mechanical object. Positioning applications involve numerical control machinery, process control indicating equipment, and robotic systems. Speed applications involve conveyor belt control units, spindle speed control in machine tool operations, and disk or magnetic tape drives for computers. As a general rule, a servo system follows a closed-loop control path, Figure 6-16.

The input of a servo system is the reference source to which the load responds. By changing the input in some way, a command is applied to the ***error detector.*** This device receives data from both the input source and the output device. If a correction is needed, a signal is amplified and applied to the actuator. The actuator is normally a servomotor that produces controlled shaft displacements. The output device is usually a synchro system that relays information back to the error detector for position comparison.

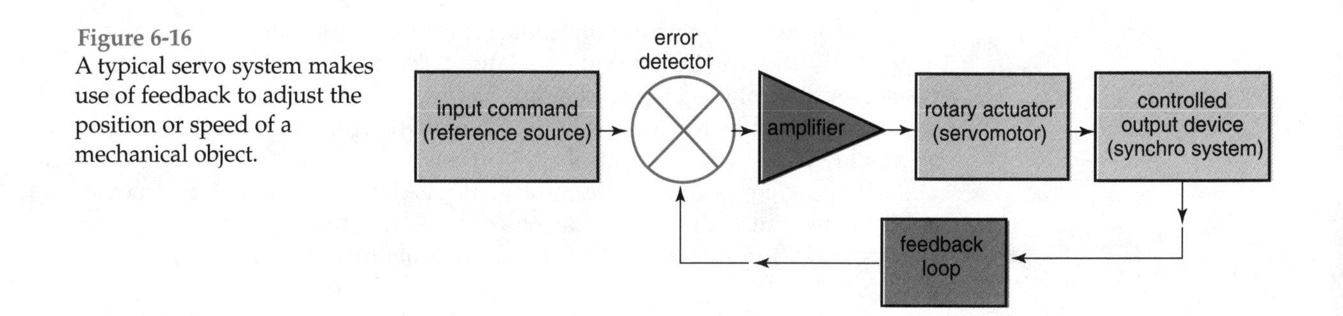

Figure 6-16
A typical servo system makes use of feedback to adjust the position or speed of a mechanical object.

Servomotors

A ***servomotor*** is used to achieve a precise degree of rotary motion. Servomotors must respond accurately to signals developed by the system's amplifier. They must also be capable of reversing direction quickly. The amount of torque developed by a servomotor must be high. There are two types of servomotors: the synchronous motor and the stepping motor.

Ac synchronous motors. An *ac synchronous motor*, Figure 6-17, contains no brushes, commutators, or slip rings. It is made up of a rotor and a stator assembly. There is no direct physical contact between the rotor and stator. For it to operate, an air gap must be carefully maintained.

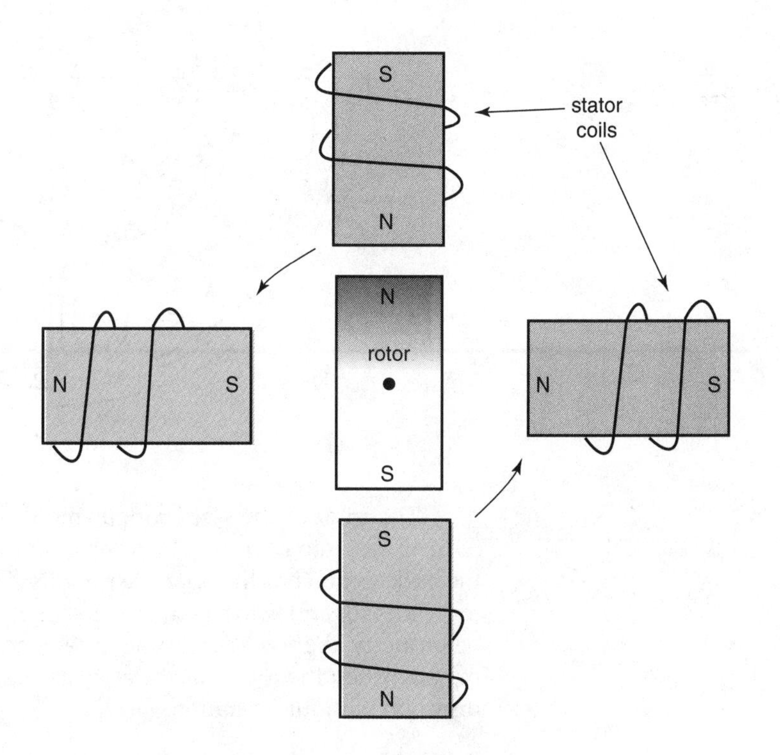

Figure 6-17
A single-phase synchronous motor is constructed simply, and is typically used in low-power applications. The rotor is a permanent magnet. Stator coils are electromagnets spaced 90° apart.

The speed of a synchronous motor is directly proportional to the ac frequency and the number of pairs of stator poles. Since the number of stator poles cannot be altered after the motor has been made, frequency is the most significant speed factor. Speeds of 28, 72, and 200 rpm are typical; 72 rpm is commonly used in digital control applications.

The single-phase ac synchronous motor is commonly used in low-power applications. Since this motor develops excessive amounts of heat during starting conditions, it is normally limited to rather low-output-power applications.

Figure 6-18 shows the stator layout of a two-phase synchronous motor with four poles per phase. Poles N_1 to S_3 and N_5 to S_7 represent one phase. Poles N_2 to S_4 and N_6 to S_8 represent the second phase. There is room for 48 teeth around the inside of the stator. One tooth per pole, however, must be eliminated to provide a space for the windings. This provides for a total of 40 teeth. The four coils of each phase are connected in series to achieve the correct polarity.

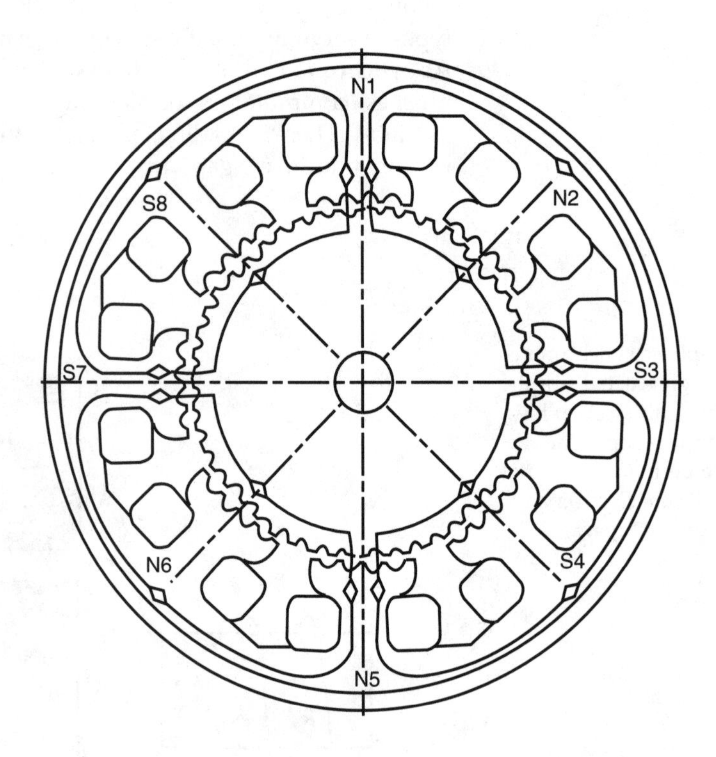

Figure 6-18
The stator layout of this two-phase synchronous motor shows the four poles per phase. It can start, stop, and reverse quickly. (Superior Electric Co.)

The rotor of the synchronous motor is a permanent magnet. There are 50 teeth cast into its form. The front section of the rotor has one polarity while the back section has the opposite polarity. The difference in the number of stator teeth (40) and rotor teeth (50) means that only two teeth of each part can be properly aligned at the same time. Since the rotor's sections have opposing polarities, the rotor has the ability to stop very quickly. It can also reverse direction without hesitation.

Because of its gears, the synchronous motor has the capability of starting in one and one-half cycles of the applied ac frequency. It can be stopped in five degrees of mechanical rotation. Synchronous motors of this type draw the same amount of current when stalled as they do when operating. This is very important in automatic machine tool applications where heavy mechanical loads are used.

Dc stepping motors. Nearly all high-power servomechanisms, Figure 6-19, use ***dc stepping motors.*** These motors are used primarily to change electrical pulses into rotary motion. They are more efficient and develop significantly more torque than the synchronous servomotor.

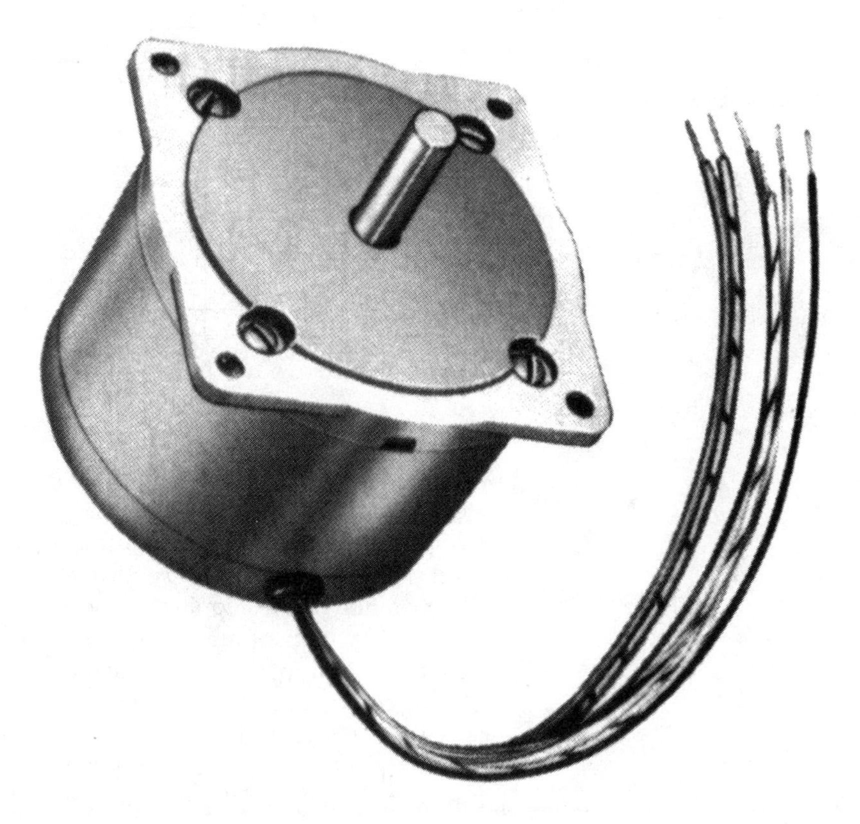

Figure 6-19
A dc stepping motor is found in nearly all high-power servomechanisms. (Superior Electric Co.)

The shaft of a dc stepping motor rotates a specific number of degrees with each pulse of electrical energy. The amount of rotary movement, or angular displacement, can be repeated precisely.

The velocity, direction, and travel distance of a piece of equipment can be controlled by a dc stepping motor. Stepping motors are energized by a dc drive amplifier that is controlled by a computer system. The movement error is generally less than 5 percent per step.

The construction of a dc stepping motor is very similar to that of the ac synchronous motor, Figure 6-20. Some manufacturers make servomotors that can be operated as either an ac synchronous motor or as a dc stepping motor. The rotor is a permanent magnet.

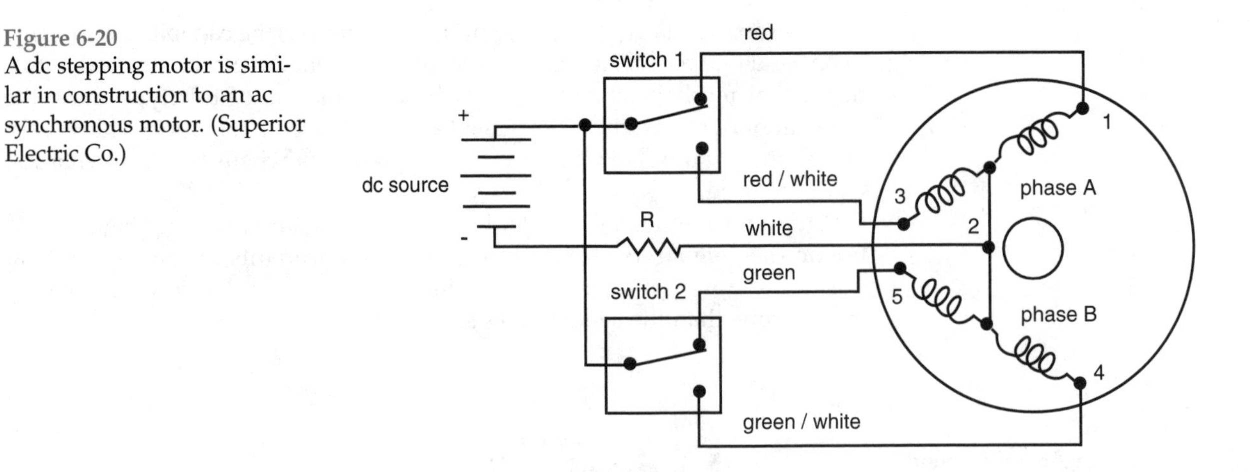

Figure 6-20
A dc stepping motor is similar in construction to an ac synchronous motor. (Superior Electric Co.)

switching sequence*

step	switch #1	switch #2
1	1	5
2	1	4
3	3	4
4	3	5
1	1	5

* To reverse direction, read chart up from bottom.

The stator coils are wound using ***bifilar construction,*** in which two separate wires are wound into the coil slots at the same time. The two wires are small, permitting twice as many turns as with a larger wire. This simplifies control circuitry and dc energy source requirements.

Operation of the stepping motor is achieved using a four-step switching sequence. Any of the four combinations of switches 1 or 2 will produce an appropriate rotor position. The switching cycle then repeats itself. Each switching combination causes the motor to move one-fourth step.

Using this rotor in the circuit shown in Figure 6-20 would permit four steps per tooth, or 200 steps per revolution. The amount of linear displacement, or step angle, is determined by the number of teeth on the rotor and the switching sequence. A stepping motor that takes 200 steps to produce one revolution will move 360°/200, or 1.8°, per step. It is not unusual for stepping motors to require eight switching combinations to achieve one step. In this case, each switching combination could be used to produce 0.25° of linear displacement, making very precise movement possible.

Overload Protection

The end effectors of industrial robots must have some type of protection against overload. Ordinarily, a feedback signal is sent to the computer system, and the manipulator is withdrawn before damage occurs. "Break-away wrists" or rapid withdrawal of a manipulator can be produced by means of mechanical fuses, detents, and preloaded springs.

Mechanical fuses are pins or tubes that will break or buckle under extreme stress. Mechanical fuses must be replaced after they perform their function, but they are not as expensive as other overload protective devices. ***Detents*** are two or more elements held in position by spring-loaded mechanisms. They move from their original positions when placed under excessive stress. ***Preloaded springs*** may also be used to prevent overload conditions. Excess stress causes the spring to release and the end effector breaks away from the work area. These devices reset automatically when the overload is removed.

Important Terms

ac synchronous motor
alternating current (ac)
armature
bifilar construction
brushes
commutator
comparator
compound-wound dc motor
control
counter electromotive force (cemf)
cycle
cycle timing
dc stepping motors
delay timing
detector
detents
digital systems
direct current (dc)
electric motors
electromechanical system
error detector
feedback
field winding
indicator
interval timing
light pipes
load
mechanical fuses
permanent-magnet dc motor
preloaded springs
rectification
rotary electric actuators
rotor
sensing systems
series-wound dc motor
servo systems
servomechanism
servomotor
shunt-wound dc motors
single-phase ac motors
single-phase induction motor
slip
squirrel cage rotor
stator
subsystem
synchro systems
synchronous speed
synthesized system
system
three-phase ac motors
three-phase induction motor
three-phase synchronous motors
timing systems
torque
transmission path
universal motors
work

Review Questions

Write your answers on a separate sheet of paper.

1. What are some systems that must be utilized in the operation of manufacturing equipment?

2. What is a subsystem?
3. What is a synthesized system?
4. How are mechanical power systems used in the operation of industrial robots?
5. How are sensing systems, timing systems, and digital systems used in the operation of industrial robots?
6. Describe how control of industrial robots is accomplished.
7. Discuss briefly the control of automated manufacturing systems.
8. What is meant by the terms "full control" and "partial control"?
9. What is "feedback" in a closed-loop system?
10. List the basic parts of an electric motor.
11. Describe the relationships of load, speed, cemf, current, and torque in an electric motor.
12. How is the horsepower of a motor determined?
13. What are some types of dc motors?
14. List the various types of single-phase ac motors.
15. Give some examples of three-phase ac motors.
16. How is the speed of an ac induction motor determined?
17. What is a synchro system?
18. Describe the purpose of a servo system.
19. What is a dc stepping motor?
20. Why is overload protection used for industrial robots?

7 Fluid Power Systems

Overview

Fluid power systems are those that use a gas (air) or a liquid (oil), or a combination of both to transfer power. Systems that use oil or a similar liquid are called hydraulic systems. Systems operated with air are pneumatic systems. The operating principles associated with these systems are similar in many respects.

In fluid power systems, electrical energy is often used to drive a fluid pump. In this way, electrical energy and mechanical motion are converted into the energy of a flowing fluid. Fluid power systems are quite reliable and flexible. Power can be transferred to any location where a pipe, hose, or piece of tubing can be placed. Pneumatic systems are used to power hand tools and to provide power for lifting and clamping during machining operations. Hydraulic systems are used for control of automated machinery and in material handling.

In this chapter, principles and applications of fluid power systems and its applications will be covered. Both hydraulic and pneumatic systems will be discussed.

Systems Models

Like an electromechanical system, a fluid power system consists of an energy source, transmission path, control, load, and indicator.

The Hydraulic System Model

The energy source for a ***hydraulic system*** is usually a pump driven by electricity. See Figure 7-1. Rotary energy is changed into fluid energy by the pump. During pump operation, fluid is drawn into the inlet port, pressurized, then ejected through the outlet port.

A typical hydraulic fluid power system is shown in Figure 7-2. It includes a number of control devices. The hand shutoff valve permits control by stopping the fluid flow. The four-way valve also has a control function. It can be positioned to restrict the amount of fluid reaching the cylinder, or to alter the flow path and reverse cylinder movement. By placing the valve's

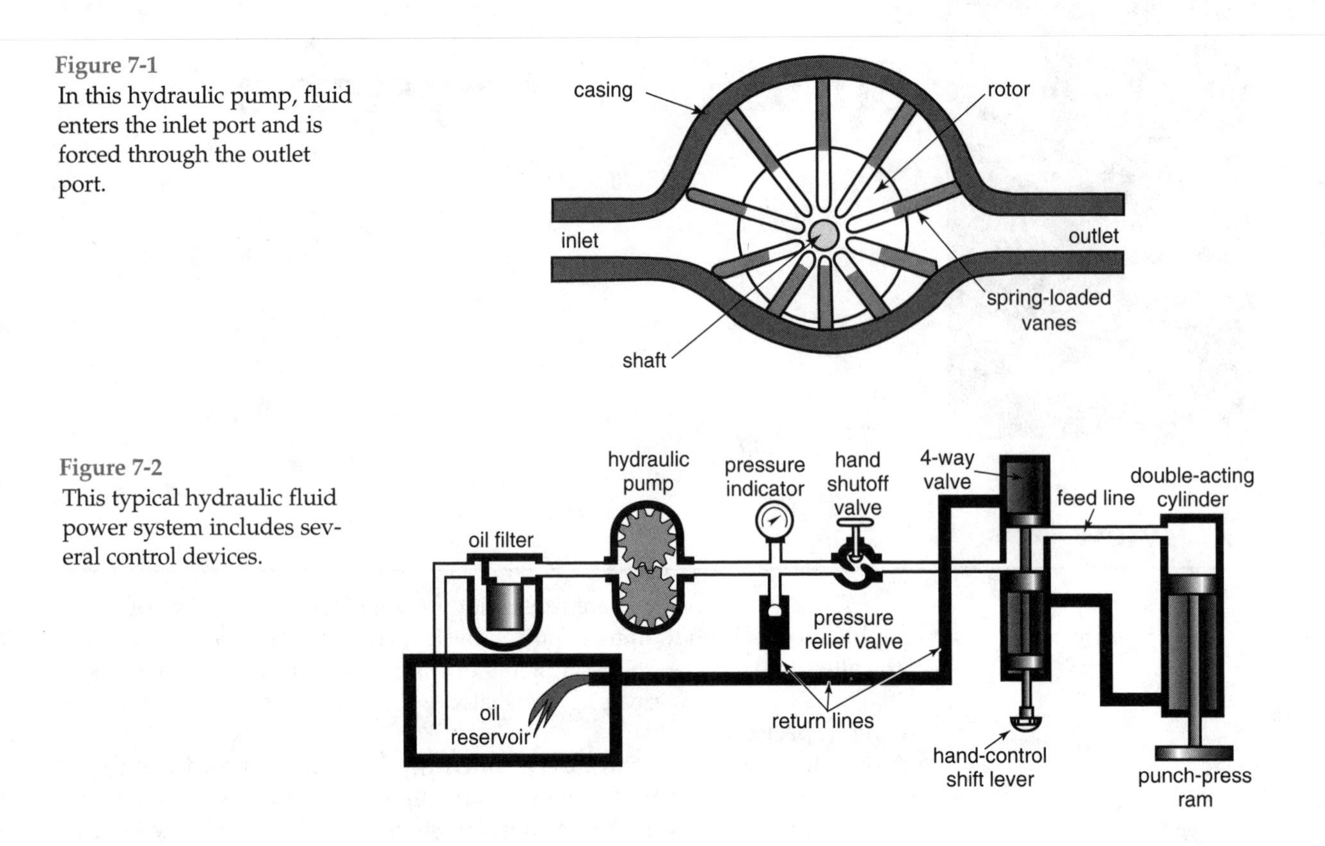

Figure 7-1
In this hydraulic pump, fluid enters the inlet port and is forced through the outlet port.

Figure 7-2
This typical hydraulic fluid power system includes several control devices.

hand-control shift lever in its off position, fluid flow can be stopped completely. The pressure relief valve is a control device that automatically protects the system from overpressure. For example, running the pump with the hand shutoff valve closed will cause the relief valve to open. High-pressure fluid is returned to the reservoir.

The double-acting ***cylinder*** serves as the system's load device. This cylinder changes the energy of hydraulic fluid flow into linear movement (in this case, the motion of the ram in a punch press).

The transmission path of a hydraulic system is often solid pipes, but may be reinforced flexible tubing. As the fluid passes through the transmission path, it encounters resistance to its flow. This builds up system pressure. The pressure inside the system can be altered by changing the operating speed of the pump. Both single-pressure and high/low-pressure hydraulic systems are used.

The pressure indicator is an optional item in the hydraulic system, as it is in other systems. Monitoring the fluid pressure and maintaining it at a constant level ensures consistent operation.

The Pneumatic System Model

The energy source of a pneumatic system is a pump or a compressor and a storage tank which holds the air. The pump or compressor may be driven by an electric motor or by an internal combustion engine.

Figure 7-3 shows a typical ***pneumatic system*** with an electrically driven air compressor. Through the action of the compressor, air is forced into a stor-

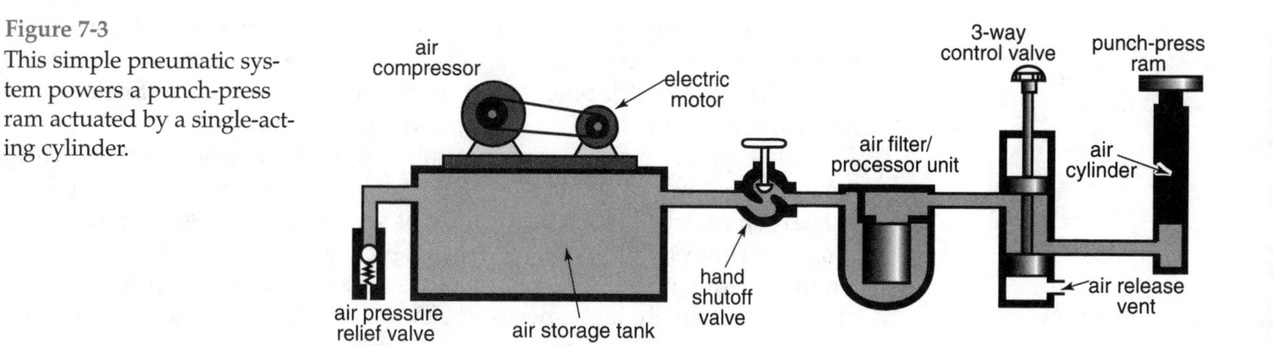

Figure 7-3
This simple pneumatic system powers a punch-press ram actuated by a single-acting cylinder.

age tank under pressure. The storage tank of compressed air serves as the reservoir for the system.

Compressed air must be conditioned to remove dirt and moisture that would damage the system. ***Conditioning*** is done with an air filter that includes a condensation trap and drain. A fine mist of oil is added to the compressed air to provide lubrication throughout the system.

Air pressure of the system must be adjusted to a specific level by an air-regulating valve. Constant pressure must then be maintained during operation. Motor-driven air compressors are designed to operate whenever the pressure in the storage tank drops below a certain level.

The transmission path of a pneumatic system consists of solid pipes, tubing, and flexible hoses used as feed lines. Unlike hydraulic systems, pneumatic systems do not use return lines to the storage tank. Return air is simply dumped into the atmosphere.

The pneumatic system in Figure 7-3 has several controls. Both the hand shutoff valve and the pressure-relief valve control air circulating through the transmission path. Air flow can also be altered by the regulator and the three-way valve.

The load device in this system is the pneumatic cylinder. It changes the mechanical energy of air into linear motion that drives a press ram. Pneumatic load devices can also be used to produce rotary motion. Some industrial tools, for example, are driven by air motors.

Indicators are usually optional items in pneumatic systems. Most commonly, one is added to monitor tank pressure. Regulator output pressure is also monitored so that exact pressures can be determined. Test indicators are frequently used to troubleshoot faulty components.

Characteristics of Fluid Power Systems

A number of recently developed industrial systems produce mechanical energy by combining fluid and electrical systems. Hybrid systems of this type play an important role in automated manufacturing, including robotics. A person working with automated manufacturing systems must be familiar with both fluid and electrical system basics.

For many applications, fluid power is the only practical solution to mechanical power transmission. These systems are useful in transferring power to inaccessible locations over moderate distances. A very fine degree of control gives a wide range of speeds and a reversing capability.

Fluid power systems have a number of characteristics that distinguish them from other power systems. For example, a very small force of a few ounces can control a much larger force of several tons. When operated under computer control, fluid-powered machines can move within tolerances of one ten-thousandth of an inch. Fluid power systems can also provide rotary motion at extremely high speeds or develop creeping speeds of a fraction of an inch per minute. They are compatible with electrical, digital, or mechanical systems. They are efficient, dependable, easy to maintain, and economical to operate for long periods of time. A high percentage of industrial machine tools are operated by fluid power.

An example of a hybrid fluid power system is the hoist used to lift cars in service stations, Figure 7-4. Both air and oil produce the power needed to lift an automobile. Adding air under pressure to the top of a long cylinder within an oil-filled tube forces the cylinder to move upward. The tube and cylinder are normally placed in the floor so that the entire unit retracts when air pressure is removed.

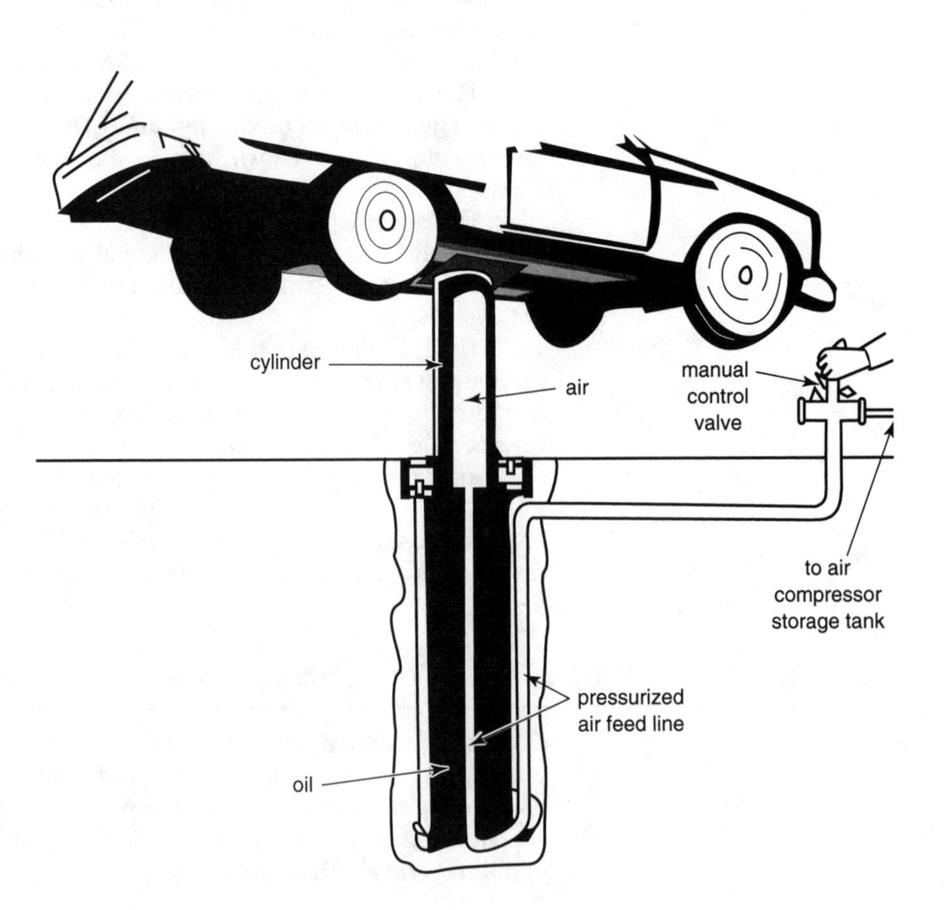

Figure 7-4
A hybrid (air and oil) fluid power system drives this automobile hoist.

Principles of Fluid Power

In 1653, the French scientist Blaise Pascal discovered that pressure applied to a confined fluid is transmitted undiminished throughout the fluid. The pressure acts on all surfaces at right angles to those surfaces. The principles set forth in ***Pascal's law*** are the basis of all fluid power systems.

Figure 7-5 illustrates Pascal's law. The force applied to the fluid at piston A is instantly transferred to all parts of the cylinder. The force exerted on piston B is equal to the force originating at A. This occurs whether or not both pistons are of the same physical size. The force acting on piston B is also applied to the walls of the cylinder. The strength of the walls must be capable of withstanding it.

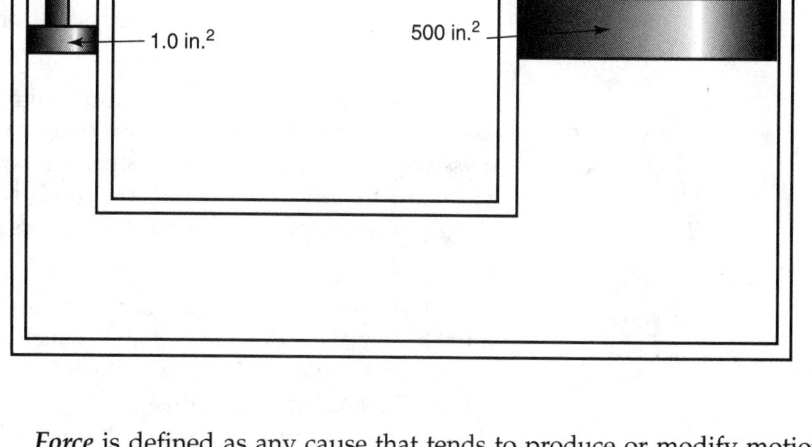

Figure 7-5
A simple hydraulic system that illustrates Pascal's law. Pressure exerted by movement of piston A acts equally on piston B and all parts of the container.

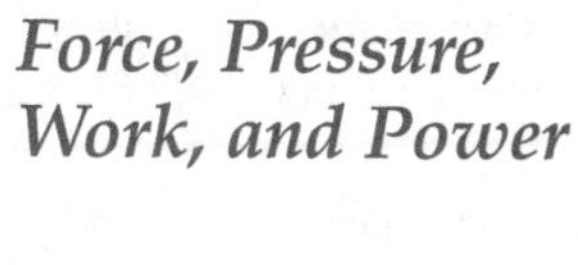

Force, Pressure, Work, and Power

Force is defined as any cause that tends to produce or modify motion. For example, to move a body or mass, an outside force must be applied to it. The amount of force needed to produce motion is based on the ***inertia*** (resistance to change) of the body. Force is normally expressed in units of weight.

In scientific terms, ***weight*** is defined as the gravitational force exerted on a body (or mass) by the earth. Since the weight of a body is a force (not a mass), we must use units of force to express both weight and force. The basic unit of force in the customary measurement system is the *pound* (lb.). In the metric system, the basic unit of force is the *newton* (N).

Pressure is a term used to describe the amount of force applied to a specific area. It is expressed in pounds per square inch (lb/in^2, or more commonly, psi) in the customary system or as newtons per square meter ($N{\bullet}m^2$) in the metric system. The *pascal* (Pa) is the basic unit of pressure in the metric system (1 Pa = 1 N•m^2). For convenience, metric pressure measurements are usually expressed in *kilopascals* (kPa). One kPa equals 1000 Pa.

At sea level, the pressure of the atmosphere on the surface of the earth is 14.7 psi (101.3 kPa). In industry, giant hydraulic presses can squeeze metals with a pressure as great as 100 x 10^6 psi. Mathematically, pressure is expressed as:

$$P = \frac{F}{A}$$

P = pressure, psi or kPa
F = force, lb. or N
A = area, in.^{2} or cm^2

An important fact about force and pressure is that they only measure effort. A measure of what the system actually accomplishes is called *work.* In Figure 7-5, work occurs when the force applied to piston A causes it to move a certain distance. Work is expressed in foot-pounds or newton-meters (joules). The mathematical formula for this relationship is

$$W = F \times D$$

W= work, ft/lb or N•m (J)
F = force, lb. or N
D = distance, ft. or m

A realistic concept of work must take into account the length of time it takes to perform. The term ***power*** is used to express this relationship. The term *horsepower* is used to describe mechanical power. Moving 33,000 pounds a distance of 1 foot in 1 minute or 550 pounds a distance of 1 foot in 1 second are descriptions of horsepower. Electric motors are rated in horsepower.

A Simple Fluid Power System

A simple static fluid power system is shown in Figure 7-6. In this system a 100 lb. force is applied to piston A, which has an area of 1 in.^{2}. This develops a pressure of 100 psi (690 kPa), which is transferred through the fluid to piston B. The area of piston B is 100 in.^{2}. Since pressure is transferred equally through the fluid to all parts of the cylinder, each square inch of piston B receives 100 lbs. of force. As a result, 100 psi x 100 in.^{2} equals 10,000 lbs. of force applied to piston B.

The distance that piston B moves is directly proportional to the *area ratio* of the two pistons. Moving the 1 in.^{2} piston 4 inches into the cylinder displaces 4 in.^{3} of fluid. This displaced volume is based on the area of the piston times the distance it moves. Therefore, 1 in.^{2} x 4 in. = 4 in.^{3} of fluid displacement. Spread over the 100 in.^{2} surface of piston B, this displacement causes piston B to move only 1/100 of the distance traveled by piston A (1/100 of 4 in. equals only 0.04 in. of motion). Piston B, therefore, receives more force because of its size, but travels only a short distance.

The amount of work done by pistons in a static system shows an unusual relationship. The work done by piston A is 100 lb. x 4 in., or 400 in./lbs. At piston B, the amount of work achieved is also 400 in/lbs This is determined by multiplying the applied force by the distance moved. Therefore, 10,000 lbs. x 0.04 in. = 400 in/lbs of work.

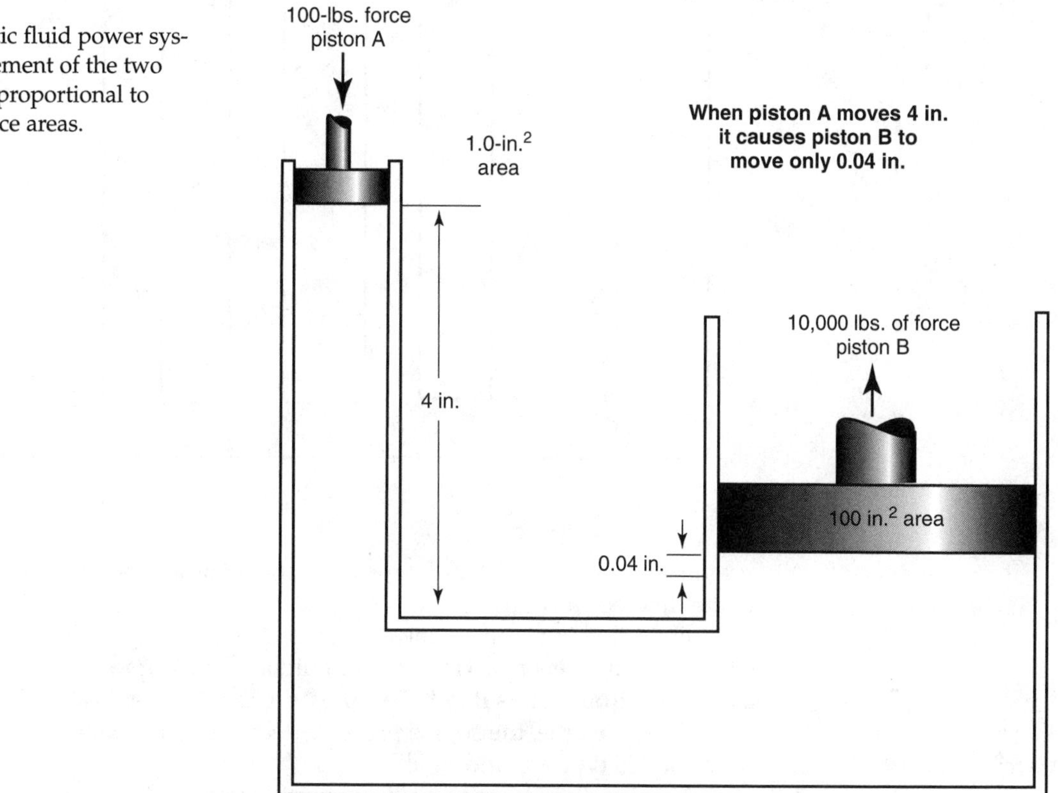

Figure 7-6
In this static fluid power system, movement of the two pistons is proportional to their surface areas.

Characteristics of Fluid Flow

In actual practice, fluid power systems do not achieve a 100 percent transfer of power from input to output. For example, fluid moving along cylinder walls encounters surface friction. The amount of power loss appears primarily as heat. In a static system, the amount of power loss due to heat is negligible. In systems that move larger volumes of fluid over longer lines, however, losses of this type are important. As a general rule, friction losses can be controlled by reducing the line length, limiting the number of bends, and by using lines of the proper size to prevent high fluid velocity. Proper design generally takes these things into account, so that a high level of efficiency can be achieved.

In a fluid system, friction is usually called ***resistance.*** System pressure develops as a result of fluid being forced against this resistance. A direct relationship, therefore, exists between system pressure and surface resistance.

Figure 7-7 illustrates the friction/pressure relationship of a static system. The pressure at point *F* is zero. A break in the system could cause such a pressure reading. Other parts of the system show varying amounts of pressure in

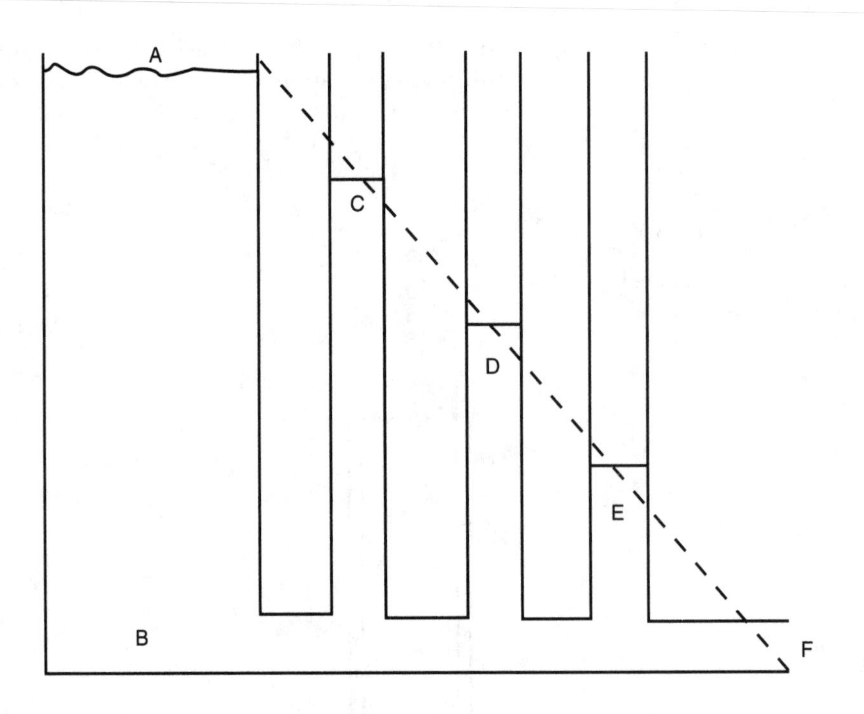

Figure 7-7
The friction/pressure relationship in a static fluid power system is shown in this graph.

response to resistance. Point *B* represents the highest pressure. The full weight of the fluid occurs at this point.

Fluid flowing from point *B* to point *F* must change all of its potential energy into heat energy. The moving fluid also undergoes a drop in pressure as it passes from points *B* to *F*. This pressure drop increases at each point after *B*. At the same time, the source pressure decreases an equal amount. Pressure drops of this type are undesirable.

Nearly all industrial systems use flowing, rather than static, fluid. Pressure drop is very important in a flowing system. Turbulence created where abrupt changes in direction occur will cause pressure drop. For example, note the corners of the system shown in Figure 7-8. System *restrictions,* such as control valves, tubing length, or reduced tubing size, are also a source of pressure drop. Smaller lines tend to increase the speed of fluid flow. This in turn causes an increase in surface friction, causing a pressure drop. Fluid also encounters more resistance when it travels long distances, with the same result. Proper system design usually minimizes this factor. In some applications, however, a drop in pressure is used to trigger the start of a second operation.

When the flow ceases, the pressure reaches a stable value throughout the system. A faulty pump or loss of electricity could cause this to occur. By comparison, a break in the system line normally causes a complete loss of pressure, or a very pronounced change in pressure value.

Compression of Fluids

Compression of fluids differs in hydraulic and pneumatic systems. All gases and liquids are compressible under certain conditions. Ordinarily, however, a hydraulic fluid (liquid) is not compressible except in extremely long

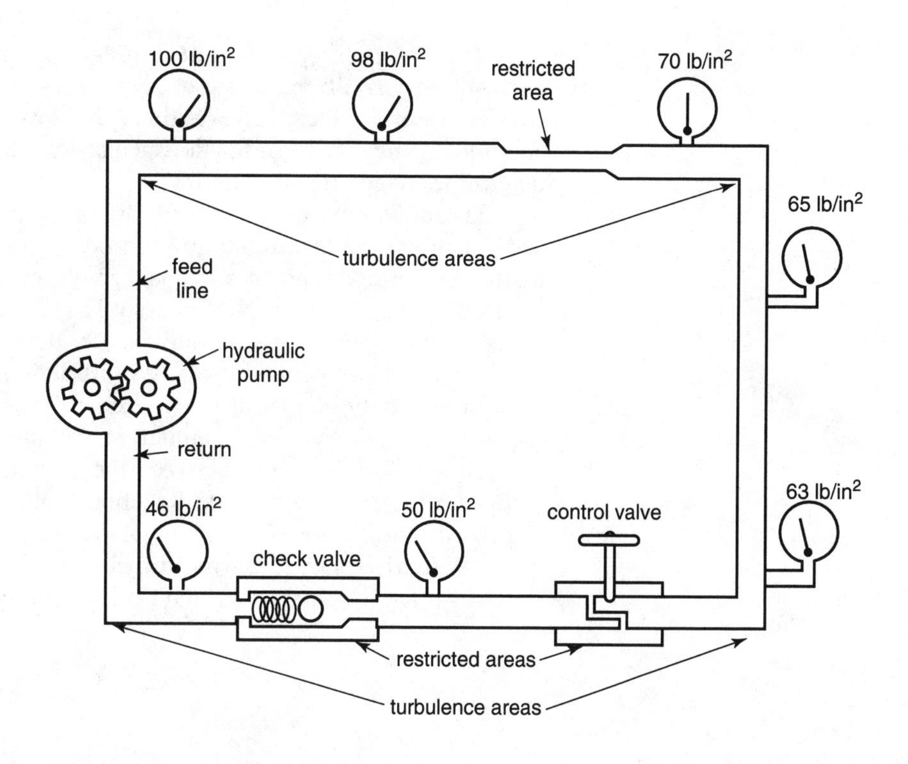

Figure 7-8
In a flowing fluid power system, pressure drops will occur where fluid flow is restricted, or where turbulence develops as a result of a change in direction of flow.

transmission lines or under very high pressures. A volume reduction of approximately 0.5 percent for every 1000 psi (6895 kPa) of pressure is typical. This is not a real problem under normal operating conditions.

In pneumatic systems, air must first be compressed before it can be effectively used. When air is compressed, its pressure increases. The air is stored under pressure in a tank and released into the system as it is needed. This difference in compressibility results in some differences between hydraulic and pneumatic system components. Typically, however, the basic function of a given component is similar for either system.

Basic Equipment

Hydraulic and pneumatic systems use similar components. In some instances, they can even be used interchangeably. However, hydraulic components are generally larger and more rugged in construction. This is necessary because oil is denser than air. In addition, hydraulic systems tend to be used at high pressure levels in heavy-duty automated operations. Aside for these differences in size, ruggedness, and pressure, hydraulic and pneumatic equipment operate on the same basic principles and respond in a similar manner.

Fluid Pumps

The pump is the heart of a fluid system. It provides an appropriate flow that will develop pressure. In this way, it is similar to the human heart.

The pump accepts fluid at an inlet port, moves it through a confined area, and expels it from an outlet port. Gases are typically compressed into a smaller volume, which increases their pressure. To increase the pressure of oil and liquids, they are forced to flow at a faster rate. The specific use for a pump determines which role it will play.

Hydraulic pumps operate continuously to keep the fluid in a constant state of motion. A pneumatic pump, on the other hand, operates intermittently. It compresses air into a smaller volume and forces it into a receiving tank for storage. When the tank pressure builds up to a certain level, the pump turns off. When pressure drops sufficiently, the pump turns on again. Air compressors are often operated for only short periods.

Pumps may be placed in two general classifications, positive displacement and nonpositive displacement. See Figure 7-9. A ***positive displacement pump*** has a close clearance between the moving member and the stationary components. As a result, a definite amount of fluid passes through the pump during each revolution. With a ***nonpositive displacement pump*** no set amount of air or fluid is moved by the impeller blades during each rotation. Flow depends upon the speed at which the blades are moving.

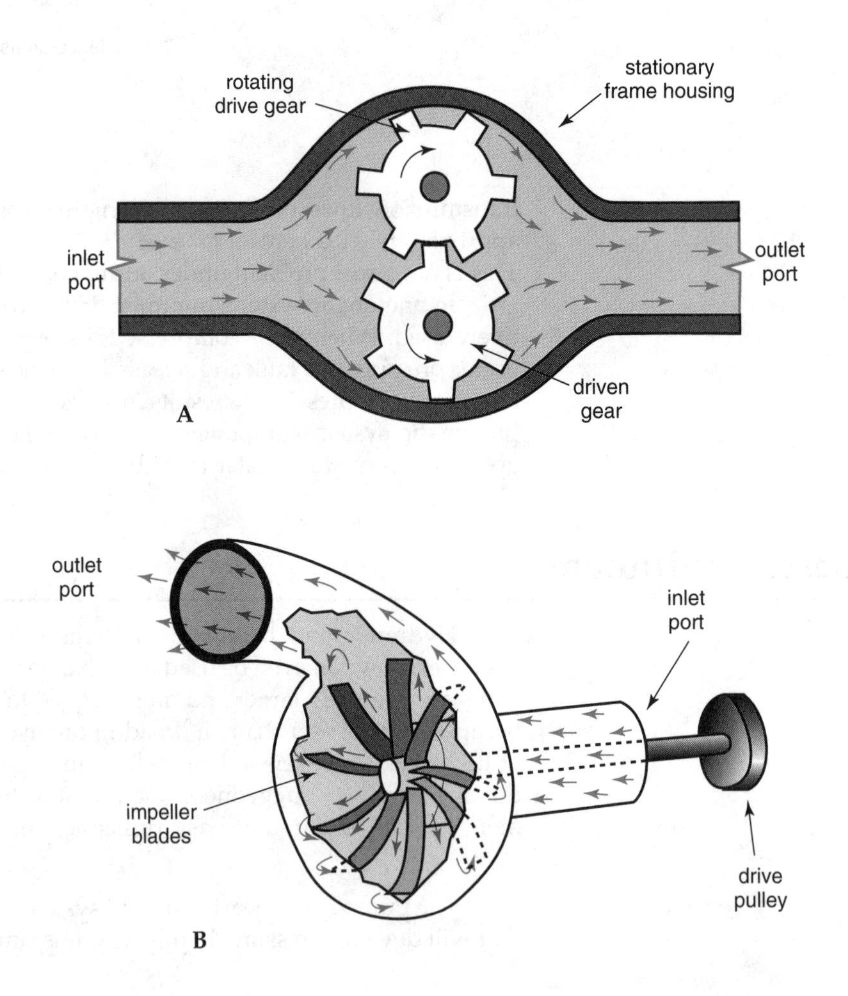

Figure 7-9
The two general types of pumps. A—Positive displacement pumps, like this gear pump, move a specific volume of fluid with each rotation or cycle. B—Nonpositive displacement pumps, such as this centrifugal pump, do not move a specific volume of fluid with each rotation. Flow volume depends upon pump speed.

The applications for fluid pumps in automated manufacturing are so varied that it is difficult to list all of the different kinds used. Four basic pumps are covered in the following sections. Three are positive displacement types, the fourth is a nonpositive displacement pump.

Rotary motion is commonly used to produce pumping action. Gear and vane pumps operate on this principle. They are the most common pump types, and are available in many sizes and styles. The centrifugal pump uses an impeller blade. It is more speed-dependent than the other rotary types. The back-and-forth movement of a piston is used by the reciprocating pump to alternately fill and empty a chamber. Reciprocating pumps are often used to compress air.

Reciprocating pumps

A ***reciprocating pump*** uses the reciprocating action of a moving piston to move a fluid into and out of a chamber. A partial vacuum is created inside the chamber by the piston as it is pulled to the bottom of its stroke, Figure 7-10A. The intake valve opens, admitting fluid (air or oil) to the chamber. The chamber fills to capacity by the time the piston reaches the end of its stroke.

As the piston reaches the bottom, the rotary motion of the drive disk causes it to change direction. The discharge valve opens and the intake valve closes. See Figure 7-10B. Fluid is forced out of the chamber.

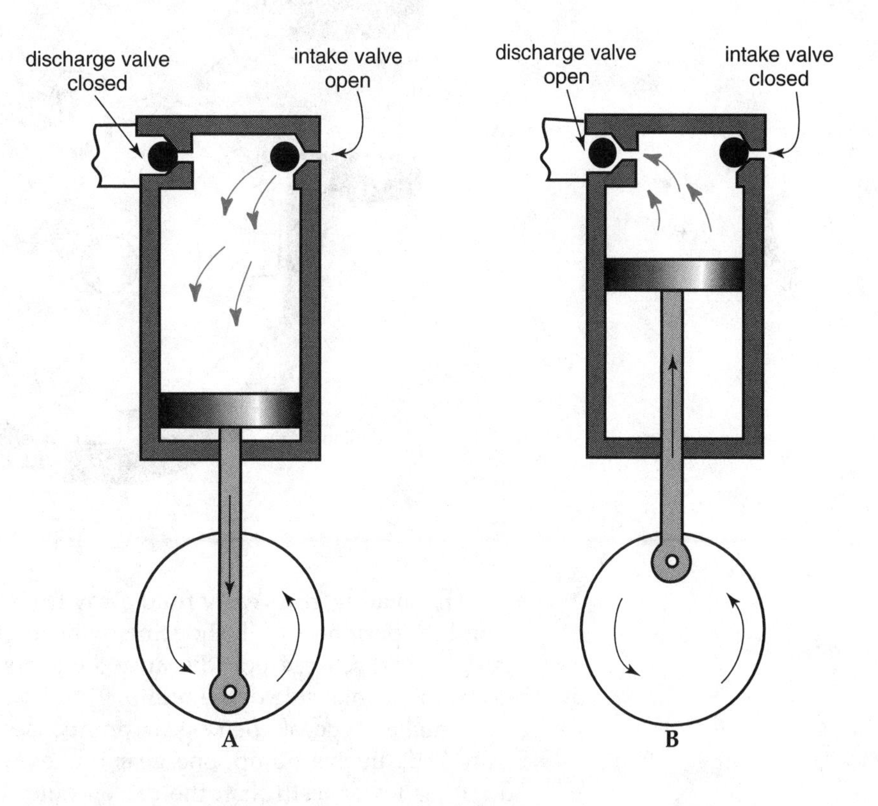

Figure 7-10
A reciprocating pump. A—During the intake stroke, the intake valve opens and the piston moves down, drawing fluid into the cylinder. B—For the discharge stroke, the intake valve closes and the discharge valve opens. The piston moves upward to force the fluid out of the cylinder.

For each revolution of the motor shaft, the reciprocating pump piston completes both an intake and a discharge stroke. Piston area and chamber volume are the key factors in determining the potential output of this type of pump. In some situations, two or more stages or cylinders may be driven by the same motor shaft.

Rotary-gear pumps

Figure 7-11 shows the basic construction of a ***rotary-gear pump.*** This type of pump, called an *external-gear* pump, contains two gears enclosed in a precision-machined housing. Rotary motion from the power source is applied to the drive gear. As it rotates, it causes the second (driven) gear to turn. The teeth of the two gears mesh in the middle of the pump.

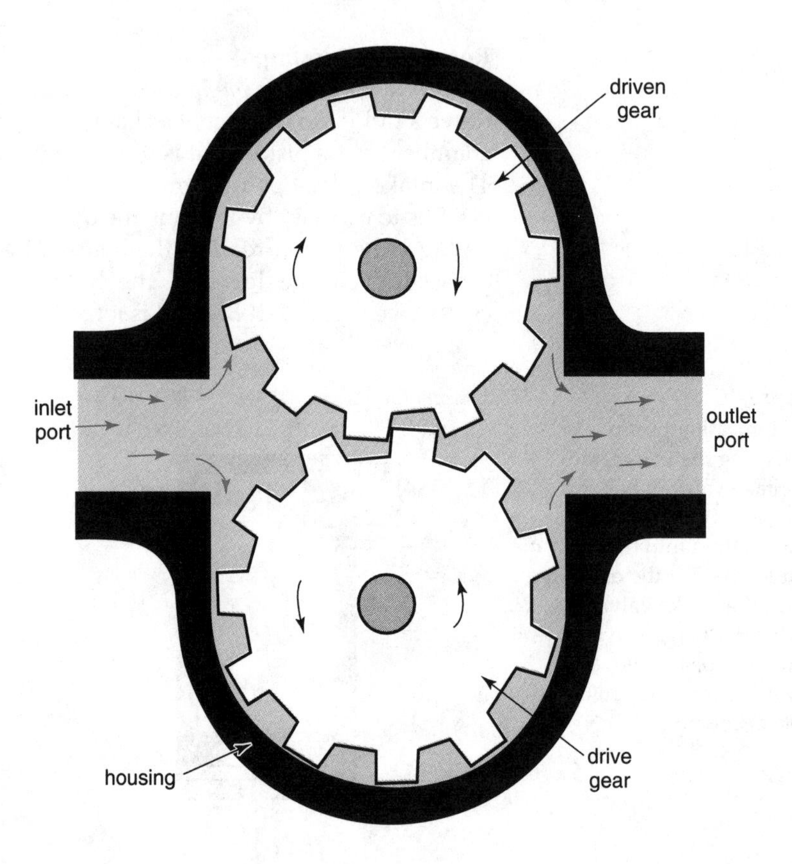

Figure 7-11
Basic construction used for the external-gear type of rotary pump. Spaces between the teeth of the rotating gears carry fluid around the inside of the housing from the inlet port to the outlet port.

The rotating gears carry fluid away from the inlet side of the pump and around the periphery of the housing to the discharge side. The fluid is ejected through the discharge port. Because the gears mesh tightly, very little fluid returns to the inlet side of the pump.

Another type of rotary-gear pump uses internal gears. As shown in Figure 7-12. In this pump, one gear rotates within another. The inner gear (idler) has fewer teeth than the driven outer gear (rotor). As the idler gear rotates within the rotor gear, the gear teeth unmesh at the inlet port and remesh at the discharge port. Fluid is drawn through the inlet port, filling the spaces between the teeth. The fluid moves smoothly around the head crescent.

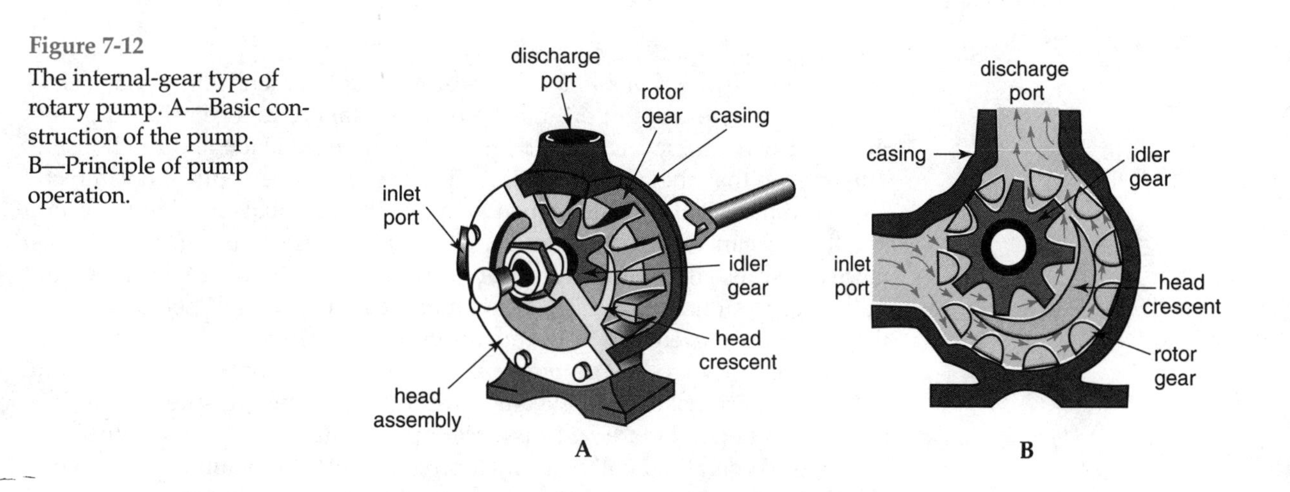

Figure 7-12
The internal-gear type of rotary pump. A—Basic construction of the pump. B—Principle of pump operation.

It is expelled at the discharge port by the remeshing of the teeth. An internal-gear rotary pump operates equally well in either direction. Output usually ranges from 0.5 gpm to 1100 gpm (gallons per minute).

Rotary-vane pumps

The ***rotary-vane pump*** uses a series of sliding vanes placed in slots around the inside of the rotor to move fluids. As the rotor turns, centrifugal force or spring action forces the vanes outward. These vanes capture the fluid as it passes by the inlet port. As the rotor turns, the fluid is moved to the outlet port.

Figure 7-13 shows the construction of an unbalanced (offset), straight-vane rotary pump. The rotor is offset toward the bottom of the housing. As a result, large volumes of oil or air can move across the top with little or no return through the bottom.

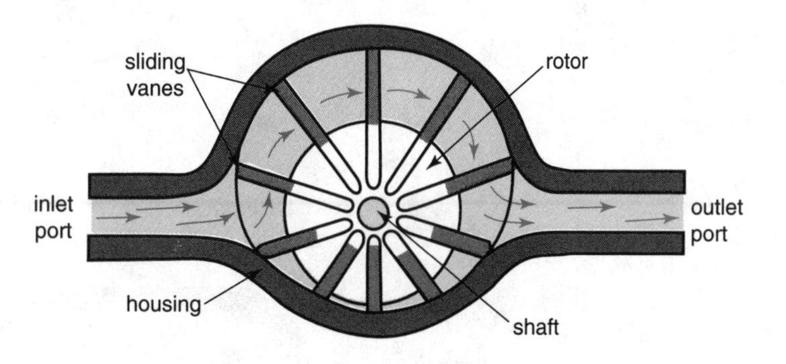

Figure 7-13
An unbalanced straight-vane rotary pump moves a large volume of fluid across the top of the chamber.

The *balanced-vane* type of pump is designed with the rotor in the center of its housing. Large areas for fluid are located in both the top and bottom of the housing. Separate inlet and outlet ports on each side make double pumping possible. As a result, the flow of air or oil is smoother than with an unbalanced type of pump.

Centrifugal pumps

The ***centrifugal pump*** is a nonpositive displacement pump that moves an indeterminate amount of fluid with each rotation. In a centrifugal pump, there is always some clearance between the impeller blade and the housing. This means that the amount of fluid leaving the outlet port is not directly related to the input. Volume depends upon the rotational speed of the pump and the resistance in the feed line connected to the outlet port. Increased resistance may cause fluid flow to slow or even come to a complete stop. Any fluid in the pump simply rotates inside without being expelled. When this occurs, the operating efficiency of the pump drops to zero. An increase in pump speed can be used to solve this problem. As a general rule, pumps of this type are only used for transferring large amounts of fluid at low pressure.

The pump in Figure 7-14A is a *volute* pump. Its housing has a volute (spiral) shape. When fluid enters the inlet port, it is set into rotation by the revolving blades of the impeller. Centrifugal force causes the fluid to move outward toward the wall of the housing. The fluid circulates in a spiral path toward the outlet port. To keep the flow continuous, the inlet port continually replaces fluid expelled from the outlet port.

In the *axial-flow* pump illustrated in Figure 7-14B, the blades of the impeller maintain fluid flow along the axis of rotation of the drive shaft.

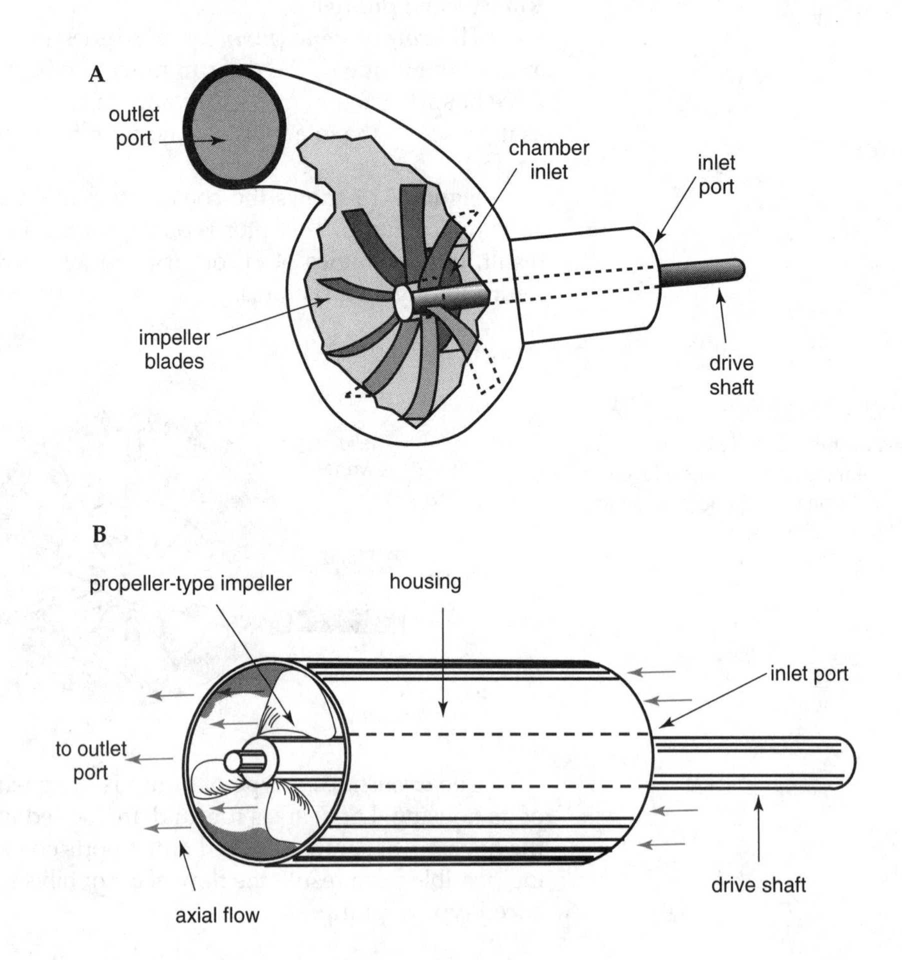

Figure 7-14
Two types of centrifugal pumps. A—The volute pump has a spiral housing that directs the flow of fluid to the outlet port. B—The axial-flow pump has an impeller that moves fluid along the axis of the drive shaft.

Fluid Conditioning Devices

Air or hydraulic fluid must be conditioned before it is processed through the system. Conditioning devices prolong the system's life by removing foreign particles or moisture, or both. The number of conditioning devices used depends upon the type of system. A simple hydraulic system may have just a line filter or strainer. A more sophisticated system may require filters, strainers, and heat exchangers. Conditioning in pneumatic systems is more complex. Devices in these systems filter the air to remove dirt and water, regulate the pressure to the proper level, and add oil as a lubricant.

Hydraulic conditioning

The number of components, types of control devices, and the operating environment are the major factors to consider in hydraulic fluid conditioning. For systems with manually operated control valves in a clean environment, a simple intake strainer may be enough. For systems with precision control valves that operate for long hours in a dirty environment, however, micron filters and several strainers are a necessity.

Strainers are inline devices that capture larger particles of foreign matter. They contain a stainless steel screen with 60 to 200 wires per square inch. Strainers are frequently placed in the reservoir filler opening, air breather, and pump inlet feed line.

Filters provide a finer grade of fluid conditioning. Typically, they are made of some porous medium such as paper, felt, or very fine wire mesh. Filter ratings range from 1 to 40 microns. A micron is a measurement equal to one-millionth of a meter or 0.00003937 in. The micron rating refers to the particle size that is permitted to pass through the filter.

Inline and T filters are commonly used in hydraulic systems. The T filter has a removable bowl or shell that contains the filter element. A bypass relief valve opens when the filter element becomes clogged and restricts flow. Inline filters must be removed when the element is cleaned or replaced. Bypass relief valves are optional with this type of filter, depending on its application.

Some hydraulic systems use ***heat exchangers*** to maintain the temperature of the fluid at a desired level. Hydraulic machinery that operates near a furnace or that is used near hot metal often requires heat exchangers to cool the fluid. Heat exchangers may be forced-air fan units, water-jacket coolers, or gaseous cooling units. Cooling is needed more often than heating, since a hydraulic system produces heat during normal operation. Heating is required only for portable systems during cold starts.

Pneumatic conditioning

In pneumatic fluid conditioning, several different types of devices are used. Filtering must remove moisture as well as foreign particles. Inline and T filters have chemical elements made of a ***desiccant.*** This is a very dry substance designed to attract moisture. Desiccant elements often require periodic *recharging,* a heating process that dries the element. Some T filters contain a glass-bowl moisture trap at the bottom of the bowl. If the bowl shows an accumulation of moisture, it must be cleared by opening a drain valve.

Pressure regulation is also necessary in pneumatic systems. After air passes through a T filter, it goes to a regulator valve. The movement of air through this valve can be controlled with an adjustment screw. The system line pressure from the receiving tank can be set to the desired operating level.

The ***pressure regulator*** shown in Figure 7-15 creates a balance between atmospheric pressure and system line pressure. Atmospheric pressure reaches the top of the diaphragm by means of the vent. System pressure is applied to the bottom of the diaphragm. Turning the adjustment screw adds mechanical pressure to the atmospheric pressure. When this combined pressure exceeds the system pressure, the diaphragm is pushed down. This action opens the poppet valve, admitting more air from the receiving tank.

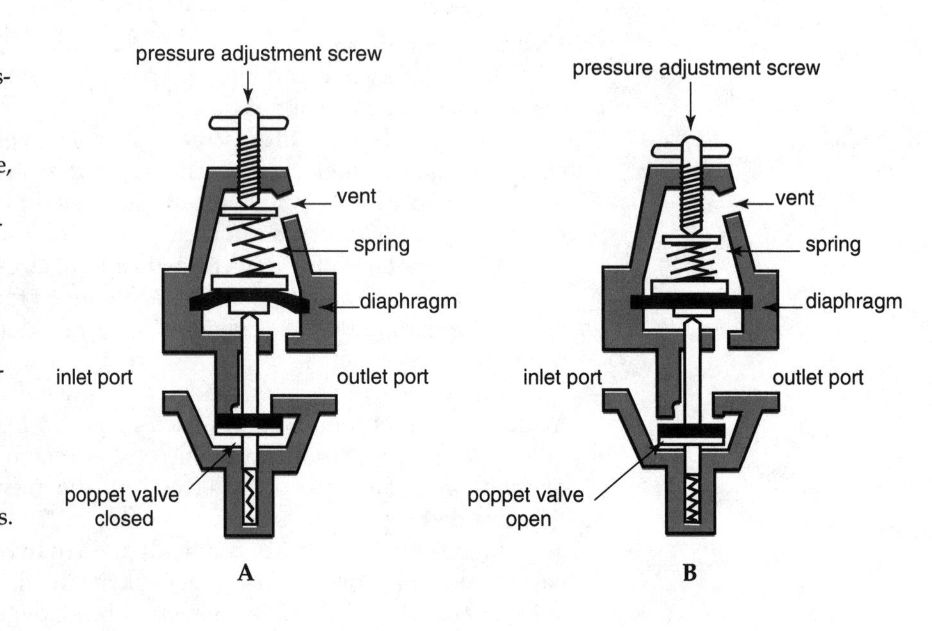

Figure 7-15
Operation of a pneumatic system pressure regulator. A—When system pressure exceeds atmospheric pressure, the diaphragm is pushed upward, allowing the spring-loaded poppet valve to close. B—If system pressure drops below atmospheric pressure, the diaphragm is pushed downward, opening the poppet valve to admit more air from the receiving tank. The pressure adjustment screw can be used to alter the pressure at which the valve opens.

When the system pressure becomes greater than the adjustment-screw pressure, the diaphragm is pushed upward. This action closes the poppet valve, maintaining pressure at a set level. Regulators may be used in several places within a system.

Lubricators are conditioning devices found only in pneumatic systems. They add a small quantity of oil to the air after it leaves the regulator. This lubrication helps valves and cylinders last longer and operate more efficiently.

Figure 7-16 shows a typical lubricator. When air enters at the inlet port, it flows into the narrowed area called the *venturi.* The air flow velocity increases and pressure decreases. Pressure in the venturi area is lower than in larger areas. As a result, oil is forced from the glass bowl into the oil tube and transported to the top of the unit. The needle valve can be adjusted to regulate the oil flow so that small droplets fall into the throat. Air velocity at the bottom of the throat breaks these droplets into a fine mist that mixes with the air. Finally, the lubricated air passes through the outlet port.

In pneumatic systems, components such as the filter, regulator, and lubricator are often placed together in an ***FRL unit***, Figure 7-17.

Transmission Lines

Transmission lines can be made of rigid metal tubing or flexible thermoplastic hose. Rigid lines are used where there is no vibration, or where move-

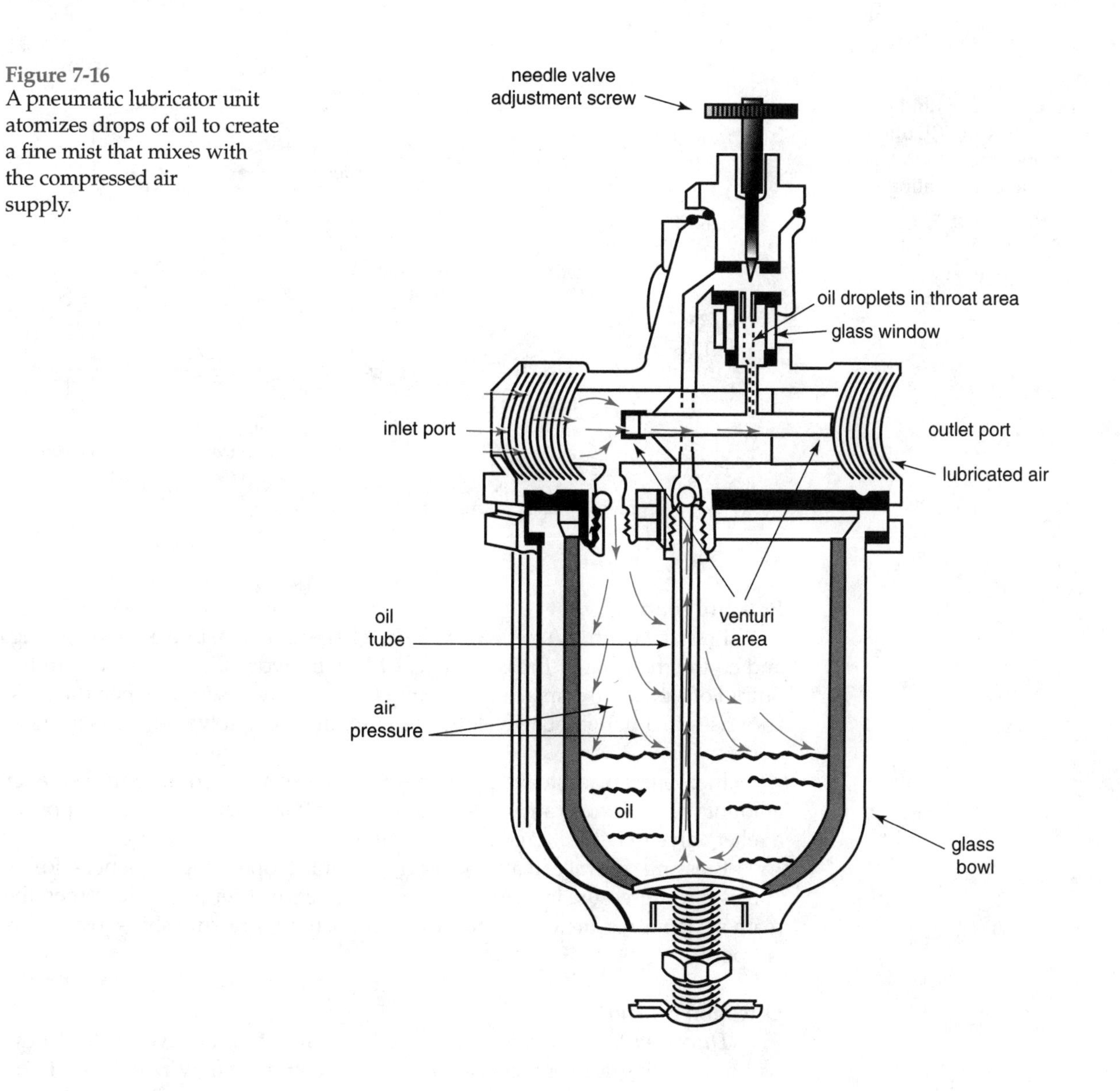

Figure 7-16
A pneumatic lubricator unit atomizes drops of oil to create a fine mist that mixes with the compressed air supply.

ment is not needed. As a general rule, they are more economical than flexible lines and less prone to trouble.

Flexible transmission lines are made in a variety of types and sizes. The system and its use determine which type of line is to be used. The tube or inner lining, the reinforcement material, and the outside cover material must all be considered. These factors determine pressure limits, temperature operating range, and resistance to exposure of the hose.

Control Devices

Control is achieved by devices that alter pressure, direction, and volume of fluid flow. Control devices are used in several different places within a system. Actual location is determined by the function of each device.

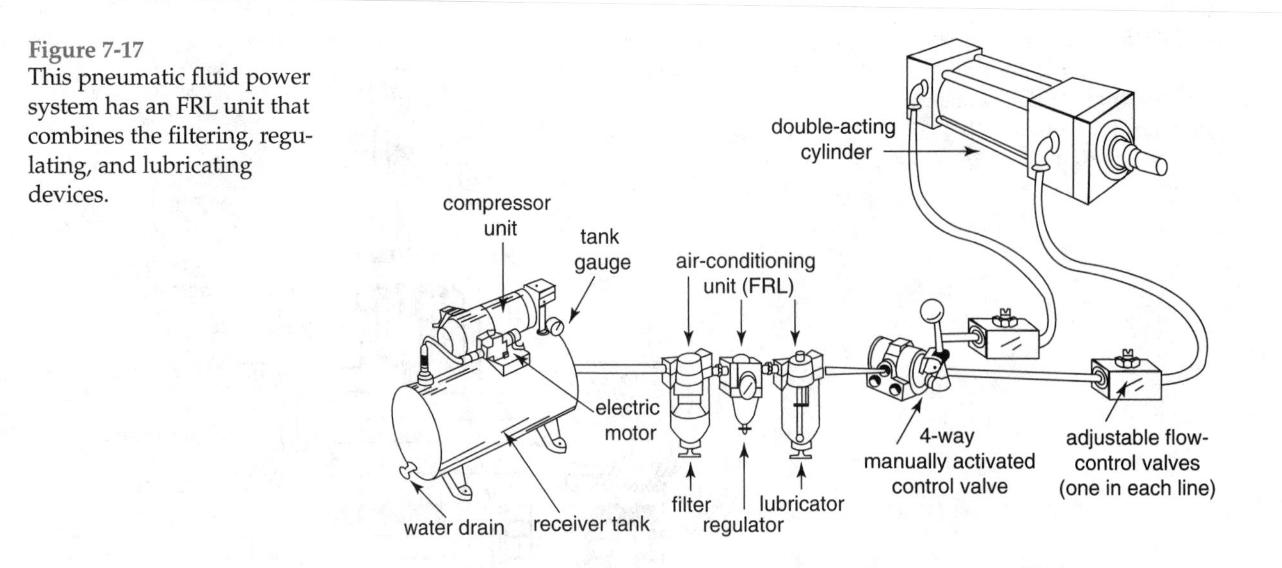

Figure 7-17
This pneumatic fluid power system has an FRL unit that combines the filtering, regulating, and lubricating devices.

Pressure control

Pressure control functions include relief, reduction, bypass, sequencing, and counterbalancing. ***Pressure-relief valves*** in hydraulic systems dump the output of a positive displacement pump back into the reservoir when the pressure rises to a dangerous level. In this case, the relief valve serves as a safety device.

In pneumatic systems, pressure-relief valves are used to control smaller amounts of air. Excess air is released into the atmosphere. The output port of a relief valve may be altered in size for this purpose.

Pressure control can also be used to establish operating sequences. Relief valves direct pressure in a predetermined sequence at certain levels. When the main system pressure overcomes the valve setting, pressure shifts to a different port.

Direction control

Direction control devices are used to start, stop, or reverse fluid flow without causing an appreciable change in pressure or flow rate. One-, two-, three-, and four-way valves are commonly used. They may be actuated by pressure, mechanical energy, electricity, or manual operation.

Control action of a directional valve occurs in different ways. One-way valves, also called *check valves,* operate on the seated-ball principle. They permit flow in only one direction.

Figure 7-18 shows the operation of a one-way valve. Pressure applied at the inlet port drives the ball away from its seat. This opens the flow path, allowing fluid to pass through the valve. Pressure applied at the outlet port, however, forces the ball into its seat. This prevents flow. Valves of this type are often used to permit free flow around controls when the flow direction is reversed.

Two-way valves are installed in along the transmission line to permit flow or shut it off. They use gates, plugs, discs, spools, or other precision-machined objects to enable or block flow. For example, the ball in a ball-type control valve can be rotated manually using an outside handle. When the handle

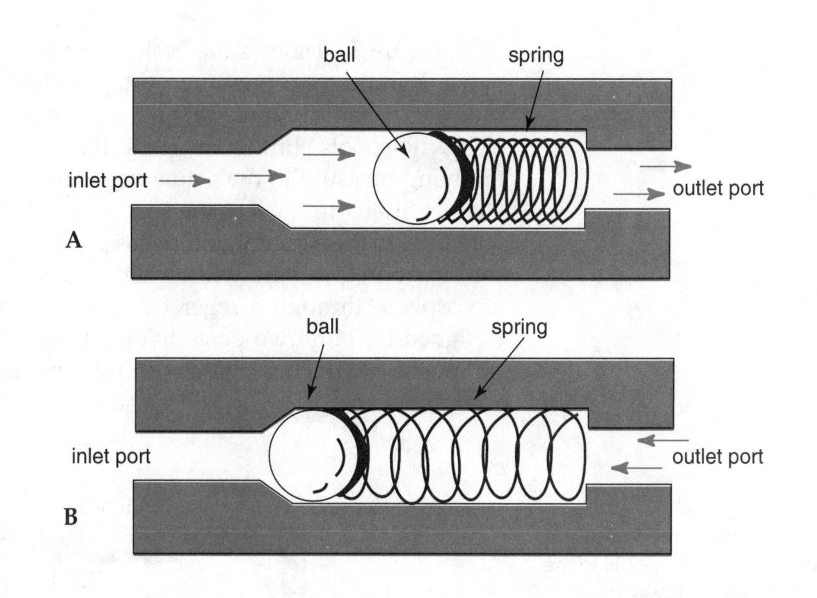

Figure 7-18
A check valve permits flow in only one direction. A—Fluid entering the inlet port pushes open the spring-loaded check ball to allow free-flow through the valve. B—If pressure from the outlet side exceeds the pressure from the inlet side, the check ball is pushed into the closed position preventing backward flow through the valve.

is in line with the transmission line, flow occurs. Turning the handle 90 degrees stops flow. This type of valve is used primarily for high-pressure applications.

Three-way valves permit shifting to two different sources of pressure or directing pressure to alternate devices. They often are used to alter cylinder operation or to control hydraulic or pneumatic motors. Valves of this type can be actuated mechanically, manually, electrically, or by pilot pressure. Basic designs, shown in Figure 7-19, include shifting spools, poppets, and sliding or rotary shear-seal plates.

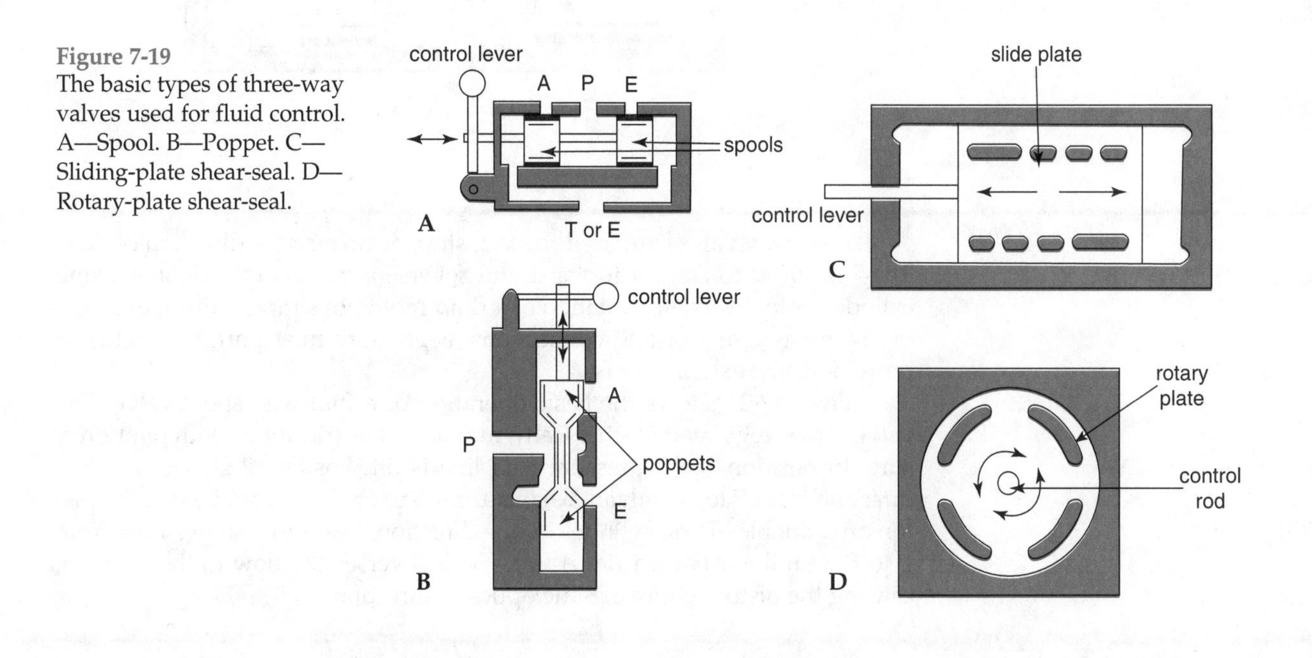

Figure 7-19
The basic types of three-way valves used for fluid control. A—Spool. B—Poppet. C—Sliding-plate shear-seal. D—Rotary-plate shear-seal.

Figure 7-20 shows the basic operation of a spool-type three-way control valve. *P* indicates a pressure port, *A* indicates an actuating port, and *E* or *T* indicates exhaust or tank. When the manual control shaft is pushed toward the right, the spools shift to the right. Flow then occurs through ports *P* and *A*, applying pressure to the actuating device controlled by the valve. When the shaft is pulled to the left, the spools shift in that direction. This cuts off pressure flow to the actuator and releases it through the *E* or *T* port. Depending on the situation, pressure will be directed to the storage tank or vented to the atmosphere through a relief valve. As a general rule, three-way valves are designed for only two-position operation. Some also have a third (neutral or off) position, which increases control capabilities.

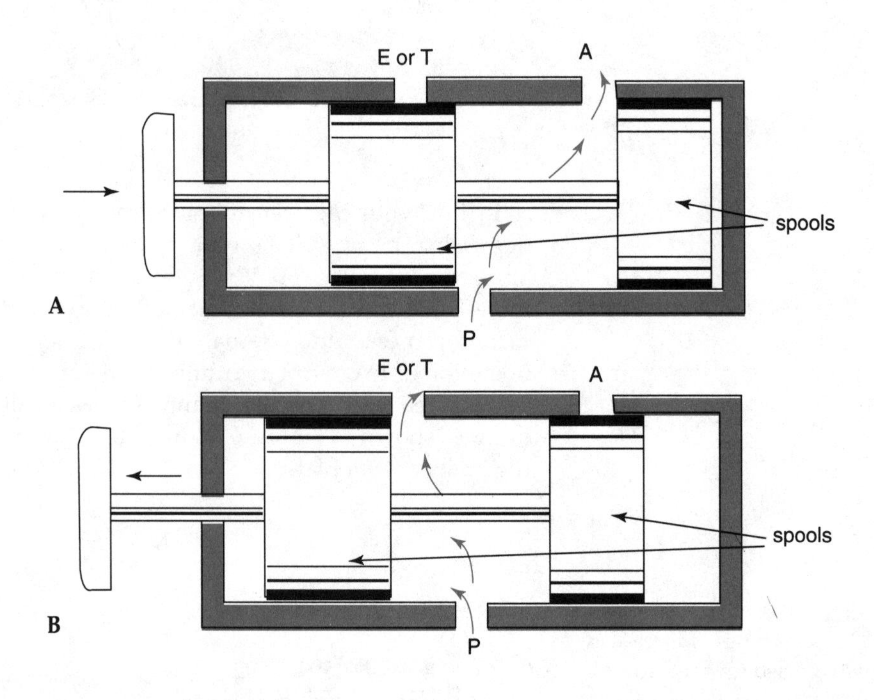

Figure 7-20
A spool-type three-way valve. A— With spools to the right, pressure flows to actuator. B—With spools to the left, pressure is directed to an air tank or exhausted to the atmosphere.

Four-way valves are used to start, stop, or reverse the direction of flow. They are used to control forward and reverse actuation of a double-acting cylinder or to reverse the rotation of a fluid motor. In simplest form, this type of valve has five working connections: a pressure inlet port, two actuator ports, and two exhaust ports.

Figure 7-21 shows the basic operation of a four-way spool valve. The valve can be regulated mechanically, manually, electrically, or with pilot pressure. In position 1, the pressure feed line is off. Position 2 shows the flow direction from *P* to *A* with an exhaust from *B* to E_2. This would move the piston on a double-acting cylinder in one direction. Position 3 shows flow from *P* to *B* with the exhaust from *A* to E_1. This reverses the flow to the cylinder, causing the piston to move in the opposite direction.

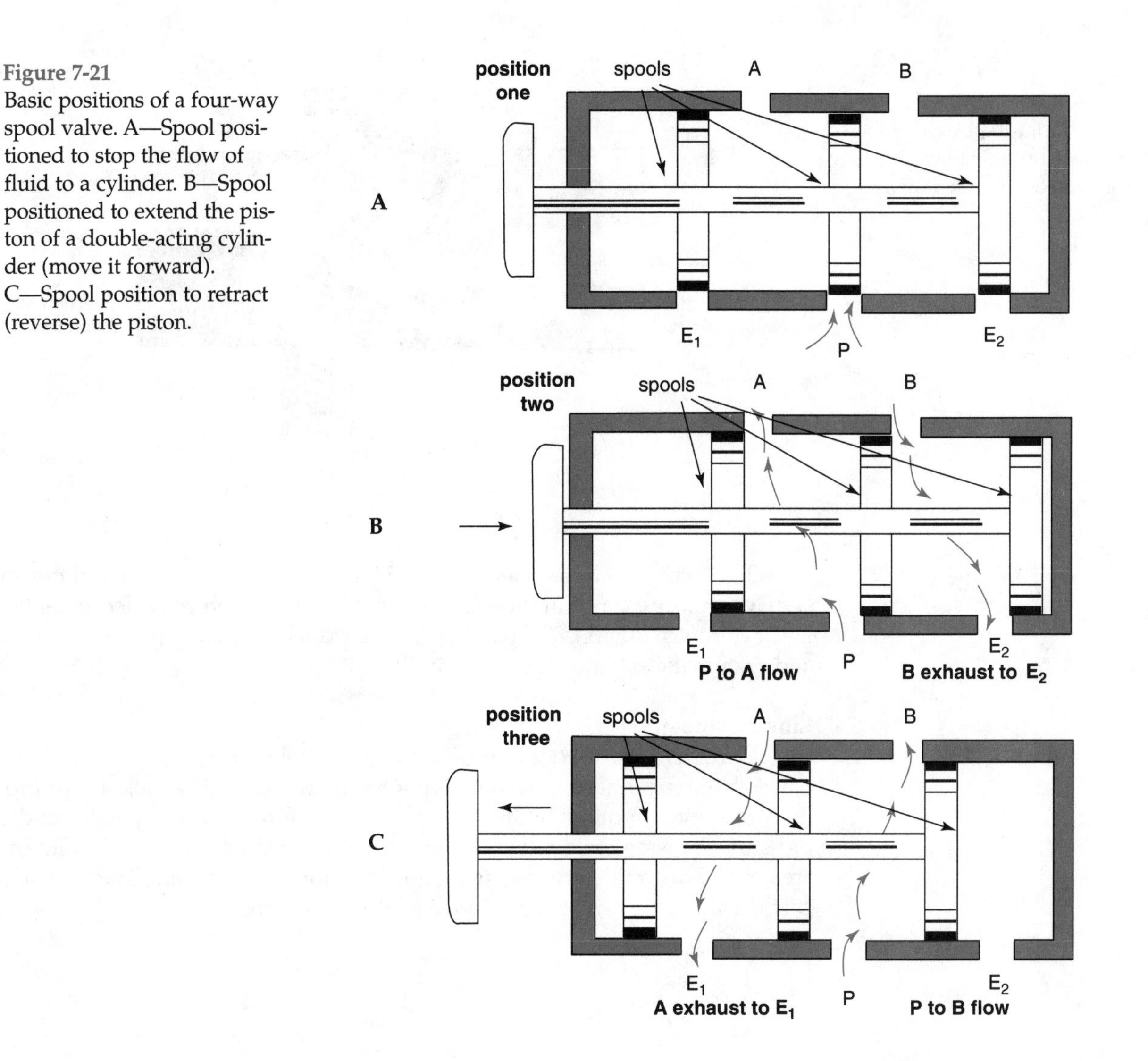

Figure 7-21
Basic positions of a four-way spool valve. A—Spool positioned to stop the flow of fluid to a cylinder. B—Spool positioned to extend the piston of a double-acting cylinder (move it forward). C—Spool position to retract (reverse) the piston.

Flow control

Flow control devices alter the volume or flow rate of the fluid. The rate at which fluid is delivered to the load of a system determines its operational speed. The speed of an air motor for example, is directly dependent upon the fluid flow rate of the fluid. By altering this rate, the motor can be made to operate at different speeds.

Cylinder actuating speed is also controlled by fluid flow. The term *metering* is often used to describe this function. To alter the linear motion of a cylinder, fluid may be controlled at the input feed line, the return line, or a combination of both.

Figure 7-22 shows the operation of a flow control valve. The letters *P* and *F* refer to the pressure and free-flow connections. Flow is controlled from *P* to *F*. Flow level in this direction is adjusted by the needle valve. Flow from *F* to *P* forces the check valve ball to move away from its seat. Free (uncontrolled) flow results.

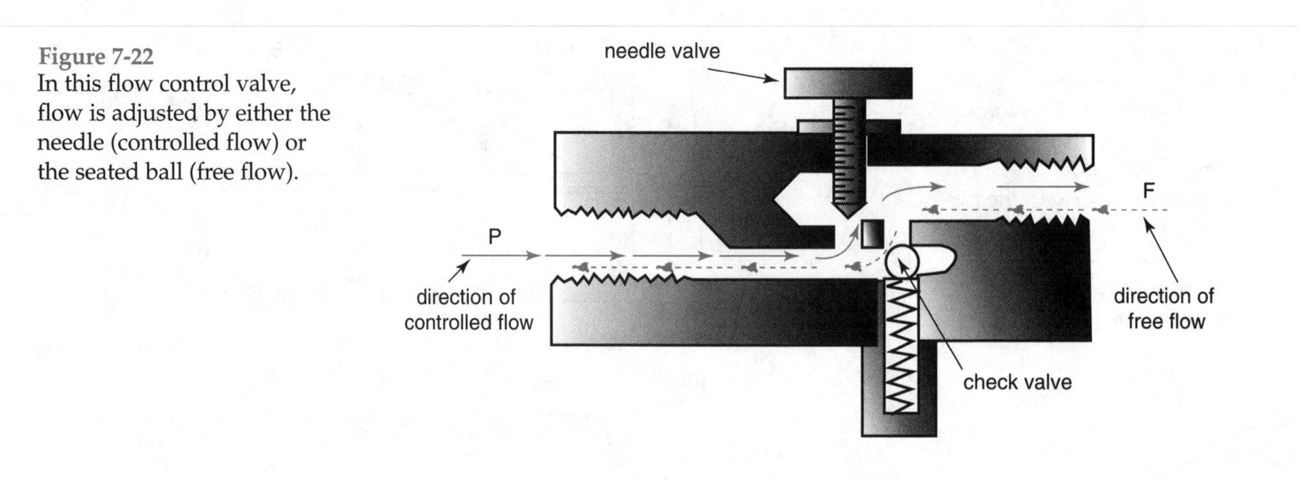

Figure 7-22
In this flow control valve, flow is adjusted by either the needle (controlled flow) or the seated ball (free flow).

Load Devices

The term *actuator* is often used to identify the load device. Fluid power systems produce work in the form of mechanical motion provided by either linear or rotary motion. The basic operating principles that apply to hydraulic and pneumatic actuators are very similar.

Linear actuators

Cylinders are used as linear actuators, as shown in Figure 7-23. They develop the force needed to lift, compress, hold, or position objects. To produce this linear motion, hydraulic fluid or air is forced into a cylinder under pressure. A piston inside the cylinder moves as pressure is applied to it. The area of the piston determines the amount of force it develops. The area of a cylindrical piston can be determined by the formula:

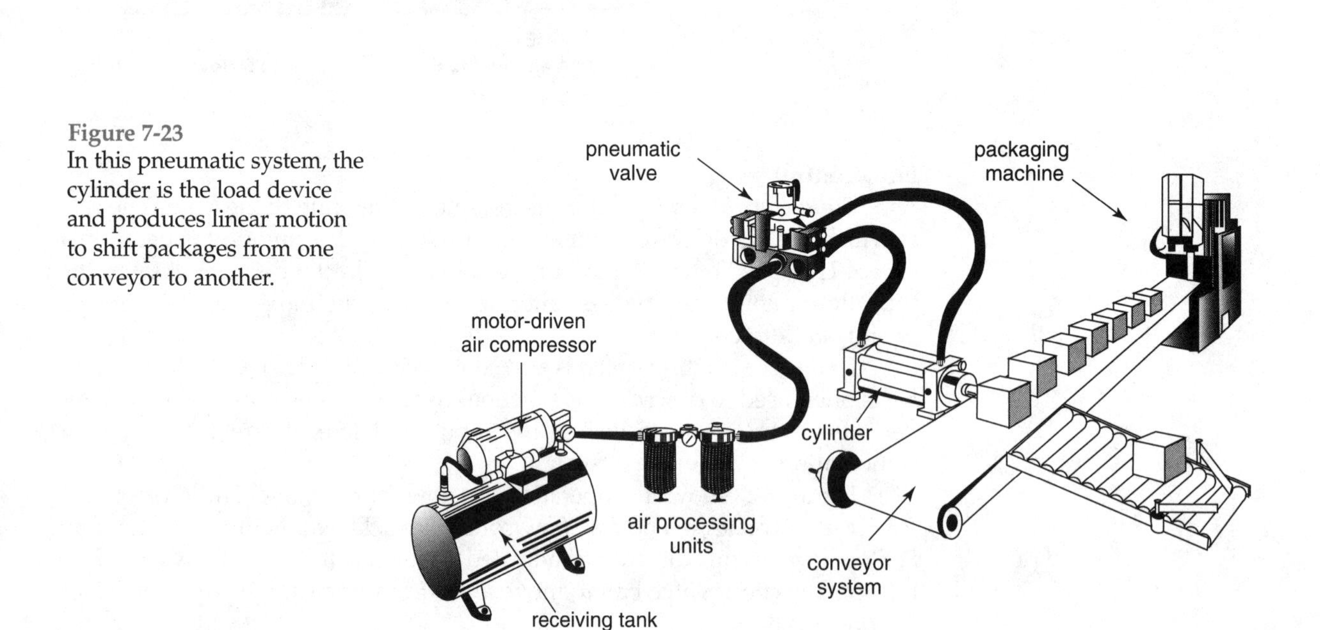

Figure 7-23
In this pneumatic system, the cylinder is the load device and produces linear motion to shift packages from one conveyor to another.

$$A = \frac{\pi D^2}{4}$$

A = area of piston, in.^{2} or cm^2

π = a constant (3.14)

D = diameter of piston, in. or cm

The force developed by this position may be determined by the formula:

$$F = PA$$

F = force developed, lb. or N

P = applied fluid pressure, psi or kPa

A = area of piston, in.^{2} or cm^2

Single-acting cylinders have only one input port, Figure 7-24. When fluid is forced into this port under pressure, the piston moves. In this illustration, a weight is lifted by the force exerted on the piston by the fluid. The combined load consists of the weight being lifted, the friction between the piston and cylinder walls, and the heat developed by that friction.

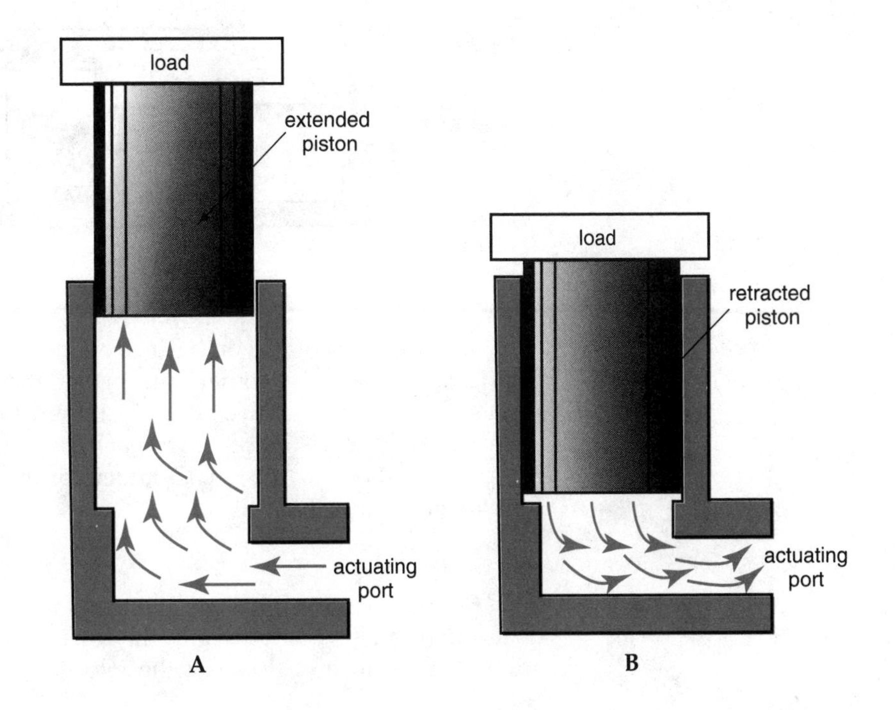

Figure 7-24
Operation of a single-acting cylinder. A— Fluid flowing under pressure through the actuating port forces the piston and its load upward. B—Weight of the piston and load retracts the piston when pressure is released.

To return the piston to its original position, the flow of fluid is stopped by closing a valve. Release of the fluid under the piston is made possible by opening the valve's exhaust port. The weight of the load then causes the piston to retract.

Double-acting cylinders can move in two directions. Two ports are needed—as fluid is supplied to one port, it is expelled from the other. Retracting of the piston is done by reversing the fluid flow. Applications for double acting cylinders include punch presses, rolling mills, machine-tool clamps, paper cutters, and robot actuators.

In Figure 7-25, fluid is applied to the right side of the piston and removed from the left side. This forces the piston to move to the left. Switching the fluid flow causes it to move to the right.

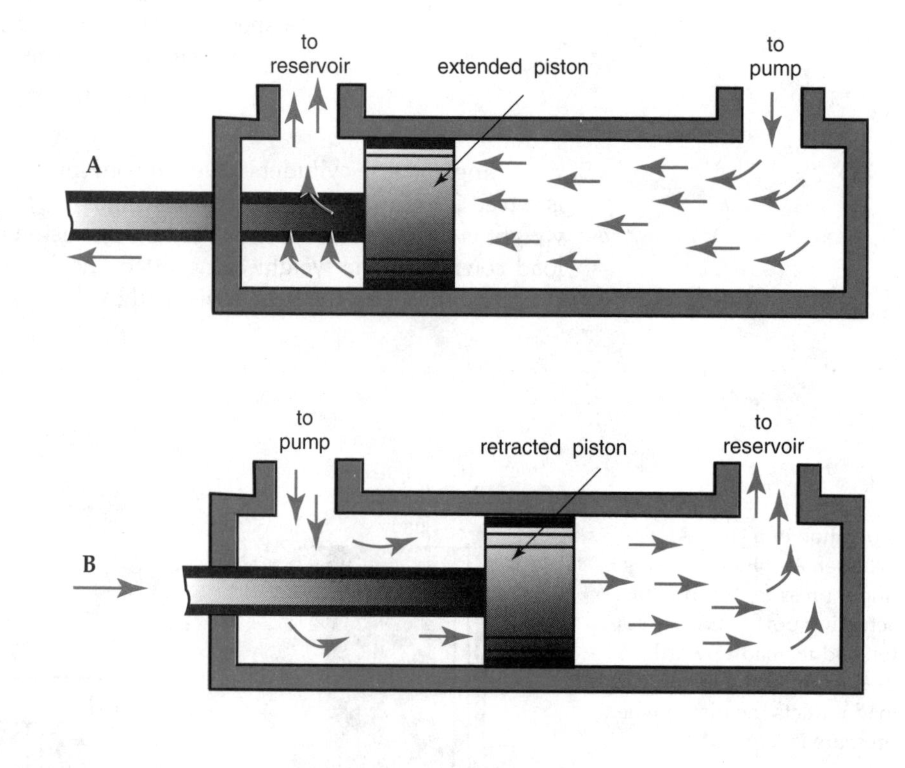

Figure 7-25
Operation of a double-acting cylinder. A—Fluid under pressure from the pump extends the piston. Fluid behind the piston is displaced and flows to the reservoir. B—Reversing fluid flow through the ports retracts the piston.

The double-acting cylinder in the illustration is of the *differential* type. The retracting force is somewhat smaller than the extending force. This results from the area of the piston being reduced on the retraction side by connection of the piston rod. A *nondifferential* cylinder has rods extending from both ends of the piston. Cylinders of this type can provide an equal force in either direction.

Rotary actuators

Rotary actuators produce a limited amount of rotary motion (twisting or turning) in either direction. In the rotary actuator shown in Figure 7-26, applying fluid to port *A* causes the rotor to move in a clockwise direction.

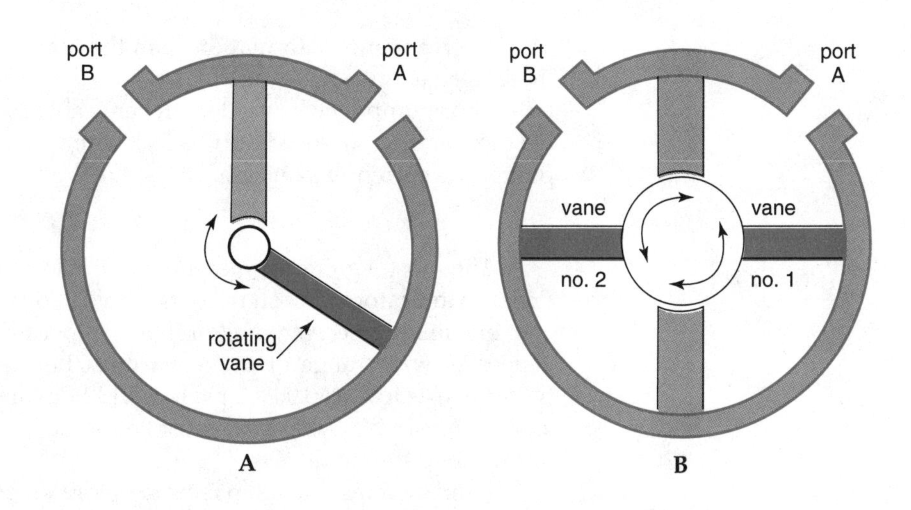

Figure 7-26
Rotary actuators. A—The single-vane type can rotate through a larger arc. B—The double-vane type has twice the turning power, but permits less movement in either direction.

Counter-clockwise rotation occurs when fluid enters port *B* and is expelled from port *A*. The single-vane rotor can be made to turn in either direction through an arc of approximately 280 degrees. The double-vane rotor has twice the turning power, but can only turn approximately 100 degrees in either direction. Actuators of this type are used for lifting or lowering, for opening or closing, and for indexing operations. One example is the reciprocating operation of a punch press.

Fluid motors

Fluid motors convert the force of a moving fluid into rotary motion through the use of vanes, gears, or pistons. Fluid motors and pumps are very similar in appearance and operation.

Fluid motors are classified according to the type of fluid displacement they use. Gear, vane, and piston motors usually have a fixed displacement. They accept and move a certain amount of fluid with each revolution or cycle. Operating speed depends entirely on the amount of fluid supplied by the source. In variable displacement motors, the amount of fluid circulated can be changed. The piston motor fits into this classification. The length of its stroke is altered to produce variable displacement. Its speed can be changed by an external adjustment. Operating speeds of up to 3000 rpm are typical.

A fluid motor's turning capability is a measure of torque. This is equal to the developed force multiplied by the radius of the rotating arm. Mathematically, this is expressed by the formula:

$$\text{Torque (in/lbs)} = \frac{\text{pressure (lb/in}^2\text{) x displacement (in.}^3\text{)/revolution}}{2}$$

The output power of a fluid motor is expressed as horsepower. Mathematically, horsepower can be determined by the formula:

$$\text{Horsepower} = \frac{\text{torque (in/lb) x speed/(rpm) x 2}}{\text{33,000 ft/lb per min}}$$

Motor performance information from the manufacturer can be used to make these calculations.

Gear pumps can be used interchangeably as fluid motors. They are capable of operating at speeds up to 5000 rpm. Both internal and external gear pumps are currently available.

Indicators

The most significant measurement made in a fluid system is pressure. Pressure indicators play an important role in overall performance. Regulators and pneumatic receiver tanks often use permanent pressure indicators or gauges. A wide range of pressures must be measured. Negative pressures (vacuums) as low as 0.00002 psi (0.00013 kPa) and positive pressures as high as 1×10^6 psi (6 895 000 kPa) must be measured. This wide range requires a number of different devices.

Fluid systems also often require measurements of flow and temperature. These indicators are valuable in analyzing system efficiency. They are only used periodically, so they are usually not permanently installed.

In ***pressure indicators,*** an element physically changes shape to show pressure changes. For example, spiral and helix coil elements uncoil when pressure is applied. The Bourdon tube element tends to straighten. The physical change causes movement of an indicator on a scale or a stylus on a paper chart.

Flow indicators are primarily used to test flow rates from pumps and at the inlet and outlet ports of actuators. By monitoring flow rates, system efficiency is measured and maintenance problems are reduced.

Important Terms

centrifugal pump
conditioning
cylinder
desiccant
direction control devices
filters
flow control devices
flow indicators
fluid motors
fluid power systems
force
FRL unit
heat exchangers
hydraulic system
inertia
lubricators
nonpositive displacement pump
Pascal's law
pneumatic system
positive displacement pump
power
pressure
pressure indicators
pressure regulator
pressure-relief valves
reciprocating pump
resistance
rotary actuators
rotary-gear pump
rotary-vane pumps
strainers
weight

Review Questions

Write your answers on a separate sheet of paper.

1. Briefly describe the characteristics of a hydraulic system. What are some industrial applications of such systems?
2. Briefly describe the characteristics of a pneumatic system and list some industrial applications.
3. Discuss Pascal's law and its application to fluid power systems.
4. Define the terms: force, pressure, work, and power.
5. Give the function of each of the following parts of a fluid power system: motor, pump, pressure relief valve, gauge, reservoir.
6. What are some types of fluid pumps? Give a brief description of each.
7. What is the purpose of the fluid conditioning components of a hydraulic system? Describe the types of conditioning components used in a pneumatic system
8. Describe the operation of a pressure regulator.
9. What are some types of control devices used with fluid power systems?
10. How is direction control accomplished in a fluid power system ?
11. What is a single-acting cylinder? A double-acting cylinder?
12. How can you determine the torque developed by the motor of a fluid power system?
13. List some types of indicators used with fluid power systems.

This robot has an air pressure gauge and air filter/lubricator installed directly on the end effector. (Deere & Co.)

8 Maintaining Robotic Systems

Overview

Robotic systems are complex because they involve many different areas of technology. These include hydraulic, pneumatic, and electrical power systems, electronic control systems, and digital computer systems. This makes their maintenance complex as well.

In this chapter, robotic system maintenance is discussed. Emphasis is placed on general servicing techniques, a well-organized approach to troubleshooting, preventive maintenance, and developing and implementing a preventive maintenance program.

General Servicing Techniques

Robots seldom require servicing of individual parts. Many repairs to robot systems take place at the subassembly level. Parts are not repaired; instead, the module (unit) is simply replaced. You should become familiar with certain general servicing techniques. They are outlined in the following seven steps.

Step 1. Initial Observation and Inspection

See if the problem is electrical or mechanical in nature. Never accept anyone's word as to what the robot is or is not doing. Check it yourself!

Inspect the power system. Determine if the proper current—single-phase ac, three-phase ac, or dc—is being used to power the equipment. Magnetic contractors, transformers, switches, circuit breakers, fuses, and other electronic controls all should be checked.

Observe the physical condition of the equipment. Look for such things as fluid or air leaks, unusual noises, odors, or poor electrical connections or loose wires.

Step 2. Gathering Information

If possible, acquire a schematic, a service or maintenance manual, and an operating manual for the robot. Such tools are essential to most equipment servicing. If these are unavailable, consult with someone who has had experience in servicing the robot. If available, an identical robotic system can be used to determine specific values (such as voltage, current, and waveforms). These values can be compared to those that exist in the robot being serviced.

Step 3. Analyzing the Equipment

After gathering the information, you should have a good idea of how the system functions. Mentally divide or "block out" the different equipment according to purpose. For example, input power systems, rectifiers, transformers, filters, feedback loops, amplifiers, interface/control circuits, and other devices can be sorted out. Then, identify the input and output of each block.

Step 4. Isolating the Problem

You are now ready to trace the malfunction to a specific subsystem within the equipment. Use a voltmeter, an oscilloscope, or a VOM (volt/ampere/ohmmeter) to read measurements or waveforms. The secret here is knowing how to use test equipment and how to interpret the measured values. Compare your readings to those indicated on the schematic or in the service manual. When you compare the test readings to the recommended values, it is usually easy to isolate the problem. Sometimes, verifying a malfunction is difficult. Generally, if a subsystem's input is correct and its output is not, you might assume the subsystem itself to be the problem. This is not true in all cases. In some cases, the output of one subsystem can become distorted because of another subsystem to which it is connected. When this is suspected, the subsystem under inspection should be disconnected and the voltages and waveforms checked again.

Step 5. Narrowing the Problem to a Component, Circuit, or Module

To isolate the malfunction of a specific component or components, use the procedure just described, making a series of measurements. Once the problem has been isolated to a specific component, circuit, or module, you must then check it to identify the problem. This can be completed in various ways with a number of different test instruments.

Step 6. Replacing a Component or Device

Any component used to replace another must be equal to, or better than, the original. Important considerations are voltage, maximum current, and power rating. *Always* replace a component with one of equal or higher maximum value. For example, a one-watt resistor may be replaced with a two-watt resistor, but never with a half-watt resistor.

Desoldering is also an important step in replacing faulty components. Use care in selecting the proper iron and soldering tip. Additionally, care must be taken not to damage components located nearby. A ***solder sucker*** must be used, when working with a printed circuit board, to properly remove a defective component without damaging the board. Care must also be taken when soldering a new component into the circuit. The replacement part must not be damaged by overheating. You must also guard against "cold" solder joints. ***Resin-core solder*** (*never* acid core) should be used for *all* electronic circuit work.

Step 7. Checking Repairs and Keeping Records

After the system has been repaired, operate it for several minutes to be sure it is working properly. When you are satisfied that operation is correct, document the problem and the solution. Make a record of the original problem, what caused it, and what you did to correct it. This information should be kept with the service manual or in a special file so that it is on hand for future reference.

Troubleshooting

Troubleshooting is the process of finding out why something doesn't work properly. If you follow the steps outlined earlier, you will be able to locate most problems. Some problems are easy to solve and require little time. Others are more difficult: they occur intermittently or are so complex that they require many hours of concentration and work.

Methods that most technicians find helpful in troubleshooting include:

- Δ Using common sense.
- Δ Following a logical sequence.
- Δ Knowing how robotic systems work.
- Δ Knowing how to use test equipment.
- Δ Knowing how to read and use schematic diagrams effectively.

To begin any kind of troubleshooting, organize your thoughts and identify possible courses of action. Without a well-organized approach, troubleshooting can become a time-consuming guessing game. You must use a planned approach in order to save time. As you become more experienced and more familiar with the equipment, less time is required. Remember, no system is perfect.

An important part of troubleshooting is the initial inspection. When this inspection is done properly, many problems can be diagnosed without a lot of unnecessary steps. This means looking for the obvious. Use your senses of sight, touch, smell, and hearing. If you suspect a specific part, turn off the equipment, then examine it carefully—smell it, touch it, listen to it, look closely at it.

A good visual inspection should be done before you get into actual circuit or system testing. Look for such things as:

- Δ Burned parts. These are often obvious. They may be charred, blistered, or discolored, and may even have holes.
- Δ Broken parts. Breaks may appear in the form of cracks, wires pulled out of parts, or parts that have been completely destroyed.
- Δ Broken wires and poor connections.
- Δ Smoke or heat damage. Parts may smoke when equipment is turned on. This identifies a damaged part, but not the cause of the damage.
- Δ Oil, air, or water leaks.
- Δ Loose, damaged, or worn parts. These can be located visually or by touch.
- Δ Noisy parts, such as motor bearings. Uncommon noises often indicate defective parts.

Keep in mind that most problems are component failures. If you know what each component is supposed to do, you will be aware of the troubles it can cause. Use the proper tests. To avoid duplicating effort, make a list of circuit or system operations you have already tested.

Good troubleshooting involves knowing how the system operates and how to use test equipment. Most of all, be patient and systematic.

Troubleshooting Charts

Most manufacturers of robotic equipment provide troubleshooting manuals with charts and diagrams to aid in repair. Use this information to help locate the faulty module or part. A typical symptom list is shown in Figure 8-1. This list itemizes problems associated with a specific robotic system. It then directs the maintenance person to detailed information, such as the material shown in Figures 8-2 and 8-3. Such detailed instructions help to specifically identify the problem. Some manufacturers use a flow chart, Figure 8-4, to guide you to the problem.

Figure 8-1
This symptom list has page references to indicate where detailed information can be found.

Symptom List	
Applications 1 Through 5 on at the same time	10-10
Axis does not move	10-11
Axis moves erratically or runs away	10-13
Control panel failures	10-15
Data Error (DE) LED on or blinking	10-17
DI/DO failure (Standard)	10-19
DI/DO failure (Extended)	10-21
Does not home	10-23
Manipulator power does not work (No other indications)	10-25
MTCB indicator failure	10-26
MTCB switches do not work	10-27
Overtime error (OT) LED on	10-28
Overrun (OR) LED on	10-29
Overrun detection failure	10-32
Overrun reset failure	10-33
Power failure (PF) LED on	10-34
Remote operator panel interface failure	10-35
Repeatability varies	10-36
Servo Error (SE) LED on	10-39
Servo Error (SE) LED blinks and operator panel beeps	10-40
Transmission Error (TE) LED on or communication error	10-42

If you understand how the circuit, device, or system functions and know how to use test equipment, troubleshooting and testing are relatively easy to accomplish. This is true for the simplest circuit or the most complex system. Practice is the best way to learn the procedures.

Preventive Maintenance

Unfortunately, many companies today practice management by crisis: they avoid thinking about a problem until one occurs. In

Figure 8-2
Detailed troubleshooting information helps to identify the actual problem.

Defective Battery

Note: Do not dispose of the battery in a fire.

With the system off, check the voltage across the battery. The voltage should be between +3.2 and +4.2v dc.

A discharged battery may load down the power supply while charging. To test for this, unplug the battery and reload the applications.

If the battery is suspected of only being discharged, leave the controller powered up for eight hours to sufficiently charge the battery. If the voltage is still low after charging, replace the battery card.

Low Power Supply Voltages

Check for 24v dc, -12v dc, 12v dc, 5v dc at the power supply. If any of the voltages differ by more than 5%, adjust or replace the power supply. See power supply adjustment (4009).

Note: The power supply may be left powered on with no load attached.

Defective Motor Control Board

If the battery is good and the voltages are good, suspect the Motor Control Board.

Figure 8-3
Manufacturers usually provide complete directions for diagnosing and solving problems. Note the safety warning included as part of a diagnostic procedure.

AXIS DOES NOT MOVE

Defective Servo Pack Fuse

Swap the servo pack fuse for the failing axis. Do not rely on the indicator.
Theta 1 is labeled SV1 in the controller. (15 amp.)
Theta 2 is labeled SV2 in the controller. (10 amp.)
Z-Axis is labeled SV3 in the controller. (7.5 amp.)
Roll is labeled SV4 in the controller. (7.5 amp.)

Defective Z-Axis Brake

The brake should remain locked until manipulator power is brought up. Then the Z-Axis servo pack takes over by controlling the power to the servo Z-Axis.

If the brake remains locked after manipulator power is up, then suspect the magnetic brake or circuit to the brake (brake is approximately 85 ohms when good). Also the brake can be tested by disconnecting connector CN16H with manipulator power on, this makes a clicking noise.

Out of Adjustment Zero

Perform the "Servo Pack Zero Adjustment."

Defective Motor Control Board

1. If the Z-axis only does not move, set controller power to 0 (off), then ground CN16H-1 (WD16) to release the brake. Set the power switch to I (on). Move the axis to check for binds. If the axis cannot be moved, isolate the bind. Remove the jumper.

   For any other axis, set the power switch to 0 (off). Move the axis to check for binds. If the axis cannot be moved, isolate the bind.

2. Set the controller power switch to I (on). Press the Manip Power key on the control panel.

   DANGER:
   THE NEXT STEP REQUIRES YOU TO PUSH ON THE AXIS WITH MANIPULATOR POWER ON. DO NOT ALLOW YOUR BODY TO ENTER THE WORK SPACE.

3. Attempt to push or turn the axis. If the axis moves easily, then suspect the Servo Pack or wiring from the Servo Pack to the motor (WD07-WD10).

   >>> Continued on the next page.

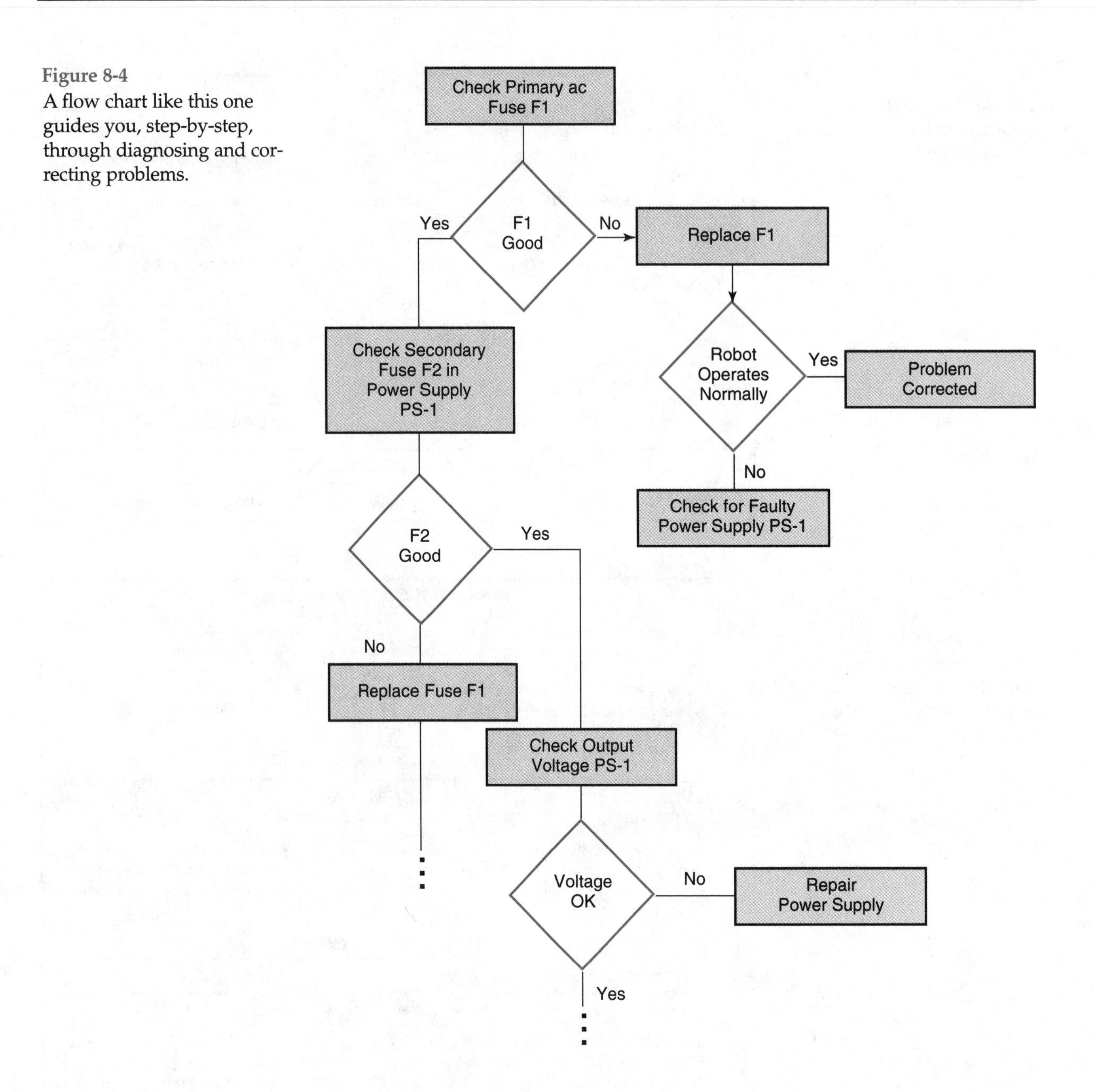

Figure 8-4
A flow chart like this one guides you, step-by-step, through diagnosing and correcting problems.

terms of maintenance, this means not fixing equipment until it breaks down. This leaves the production line always vulnerable to equipment failure.

Equipment failures are costly, since they can create production bottlenecks. When work in process backs up at a workstation shut down by equipment failure, it means the operation must later run additional hours to "catch up." This causes the operations that follow to need additional production time, as well. Idle time results when operators or machines are not being utilized. If the order is not produced in time, the customer may go to another source.

Equipment failure also causes increased rework and scrap because the equipment had excessive wear prior to breakdown. ***Rework*** is the process of fixing parts that do not meet product specifications. ***Scrap*** is a part that cannot be fixed. Inventory costs are also increased when materials must be held beyond the normal time it takes for them to cycle through the manufacturing process.

Preventive maintenance is the best solution for avoiding the costly results of equipment failure. It is the process of regularly checking equipment, cleaning and maintaining it, and replacing worn parts before breakdowns occur. It can play an important part in an organization's management strategy.

Preventive maintenance is essential to robots and all other types of industrial equipment. It can increase flexibility, help maintain production flow, and allow for a continuous analysis of equipment. It decreases the need for a large on-call maintenance crew. Used in combination with mathematical analysis, it aids in predicting when equipment needs to be adjusted or replaced to avoid breakdowns. Preventive maintenance can be performed at convenient times, such as when a robot is not being used. This reduces overtime for skilled operators and balances the workload among shifts. When performed conscientiously, preventive maintenance ensures consistent quality, reduces product costs, and reduces downtime of critical production equipment.

Developing a Maintenance Program

To develop a preventive maintenance program, a few basic steps should be followed:

Step 1. Establish a schedule

Each shift must share in maintenance responsibilities. An example of a maintenance schedule is shown in Figure 8-5. Such a schedule can be altered to accommodate the appropriate number of shifts. However, it must be followed carefully for maximum effectiveness. If one shift fails to follow the schedule, it can cause problems for the entire program. Usually, the schedule is easy to maintain and should be posted on the machines.

Step 2. Use an assignment sheet

An assignment sheet describes tasks that must be performed on a regular basis, Figure 8-6. Each assignment sheet identifies the unit to be maintained, the maintenance intervals, and the type of action required. The steps may be carried out in any order, but each should be done completely. Assignment sheets should be posted on the machines.

Step 3. Keep work areas clean

Once a preventive maintenance program is begun, machinery and work areas must be kept very clean. This will allow any leaks or visible damage or wear to be noticed more easily by the operator.

Step 4. Make the operator part of the program

Preventive maintenance places a great deal of responsibility on the operator. That person must be dedicated to the program for it to work properly. This does not mean that the operator is personally responsible for mainte-

Daily-Weekly Preventive Maintenance Schedule									
Machine Type: IBM-7547 Department: 179								Serial Number: 257698ZMW Building Number: 5	
Shift	Week	Check as completed Mon	Tues	Wed	Thurs	Fri	Weekly	Operator	Special Comments
Shift 1	1	X	X	X	X	X	X	John Wasson	Air leak
Shift 2	2	X	X	X	X	X	X	Robert Towers	Bad Relay
Shift 3	3	X	X	X	X	X	X	Stephen Fardo	No Problems
Shift 1	4	X	X	X	X	X	X	John Wasson	No Problems
Shift 2	5	X	X	X	X	X	X	Robert Towers	Air Leak
Shift 3	6	X	X	X	X	X	X	Stephen Fardo	Bolt Loose
Shift 1	7	X	X	X	X	X	X	John Wasson	Motor Noise
Shift 2	8	X	X	X	X	X	X	Robert Towers	No Problems
Shift 3	9	X	X	X	X	X	X	Stephen Fardo	Bad I/O Port
Shift 1	10	X	X	X	X	X	X	John Wasson	Loose Clamp
Shift 2	11	X	X	X	X	X	X	Robert Towers	Air Leak
Shift 3	12	X	X	X	X	X	X	Stephen Fardo	No Problems
Shift 1	13	X	X	X	X	X	X	John Wasson	No Problems
Shift 2	14	X	X	X	X	X	X	Robert Towers	Bolt Loose
Shift 3	15	X	X	X	X	X	X	Stephen Fardo	No Problems
Shift 1	16								
Shift 2	17								
Shift 3	18								
Shift 1	19								
Shift 2	20								
Shift 3	21								
Shift 1	22								
Shift 2	23								
Shift 3	24								
Shift 1	25								
Shift 2	26								
Shift 3	27								
Shift 1	28								
Shift 2	29								
Shift 3	30								
Shift 1	31								
Shift 2	32								
Shift 3	33								
Shift 1	34								
Shift 2	35								
Shift 3	36								
Shift 1	37								
Shift 2	38								
Shift 3	39								
Shift 1	40								
Shift 2	41								
Shift 3	42								
Shift 1	43								
Shift 2	44								
Shift 3	45								
Shift 1	46								
Shift 2	47								
Shift 3	48								
Shift 1	49								
Shift 2	50								
Shift 3	51								
Shift 1	52								

Figure 8-5. A preventive maintenance schedule helps find and eliminate small problems before they become large ones.

Figure 8-6
Assignment sheets should be posted on the machines to identify maintenance tasks and the interval at which they should be performed.

IBM-7547 Preventive Maintenance Assignment Sheet			
Unit	**Action**	**Interval**	**Lubrication**
Manipulator	Clean Check Oil Level Check for Air Leaks	Daily Daily Daily	— — —
Controller	Check Air Filters	Monthly	—
Roll Axis Belts	Check Tension	Monthly	—
Roll Axis Gears	Lubricate	Monthly	Molycoat G
Theta 1 Axis	Change Oil	Semi-Annually	Number 10 Oil
Theta 2 Axis	Change Oil	Semi-Annually	Number 10 Oil
Bearings	Pack Bearings	Every Replacement	Number 23 Grease

nance. It *does* mean, however, he or she must keep maintenance personnel informed of any problems.

While a good preventive maintenance program will definitely reduce the rate of unexpected equipment failure, such programs are not without their drawbacks. These include:

- Δ Increased need for dedicated workers.
- Δ Increased responsibility and paperwork for operators.
- Δ Access time needed to work on equipment.
- Δ Initial cost.

Implementing a New Program

Implementing a new preventive maintenance program involves a number of steps, Figure 8-7. Because operators are so important to preventive maintenance, they should be involved early in the planning process.

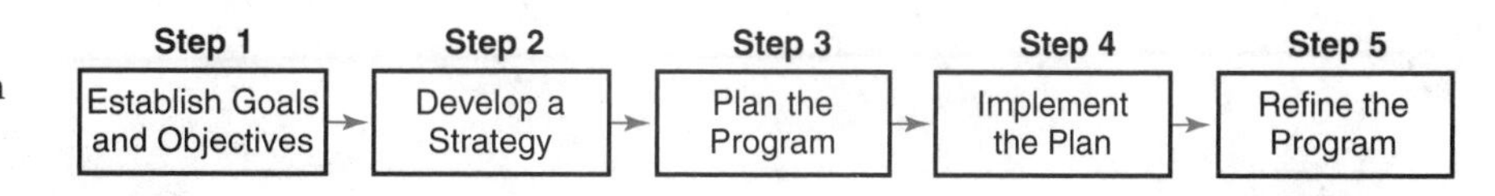

Figure 8-7
These steps should be taken when implementing a preventive maintenance program.

Important Terms

preventive maintenance
resin-core solder
rework
scrap
solder sucker
troubleshooting

Review Questions

Write your answers on a separate sheet of paper.

1. Identify several areas of technology used by robotic systems.
2. List the seven steps involved in servicing a robotic system.
3. During the initial observation of equipment to be serviced, what should you look for?
4. How does a technician proceed in gathering information about equipment?
5. What are some types of measurements that can be taken to check equipment?
6. How do you narrow down the cause of an equipment problem?
7. What factors should be considered when replacing a component?
8. Why is keeping good records important in servicing equipment?
9. What is meant by "troubleshooting?" Which of your senses are used in troubleshooting equipment?
10. What are some methods that technicians find helpful in troubleshooting?
11. Why are troubleshooting manuals important?
12. Why is preventive maintenance important?
13. What is meant by the terms "scrap" and "rework"?
14. List the steps to follow in developing and implementing a preventive maintenance program.
15. What is a PM assignment sheet?
16. What are some disadvantages of preventive maintenance?

Although this robotic workstation appears to be fairly simple in design, it is a complex system that requires regular inspection and maintenance. (Wes-Tech Automation)

SENSING AND END-OF-ARM TOOLING

The control of an industrial robots often depends upon a sensing system which use devices called transducers to convert light, heat, or mechanical energy into electrical energy. The signal output of the transducer is used to affect the operation of the robot's end effector (end-of-arm tooling). End effectors, attached to the wrist of a manipulator, can grasp, lift, transport, maneuver, or perform operations on a workpiece.

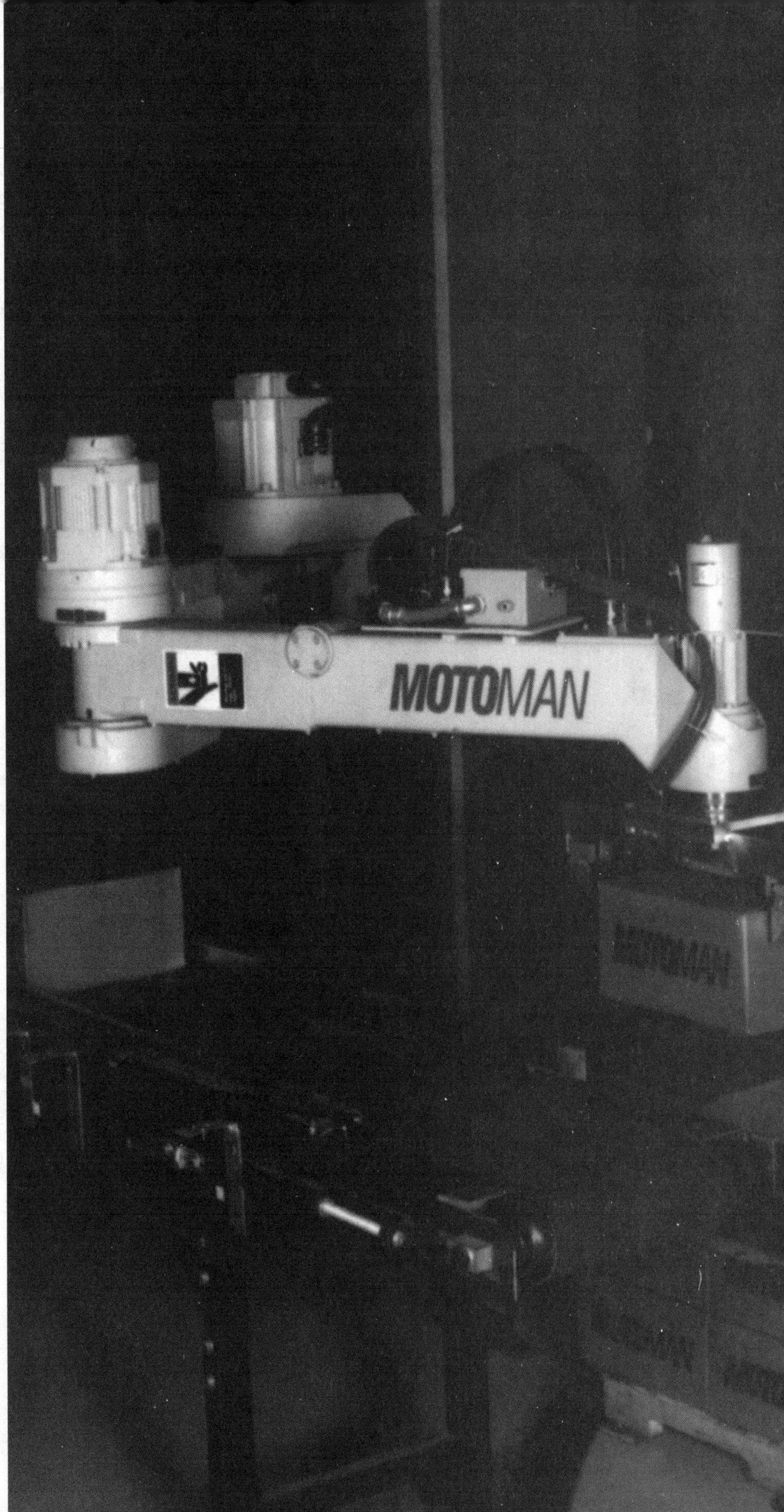

Motoman, Inc.

9 Sensing Systems

Overview

The control of many industrial robots depends upon various types of sensing systems. A sensing system uses a device called a transducer to convert light, heat, or mechanical energy into electrical energy. The signal output by the transducer is used to affect the operation of the machine. Sensors give robots a higher level of intelligence by improving decision-making capability. This chapter discusses some of the sensors used with robotic systems in industry.

Kinds of Sensing

Sensing systems allow robots to interact with their environment. It helps them deal with changes and uncertainties. A sensing system allows robots to determine their own actions. If a robot lacks sensing ability, support fixtures and tools must be built for it. Sensors help robots deal with:

- Δ Different positions and orientation of parts.
- Δ Variations in the shape and dimensions of parts.
- Δ Unknown obstacles.

Simple sensor control is accomplished by the use of contact switches. These switches can stop the motion of the robotic arm or open and close grippers. More complex robotic systems use touch, force, or torque sensing. Some robots have vision sensors that can recognize patterns or react to motion.

Types of Sensors

Many different sensing techniques and kinds of transducers are used with robot systems. The type chosen depends on what is to be sensed, the accuracy required, and the environment.

Proximity Sensors

Most ***proximity sensors*** detect the absence or presence of an object within a certain distance. Others provide feedback about the distance between the sensor and an object, such as an end effector. ***Optical proximity sensors*** measure the amount of light reflected from an object. They can respond to either visible or infrared light. Incandescent lights can be used as the light source. However, ***light-emitting diodes (LEDs)*** are generally preferred, since they are more reliable than lamps and are not sensitive to shock and vibration.

LEDs, or solid-state lamps, are small, lightweight opto-electronic devices. They are easily used with digital and other miniaturized systems. The semiconductors used in LEDs produce light when an electric current is applied. The response varies according to the type of semiconductor used. LEDs are made to produce different colors of light. The wavelength of radiated energy from an LED is beyond the visible range.

Eddy current proximity sensors produce a magnetic field in the small space of a detector unit, which can be mounted in a probe. The magnetic field induces eddy currents into any conductive material that is near the probe. A pick-up coil senses a change in magnetic field intensity when an object enters the field.

Reed switches

Reed switches can either make or break contact in response to a magnetic field. Two flat metal strips, or "reeds," are housed in a hollow glass tube filled with an inert gas. When the reeds are exposed to a magnetic field, they are forced together, completing an electrical circuit. When the magnetic field is turned off, the reeds spring open, breaking the circuit.

Because the contacts are housed inside a hermetically sealed glass tube, contact sparks are isolated from the outside. This type of construction is useful in explosive environments. Contacts are also isolated from outside dust and corrosion, which means improved life expectancy for the switch.

Touch-sensitive proximity detectors operate on capacitance developed by a large conductive object (such as the human body). This capacitance changes the frequency of an electronic circuit. A conductive plate or rod can be used to sense contact.

Acoustical proximity detectors react to sound. Standing sound waves are generated within a cylindrical, open-ended cavity inside the detector. The presence of a nearby object interferes with these waves, altering the detector's output. A microphone may be used to detect a change in sound pressure and measure the distance of the object from the detector.

Range sensors

Range sensors determine the precise distance from the sensor to an object. Such devices are useful for locating objects near a workstation or for controlling a manipulator. One range sensing system, called a *laser interferometric gauge,* is sensitive to humidity, temperature, and vibration. Another range sensing system is a television camera which operates on the sonar (sound) principle.

Tactile sensors

Tactile sensors indicate the presence of an object by touch, while stress sensors produce a signal that indicates the magnitude of the contact made. A simple type of touch sensor is a microswitch. Limit switches also respond to contact with an object. Devices called strain gauges are often used as stress sensors.

Visual Sensors

Visual sensors can be used to recognize objects or to measure their characteristics. Camera-equipped computer systems are an example. These identify a part by means of a television camera and can distinguish that part from any other part. An object may be identified by its shape, outline, or area, regardless of its orientation to the camera.

Computer vision systems are used to sense spatial relationships. Computer vision provides depth information by means of stadimetry and triangulation. ***Stadimetry*** determines the distance to an object based on the apparent size of a camera image. ***Triangulation*** involves measuring angles and the base line of a triangle to determine the object's position.

Position detection can also be done by solid-state TV cameras. A camera is placed in a robot's end effector, and feedback is used to guide the effector to a specific location. This process, which is used for moving material from one spot to another, is referred to as *visual servoing*. Servo movement may be used with either stationary or moving objects.

Light Sensors

Light sensing systems respond to changes in light energy, using various opto-electronic devices. The term ***opto-electronic*** refers to the combination of optics and electronics. These devices, along with lasers and X-ray devices, are becoming increasingly important in control circuits.

Light is a visible form of radiation. It occupies a narrow band of frequencies along the vast ***electromagnetic spectrum***. As shown in Figure 9-1, the electromagnetic spectrum includes frequencies for radio, television, radar, infrared radiation, visible light, ultraviolet light, X-rays, and gamma rays. The different types of radiation, such as light, heat, radio waves, and X-rays, differ only in frequency or wavelength.

The human eye responds only to light in the visible frequencies. See Figure 9-2. Each color of light has a different frequency and wavelength. In order of increasing frequency and decreasing wavelength, colors range as follows: red, orange, yellow, green, blue, and violet.

The wavelengths of light are measured in ***nanometers (nm)***. A nanometer (nm) is 1×10^{-9} m. Visible light wavelengths are in the 400-nm (violet) to 700-nm (red) range. *Angstrom* units are also used for light measurement. An ***angstrom Å*** unit is one-tenth of a nanometer. Thus, visible light ranges from 4000 to 7000 Å. The human eye is most sensitive near 5500 Å. It is least sensitive around 4000 Å on the lower wavelengths and 7000 Å on the higher wavelengths. The human eye perceives various degrees of brightness because of its response to the wavelengths of light. The normal human eye cannot see a wavelength of less than 4000 Å or more than 7000 Å (400-700 nm).

Opto-electronic devices fall into three categories: photoemissive, photoconductive, and photovoltaic. ***Photoemissive devices*** emit electrons in the presence of light. Phototubes are a type of photoemissive device.

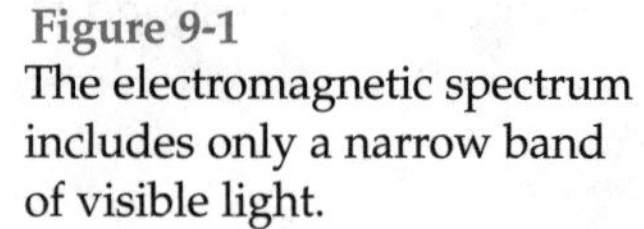

Figure 9-1
The electromagnetic spectrum includes only a narrow band of visible light.

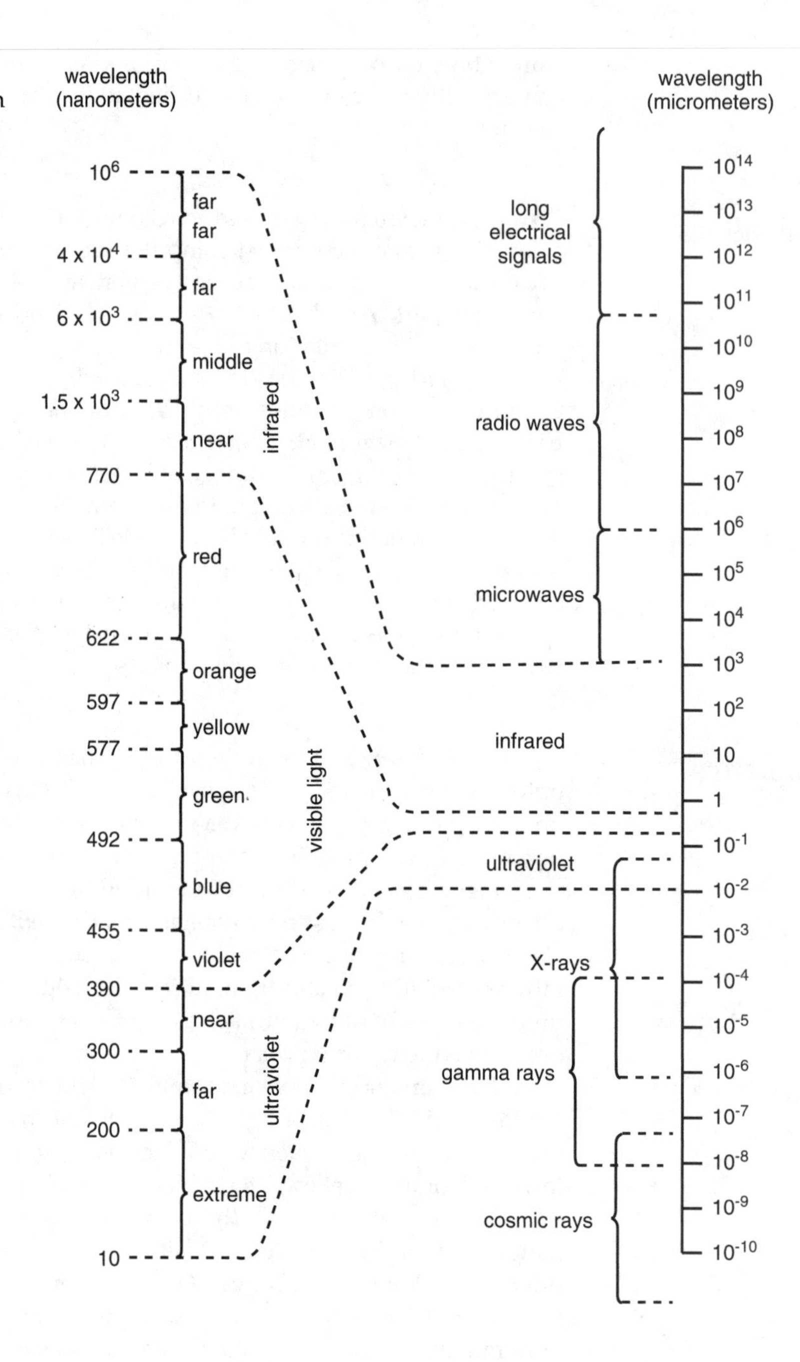

Photoconductive devices vary in conductivity according to fluctuations in light. Their electrical resistance decreases when light is more intense and increases when light intensity decreases, Figure 9-3. ***Photovoltaic devices,*** or *solar cells,* convert light energy into electrical energy. When light energy falls on a photovoltaic device, it creates an electrical voltage.

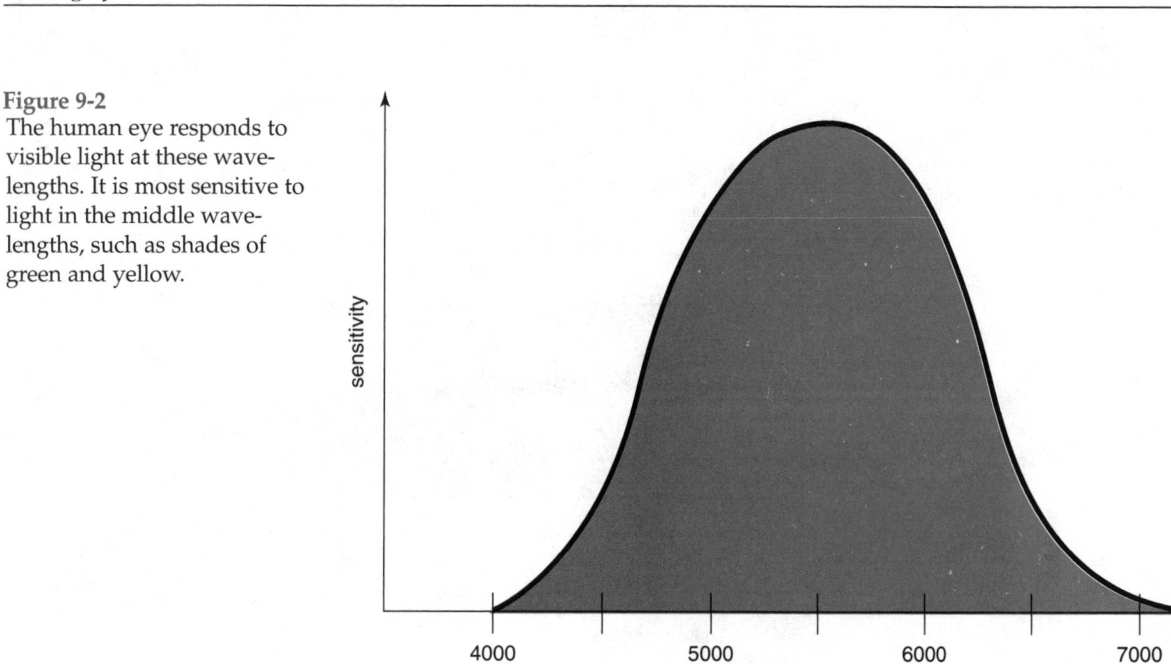

Figure 9-2
The human eye responds to visible light at these wavelengths. It is most sensitive to light in the middle wavelengths, such as shades of green and yellow.

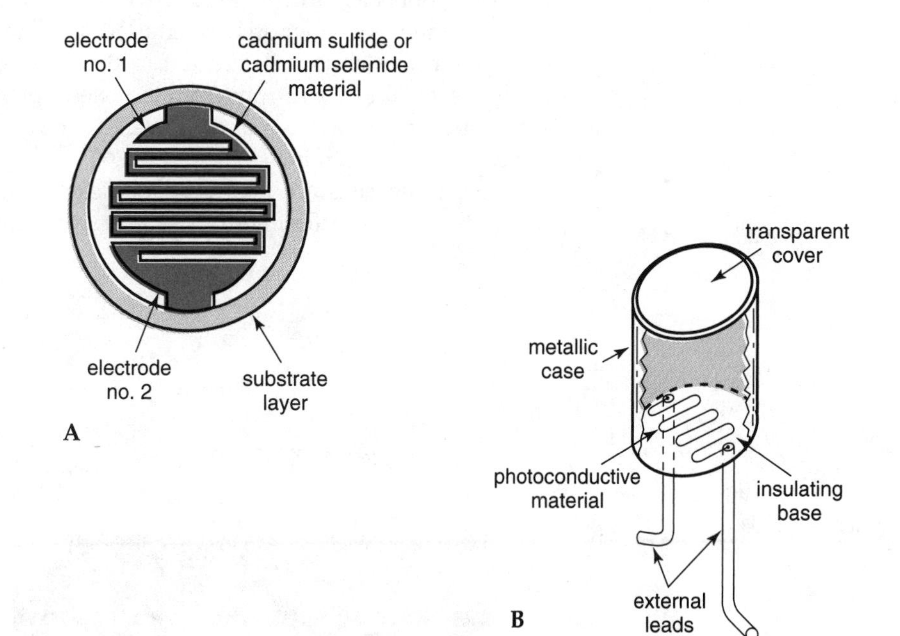

Figure 9-3
This cadmium sulfide cell will vary in conductivity with the amount of light energy striking it. A—Top view. B—Cutaway view.

The electrical output of a solar cell is proportional to the amount of light energy falling on its surface. The cell in Figure 9-4 has layers of selenium and cadmium deposited on a metal base. A layer of cadmium selenide and another layer of cadmium oxide are produced. A transparent conductive film and a

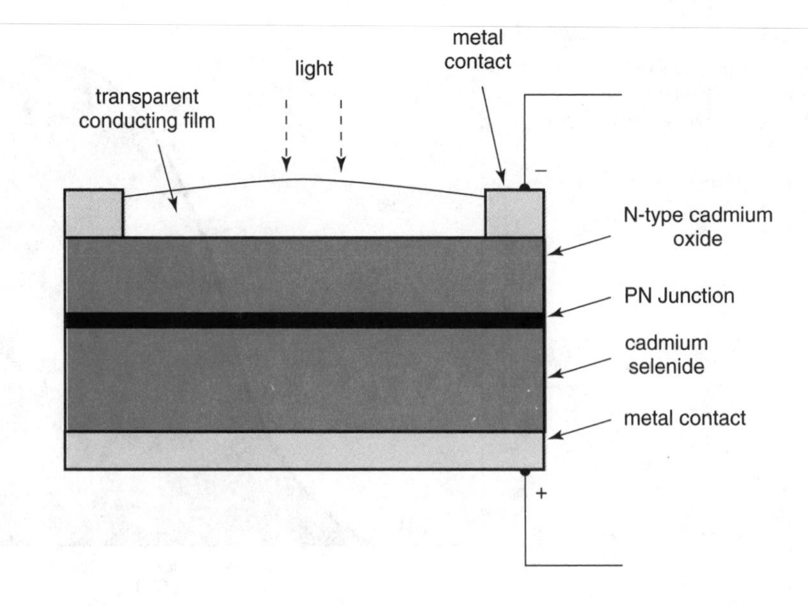

Figure 9-4
Electrons in this selenium photovoltaic cell move from layer to layer, generating electrical energy.

section of conductive alloy are placed over the cadmium oxide. External leads are then connected to the conductive material. When light strikes the cadmium oxide layer, electrons are emitted and move toward the leads. A deficiency of electrons is now created in the cadmium oxide. This deficiency is compensated for by electrons from the selenium material. Electrons then move from the metal base into the selenium. As a result, a voltage develops between the two external terminals.

Selenium cells are relatively inefficient in producing electricity; silicon cells are now more frequently used. A silicon photovoltaic cell is shown in Figure 9-5.

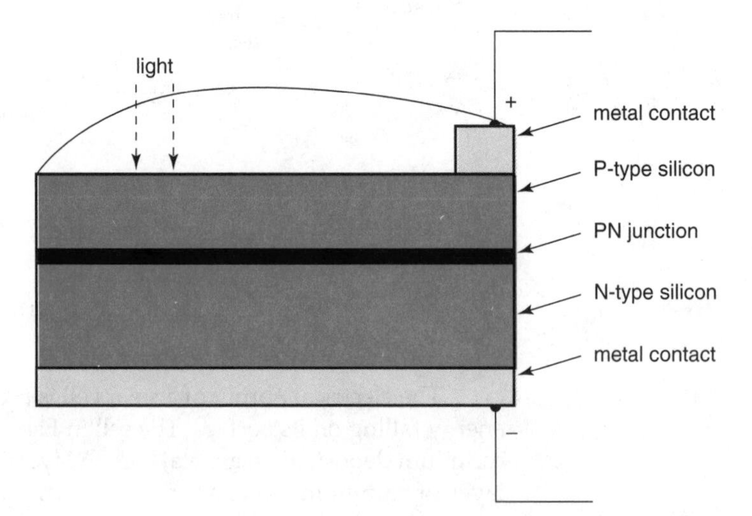

Figure 9-5
Silicon photovoltaic cells like this one produce electrical energy more efficiently than selenium cells.

Photovoltaic cells have a variety of applications. Although their electrical output is low, they are used with amplifying devices to drive a load.

Infrared sensors

Another type of device that responds to radiant energy is an ***infrared detector.*** These devices pick up radiation in the infrared region of the electromagnetic spectrum. All objects emit infrared thermal radiation. Infrared camera systems can detect such radiation even in darkness. Industrial uses include heat-sensitive control systems, optical pyrometers, and infrared spectroscopy for gas analysis.

Ultraviolet sensors

Ultraviolet sensors respond to electromagnetic radiation in the ultraviolet range. However, design problems have been encountered with such sensors. They are thus used less frequently than detectors of visible light or infrared energy.

Opto-electronic position sensors

Various types of opto-electronic devices have been designed to sense the position of a light beam. They are used with digital control systems to produce electrical outputs, Figure 9-6. The output is based on patterns created by the photoconductive material onto which the light beam is focused.

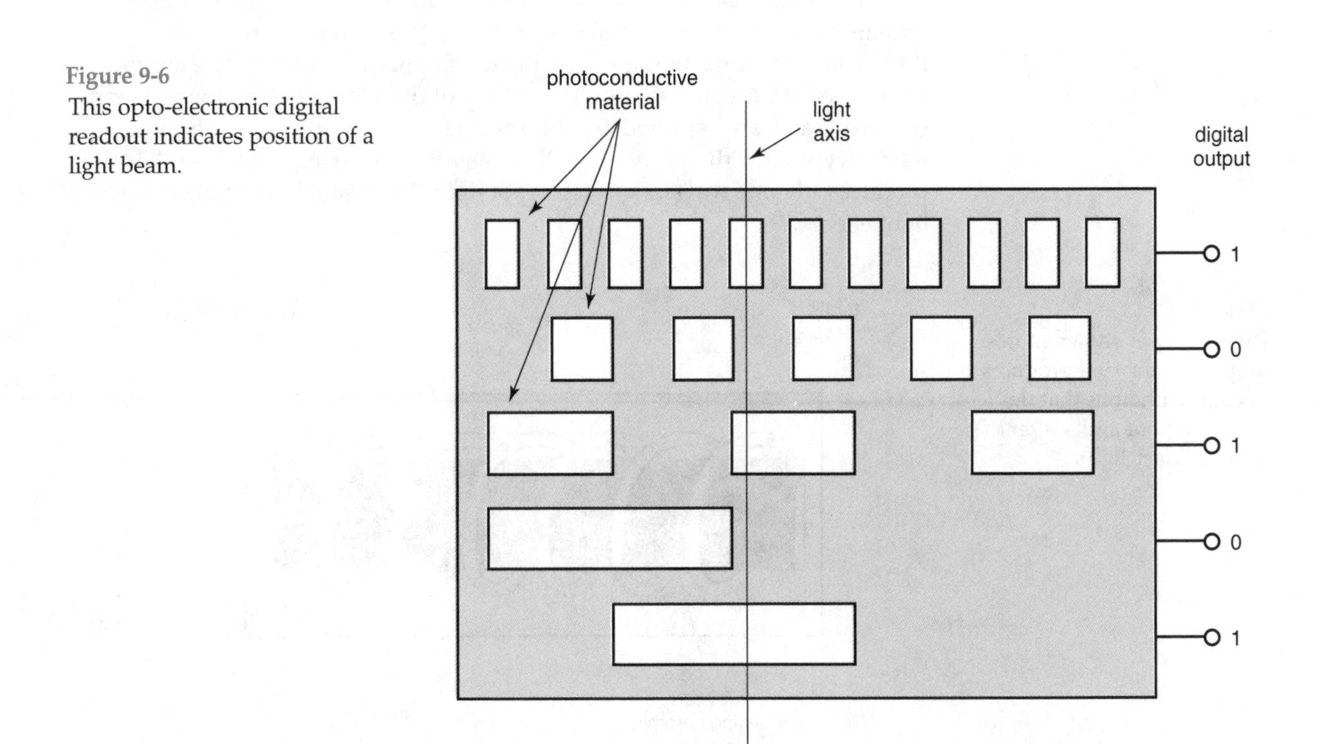

Figure 9-6
This opto-electronic digital readout indicates position of a light beam.

Fiber-optic sensors

Optical fibers made of glass or plastic can transmit light from one point to another, Figure 9-7. The light can travel around corners, within a limited space, or over long distances. The light travels through the fiber optic material regardless of how the material is bent or shaped. The cladding (cover) material is reflective, so the light bounces from side to side within the optical fiber "light pipe" as it moves. Fiber optics are used for numerous applications.

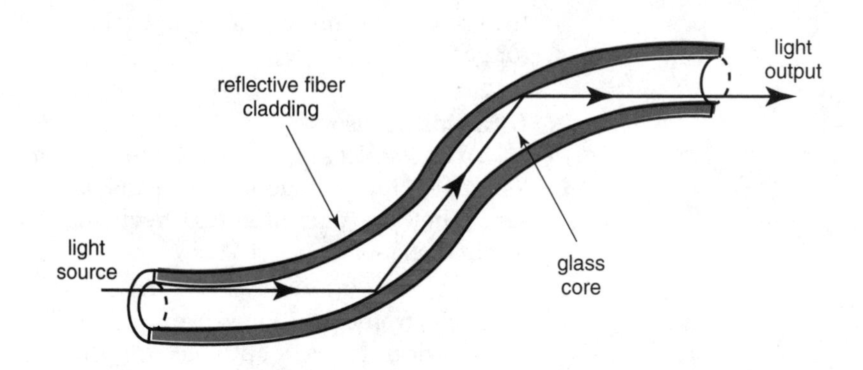

Figure 9-7
Light can be "bent" around corners through use of an optical fiber, sometimes called a "light pipe."

Laser sensors

The development of the laser has had a significant continuing impact on industrial control systems. The major advantage of the laser is that its tightly-focused beam can travel long distances with relatively little spreading.

The term ***laser*** means *L*ight *A*mplification by *S*timulated *E*mission of *R*adiation. A ruby laser is shown in Figure 9-8. When the xenon tube flashes, the chromium atoms in the ruby rod absorb photons of light. The chromium atoms then emit their own photons. Many of these travel along the axis of the ruby rod, and are reflected by the mirrors on each end. This amplifies the light. Eventually, the photons "fall into step" with one another so that they produce only one wavelength (color) of light. The result is an intense beam of light energy.

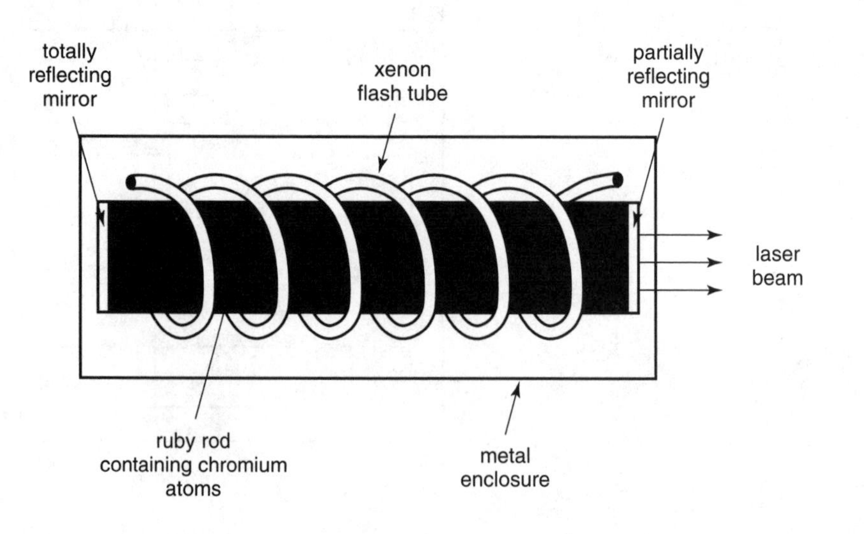

Figure 9-8
The ruby laser shown in this simplified drawing produces a beam of photons that do not spread out and scatter as does ordinary light.

Gas lasers. Gas lasers are often used with sensing and control systems. A popular type is the helium-neon laser, Figure 9-9. Many other types are available; all use basically the same operating principle, as described below.

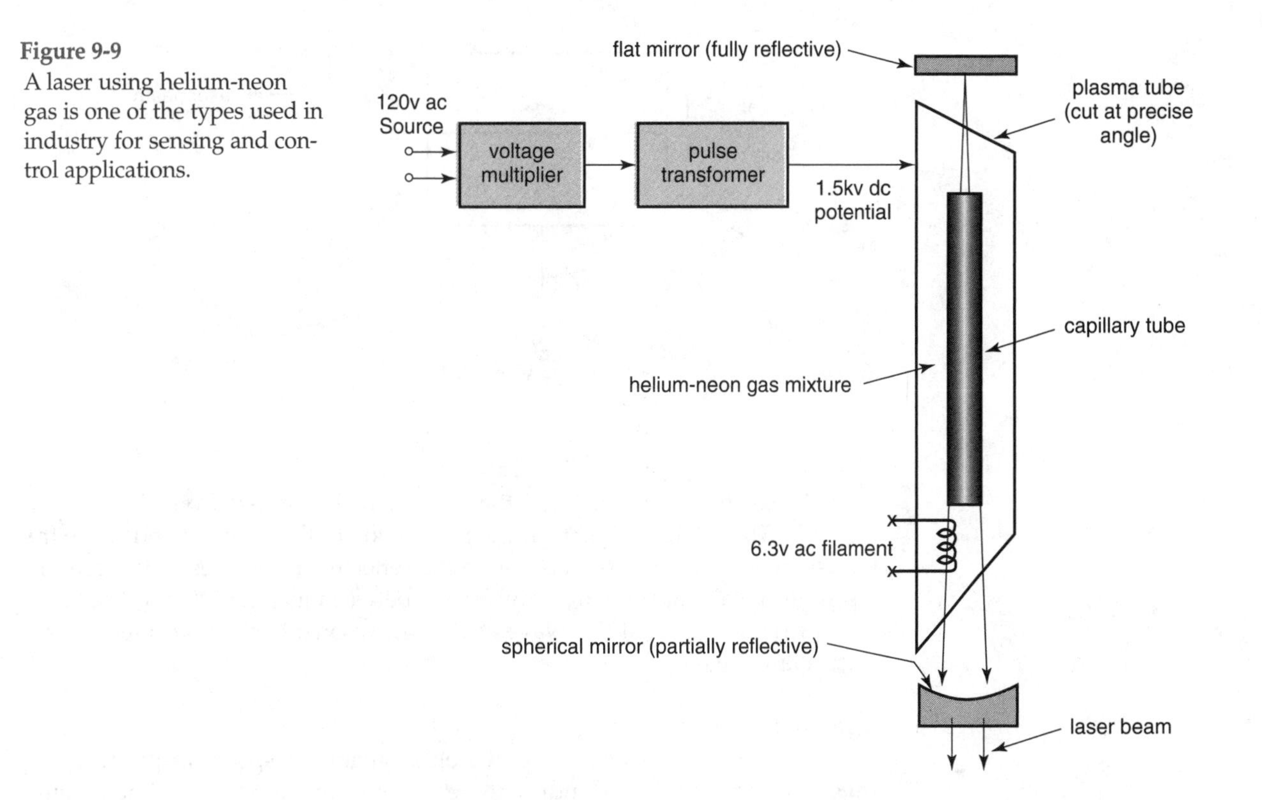

Figure 9-9
A laser using helium-neon gas is one of the types used in industry for sensing and control applications.

A high dc potential is applied to the plasma tube by means of a voltage multiplier circuit and a pulse transformer. The filament within the plasma tube is heated by a 6.3v ac source. As electrons from the filament are accelerated, they strike helium-neon gas atoms in the tube and cause them to ionize. The ionized gas emits light. This is similar to the action of a fluorescent tube used for household lighting. The light reflects from a flat, fully reflective mirror at the top. The plasma tube is cut at a precise angle to control reflection. The light is reflected back and forth several times to the partially reflective spherical mirror at the bottom. Here, it is concentrated into a laser beam that is emitted through the spherical mirror.

Semiconductor lasers. It is also possible to generate laser beams by means of semiconductors. These lasers have a resonant cavity similar to other lasers except that it is formed on a chip of semiconductor material. Semiconductor injection lasers, Figure 9-10, are very efficient and extremely tiny in size compared to other lasers. The end faces of a gallium arsenide chip are made parallel and flat. Since gallium arsenide is a reflective material, no mirrors are needed. This is a distinct advantage in terms of complexity and cost.

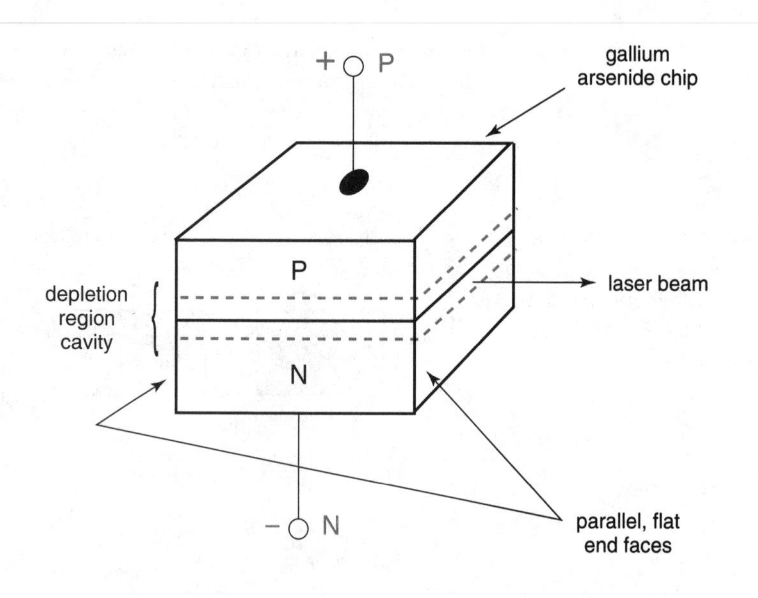

Figure 9-10
This semiconductor injection laser requires no mirrors. It is tiny and very efficient.

As current flows through the chip, light is emitted from the gallium arsenide. The atoms collide near the *pn* junction of the material and cause the release of additional photons. Due to the reflective properties of the gallium arsenide, a wave of photons is developed between its flat surfaces. The back-and-forth movement of this wave creates the resonant action required to produce a laser beam.

X-ray sensors

One band of frequencies in the electromagnetic spectrum produces ***X-rays*** Radium emits three kinds of these invisible rays: alpha, beta, and gamma rays. Some of these rays can pass through the human body, a property that makes them valuable in medical treatment and analysis.

It is possible to use a vacuum tube, Figure 9-11, to produce rays similar to those emitted by radium. This X-ray tube has a cathode that is heated by means of a filament voltage. The anode is constructed of a heavy metal. A high positive potential is applied to accelerate the electrons emitted from the cathode at a very rapid rate. The electrons strike the anode with such velocity that X-rays are created. If the potential is increased, the frequency of the X-rays also increases, while the wavelength decreases.

X-ray tubes can operate with potentials in excess of one million volts dc. The resulting X-rays are similar to the high-frequency gamma rays emitted by radium. In industry, X-rays are used to control processes that involve metals. The short wavelength of X-rays allows them to pass through metals and reveal structural characteristics.

Sound Sensors

Sound sensing systems rely upon the piezoelectric effect to convert sound to electrical energy. The ***piezoelectric effect*** occurs in certain crystals, such as quartz and Rochelle salt. When these crystals are subjected to mechanical stress, an electrical potential is developed in them.

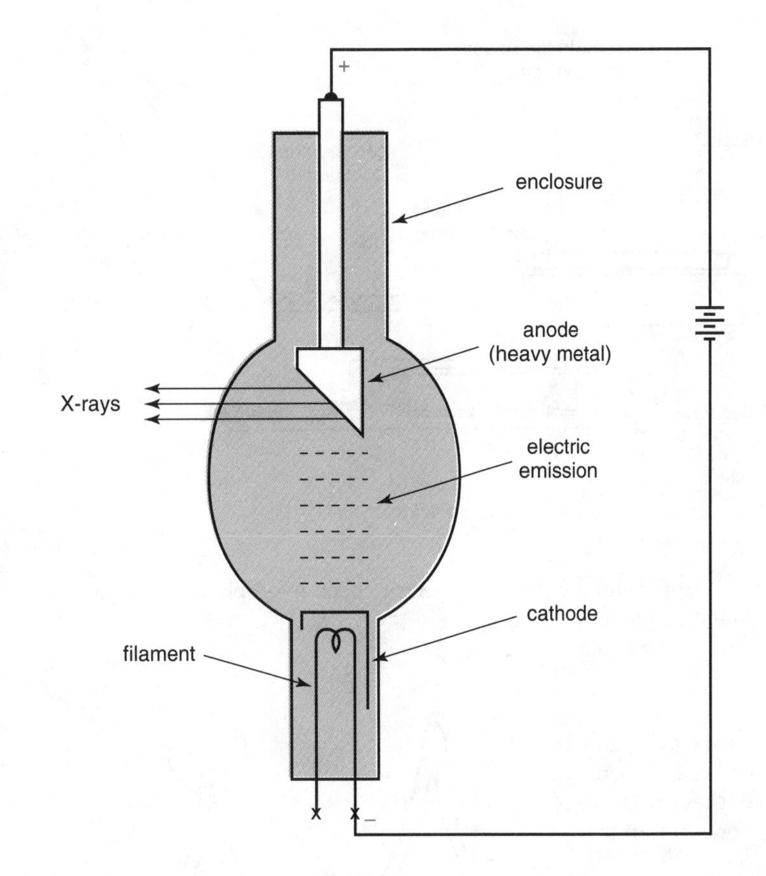

Figure 9-11
X-rays generated by tubes like this one are used to probe the structure of metals.

A common application of the piezoelectric effect is the cartridge/stylus ("needle") assembly of a phonograph, Figure 9-12A. The cartridge contains a crystal that vibrates as the stylus attached to the cartridge rides through the record grooves. This creates an electrical potential across the crystal. The small electrical variations are then amplified by a control system. Thus, mechanical energy (vibration) is converted to electrical energy.

The same principle is applied by crystal microphones, in which sound waves (vibrations in air) strike a piezoelectric crystal. See Figure 9-12B. A voltage is developed across the crystal and is amplified by the control system.

Temperature Sensors

Sensors that produce a change in electrical output due to a change in temperature are referred to as ***thermoelectric sensors.*** They are used to control processes in which temperature must be held at a given level, or not allowed to exceed a specific point.

One device that is commonly used for heat sensing is the ***thermistor.*** The resistance of this temperature-sensitive resistor decreases as temperature increases (and viceversa). This phenomenon is referred to as a "negative temperature coefficient of resistance." Various metal-oxide semiconductor materials are used to construct thermistors, Figure 9-13. Thermistors are

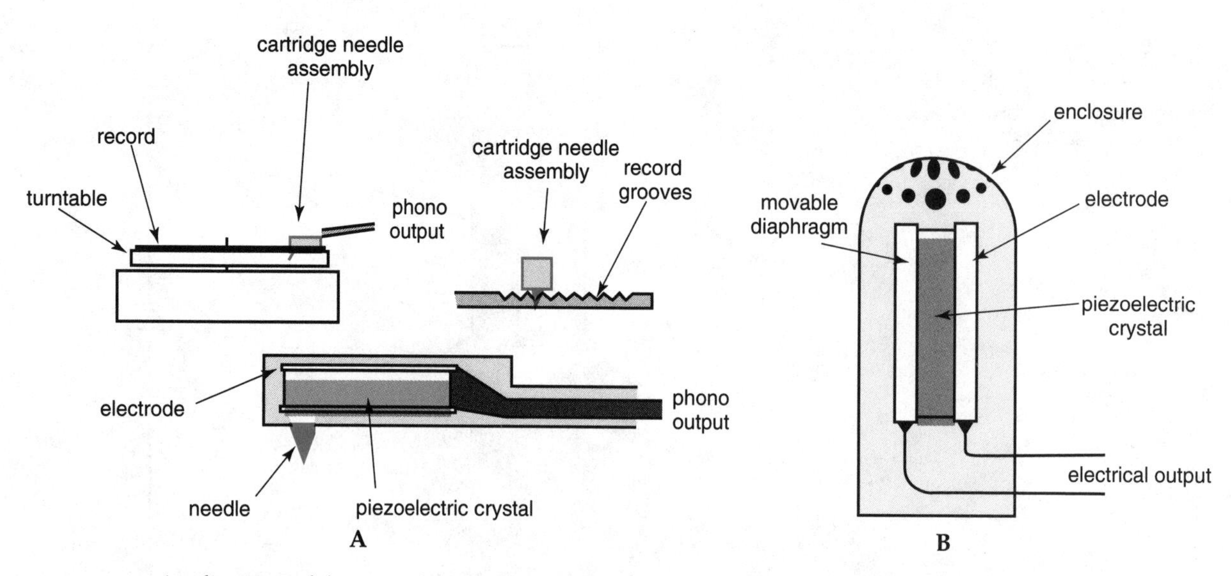

Figure 9-12 Applications of the piezoelectric effect. A—A phonograph cartridge/needle (stylus) assembly. B—A crystal microphone.

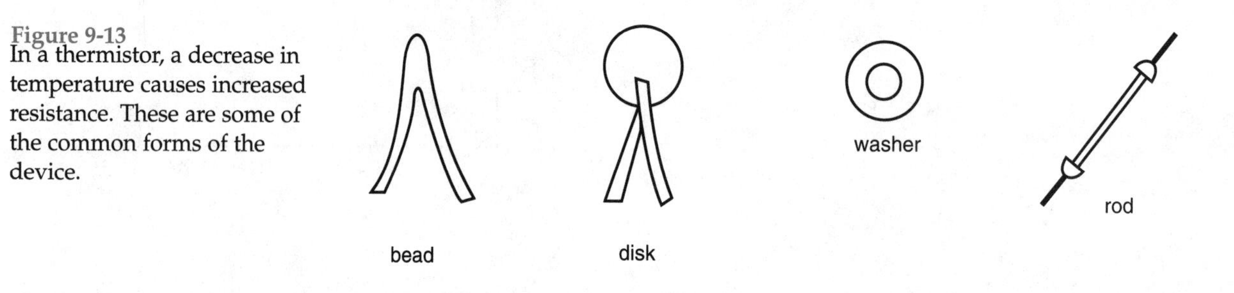

Figure 9-13
In a thermistor, a decrease in temperature causes increased resistance. These are some of the common forms of the device.

manufactured in a wide range of resistance characteristics and temperature coefficients.

Thermocouples are devices that convert heat energy into electrical energy. A thermocouple, Figure 9-14, consists of two dissimilar metal strips fused together at one end. The metals are usually combinations of iron-constantan, copper-constantan, and platinum-rhodium. Different metal combinations respond to different ranges of temperature. When the fused end is heated, a voltage develops at the unconnected ends. This voltage exists due to the differing coefficients of expansion of the two metals. The voltage produced by a thermocouple is usually in the millivolt range.

Displacement Sensors

Displacement (movement) can be detected by various types of sensors. Figure 9-15 illustrates how resistive, capacitive, and inductive sensors can be used to create electrical output that show displacement. Displacement is sensed in reference to a fixed position or to the force required to move from one position to another. It is possible to sense a displacement of less than 1 mm (0.04 in.).

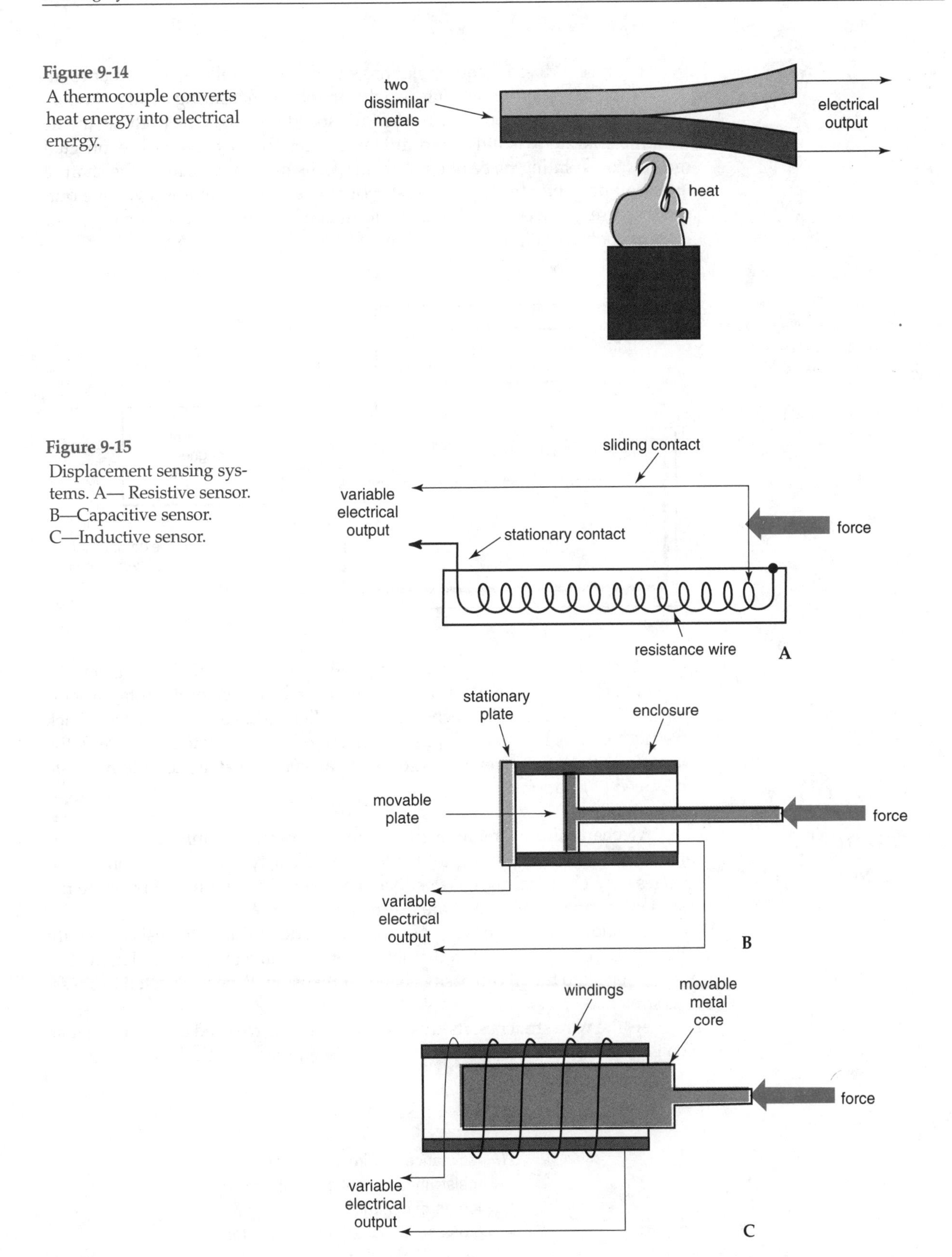

Figure 9-14
A thermocouple converts heat energy into electrical energy.

Figure 9-15
Displacement sensing systems. A— Resistive sensor. B—Capacitive sensor. C—Inductive sensor.

Speed Sensors

Because industrial equipment has shafts, gears, pulleys, and other rotating components, ***speed sensing*** usually involves measuring rotary motion. Several different methods can be used for speed sensing. One of these, the dc tachometer system, is illustrated in Figure 9-16. The tachometer is connected directly to a rotating piece of equipment. As its movement causes the shaft of the small dc generator to rotate faster or slower, the generator's voltage output increases or decreases. Voltage output is translated into speed changes or used to control equipment operation.

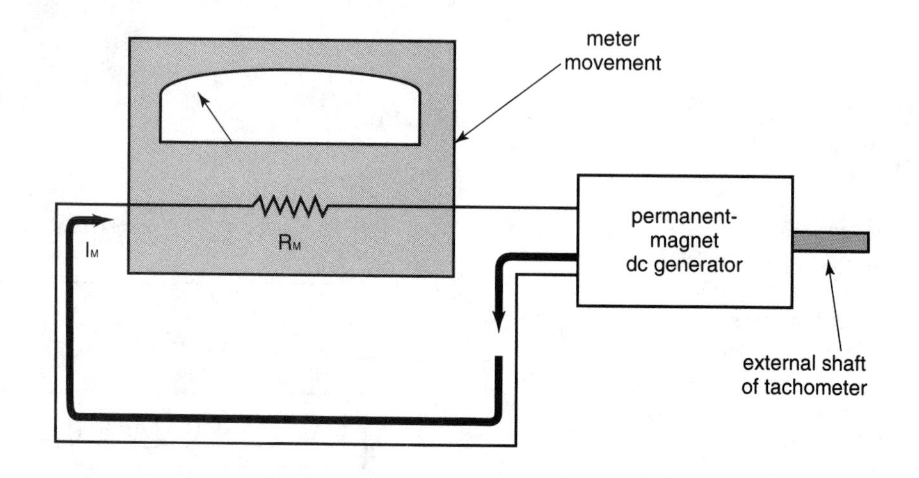

Figure 9-16
A dc tachometer senses speed changes and displays them as meter fluctuations.

Electronic tachometers are now popular because of their precision and ease of use. A reflective material is placed on the surface of the rotating portion of the equipment. The tachometer has a light source that is reflected back to a photocell as the material passes in front of it. The photocell converts the pulses of reflected light energy into electrical signals that are used to measure the speed of rotation.

Mechanical Movement Sensors

Mechanical movement (a change in dimension resulting from strain or stress) can be sensed using a ***strain gauge.*** In a typical application, strain gauges emit electrical signals (feedback) based on the amount of pressure the mechanical fingers of a robot are exerting to lift an object.

As shown in Figure 9-17, the gauge is made of fine-gage resistance wire about 0.001 in. (0.025 mm) in diameter mounted on a strip of insulation. The wire is elastic so that it can easily change dimension. When subjected to stress, it is stretched.

As the wire stretches, its cross-sectional area is reduced and its resistance changes. The wire's resistance can be expressed mathematically as:

$$R = \rho \frac{l}{A}$$

R = resistance of wire conductor
ρ = resistivity constant of conductor
l = length of conductor
A = cross-sectional area of conductor

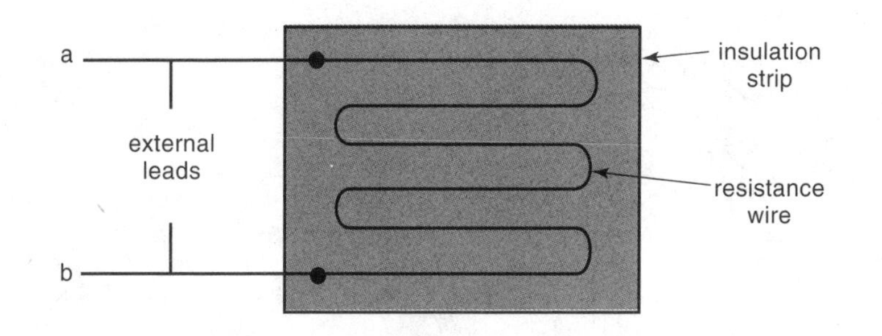

Figure 9-17
A strain gauge responds to mechanical energy by changing its resistance.

Recently, semiconductors have been used in place of traditional metal strain gauges. Their rate of change of resistance is approximately 50 times higher, and they are more sensitive to small changes. A semiconductor strain gauge is as stable as the metallic type, but has a much higher output.

Types of Transducers

The conversion of physical quantities to electrical quantities is a basic sensing function. ***Transducers*** are used to make this conversion. For example, a thermocouple is a transducer that converts heat energy into electrical energy. A microphone is a transducer that converts sound energy into electrical energy. Numerous other examples of transducers can be found in the home as well as in industry. Transducers may be resistive, capacitive, or inductive.

Resistive Transducers

Resistive transducers convert resistance variations into electrical variations. One type works on the potentiometer principle, as shown in Figure 9-18. This transducer changes resistance when the position of its movable contact is changed. By increasing the length of wire between terminals A and B, the resistance between those two points is increased. This transducer is often used to sense physical displacement. Displacement causes movement of the sliding contact, and thus a resistance change in the control circuit.

Capacitive Transducers

Capacitive transducers measure a change in *capacitance,* or ratio of charge to the potential difference between conductors. Capacitance exists when two conductive materials (plates) are separated by an insulating material. Capacitance can be increased by increasing the area of the plates or by decreasing the thickness of the insulation.

One use for capacitive transducers is for sensing fluid pressure, Figure 9-19. As shown, the transducer is placed into a fluid line. Plate 1 is a conductive diaphragm that senses any variation in fluid pressure. Plate 2 is stationary, while plate 1 varies position due to changes in pressure. When the pressure of the fluid in the line increases, plate 1 moves closer to plate 2. When the distance between capacitor plates decreases, the capacitance between terminals A and B increases. When pressure decreases, the distance between plates increases, and capacitance decreases.

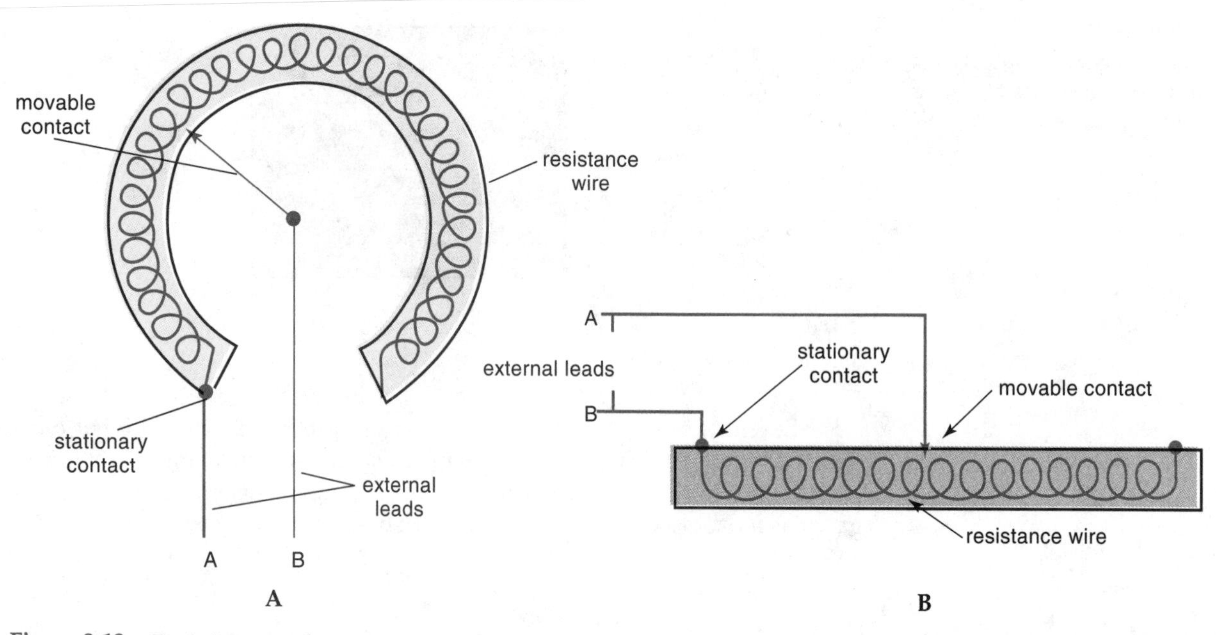

Figure 9-18 Resistive transducers.
A—Rotary potentiometric transducer. B—Flat potentiometric transducer.

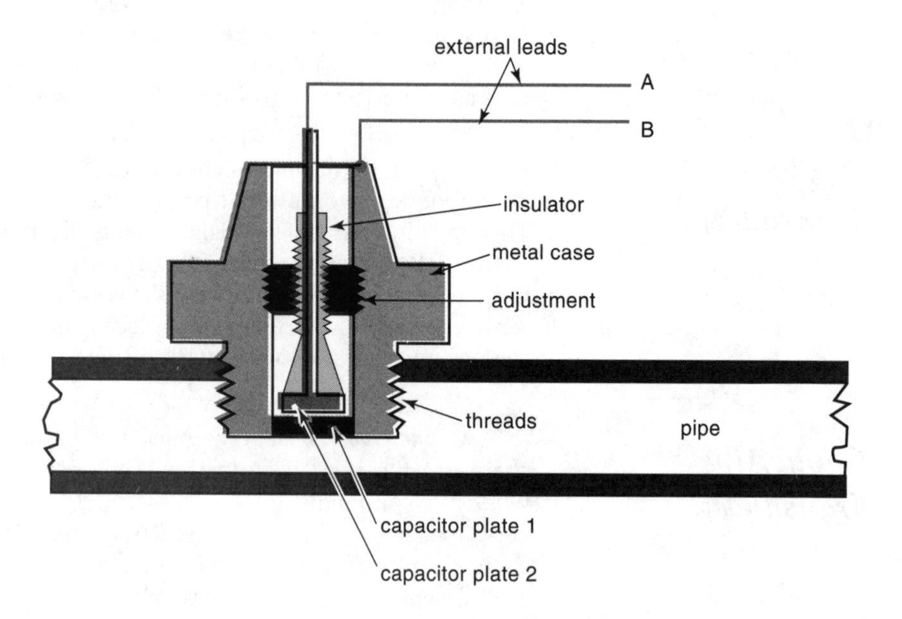

Figure 9-19
This capacitive transducer senses fluid pressure. Pressure changes cause movement of plate 1, changing the capacitance of the circuit and affecting the control signal that is emitted by the transducer.

Inductive Transducers

An ***inductive transducer***, such as the one shown in Figure 9-20, usually has a stationary coil and a movable core. The movable core is connected to an object whose movement is to be measured. As the core changes position within the coil, the inductance of the coil varies. The current flow through the coil drops as inductance increases ($X_L = 2\pi fL$ and $I = V/X_1$).

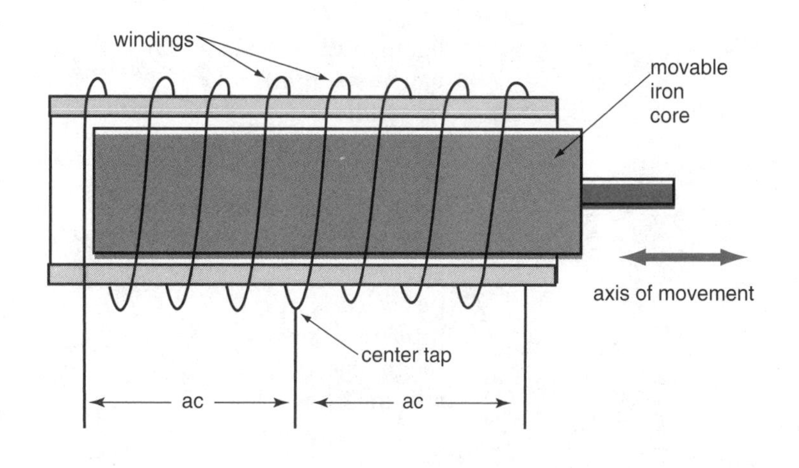

Figure 9-20
A movable-core inductive transducer permits measurement of movement.

One common type of inductive transducer, the linear variable differential transformer (LVDT) is illustrated in Figure 9-21. A movable metal core is placed in a tubular housing that has three windings. The center winding (primary) is connected to an ac source. Voltage is induced in the two outer windings by the primary winding. Initially, the voltages induced in the two outer windings are equal. Any movement of the core will cause one induced voltage to increase and the other induced voltage to decrease. The difference between the induced voltages depends upon the amount of core movement.

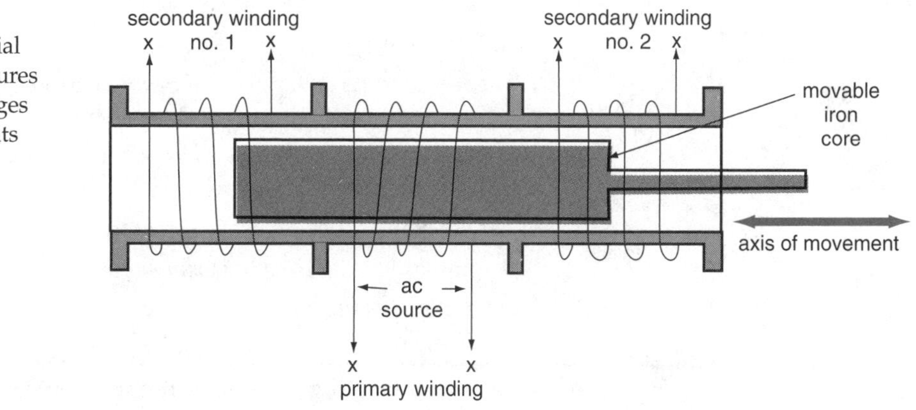

Figure 9-21
A linear variable differential transformer (LVDT) measures movement based on changes of the voltage induced in its secondary windings.

Important Terms

acoustical proximity detectors
angstrom (Å)
capacitive transducers
computer vision
displacement
eddy current proximity sensors
electromagnetic spectrum
inductive transducers
infrared detector
laser

light-emitting diode (LED)
nanometer (nm)
optical fibers
optical proximity sensors
opto-electronic
photoconductive devices
photoemissive devices
photovoltaic devices
piezoelectric effect
proximity sensors
range sensors
reed switches
resistive transducers
sound sensing systems
speed sensing
stadimetry
strain gauge
tactile sensors
thermistor
thermocouples
thermoelectric sensors
touch-sensitive proximity detectors
transducers
triangulation
ultraviolet sensors
X-rays

Review Questions

Write your answers on a separate sheet of paper.

1. What is the purpose of equipping a robot with a sensing system?
2. List some types of sensors used with robots.
3. What is a proximity sensor? What is an optical sensor? What is a LED?
4. What are tactile sensors?
5. What are visual sensors?
6. How is depth measured by a sensor?
7. What is visual servoing?
8. What are the three categories of opto-electronic devices? Define each category and give examples of devices in each.
9. What is the difference between an infrared sensing system and an ultraviolet sensing system?
10. What are some types of laser systems?
11. List some industrial uses for laser sensing systems.
12. How are X-rays used in industrial control?
13. Discuss the two types of heat sensing systems.
14. How is a strain gauge used to sense mechanical movement?
15. What is a transducer? List three types.

10 End-of-Arm Tooling

Overview

A key component in robot design is the end effector or ***end-of-arm tooling.*** End effectors are devices attached to the wrist of a manipulator. They can grasp, lift, transport, maneuver, or perform operations on a workpiece. In this chapter, the two types of end effectors will be discussed. Factors that influence end effector design are also covered.

End Effector Design

End effectors can be classified as grippers or tools. ***Grippers*** grasp an object and move it. ***Tools*** perform a specific task, such as welding or painting.

Robots used for assembly require multipurpose tooling. Two approaches can be used to give the robot the required flexibility. One is to equip it with a multifunction end effector. The other is to make it possible to easily change the end effectors. Both approaches are used, but quick-change tooling is usually preferred.

The robot's end effector is a clumsy imitation of the human hand. The human hand has the ability to adjust, grasp, pick up, and rotate different objects. It also has built-in sensing capabilities and adjusts to imperfections. The human hand can make both nonprehensile and prehensile movements. ***Nonprehensile movements*** include pushing, poking, punching, and hooking. They do not require any special dexterity or the use of the opposed thumb. ***Prehensile movements*** are used to grasp an object with the aid of the thumb. Curled fingers and the opposing thumb provide the hand with the dexterity needed for these movements. The opposed thumb is very important: it enables humans to pick up and manipulate very small objects.

The hand makes five basic prehensile (gripping) movements, as shown in Figure 10-1. They are described with the terms: palmar, cylindrical, spherical, lateral, and oppositional. As a child grows and coordination develops, so does the ability to use various grips. The ***palmar grip*** is that of a baby holding a bottle during feeding. The ***cylindrical grip*** is used for grasping a cylindrical object, such as a baby's rattle or a human finger. With the ***spherical grip*** the fingers come more into play to hold round objects. An example is holding and throwing a baseball. The ***lateral grip*** grasps larger objects in a sideways motion. The ***oppositional grip*** involves the use of the index finger and thumb, which oppose one another.

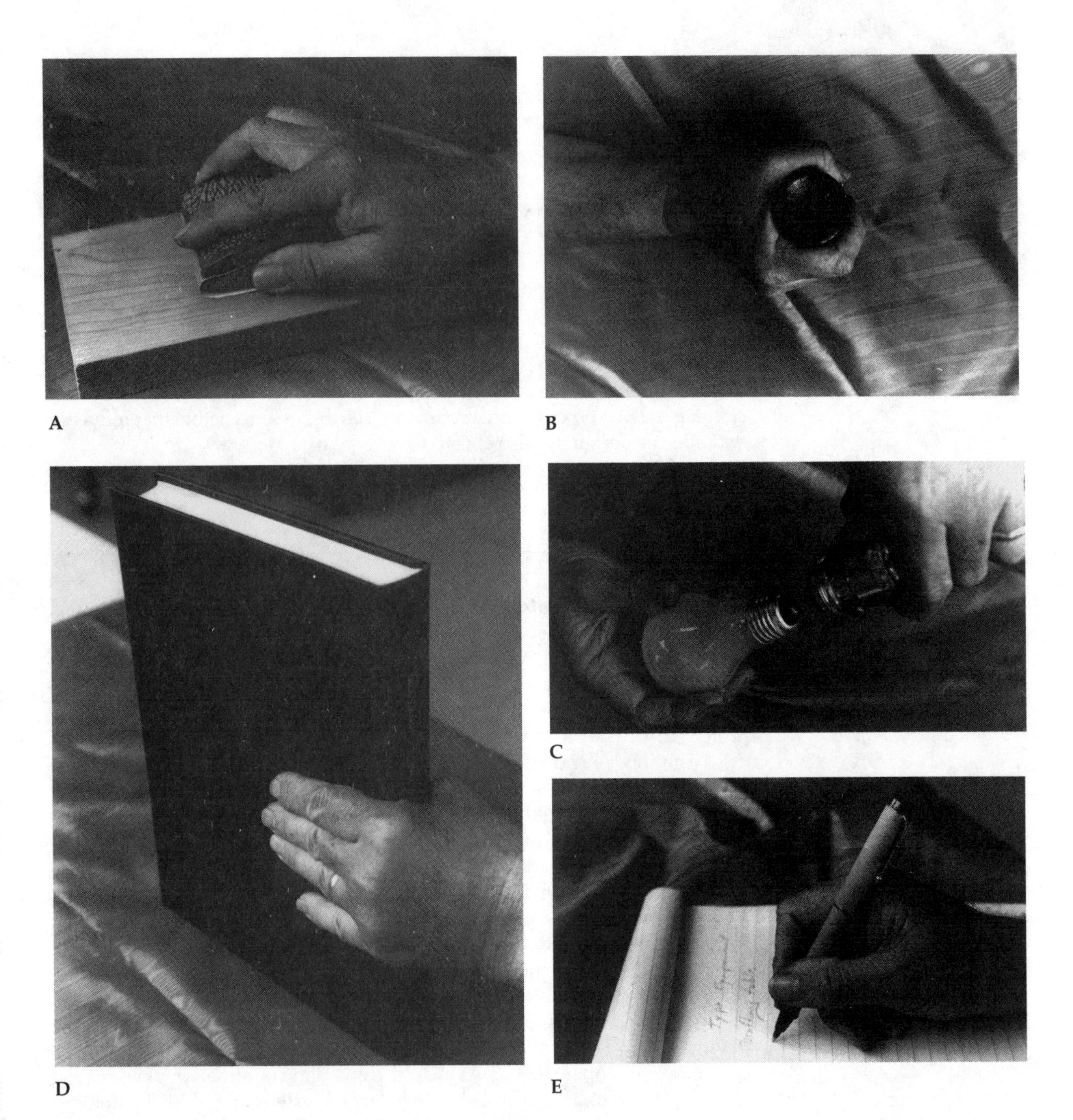

Figure 10-1 The human hand is capable of five basic prehensile grips. A—Palmar grip. B—Cylindrical grip. C—Spherical grip. D—Lateral grip. E—Opposing grip.

Two types of nonprehensile movements, hook and spread, are shown in Figure 10-2. A ***hook movement*** can be used to pull drawers open or lift objects with bail-type handles. The ***spread movement*** can be used to carry objects that have holes or openings of some kind.

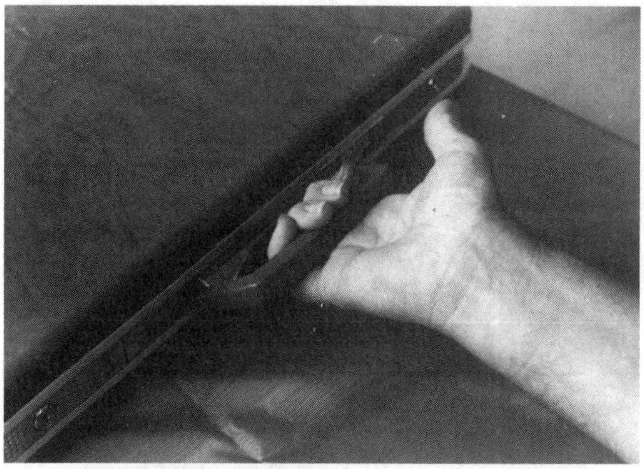

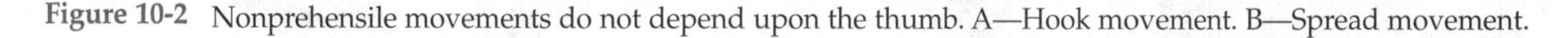

A 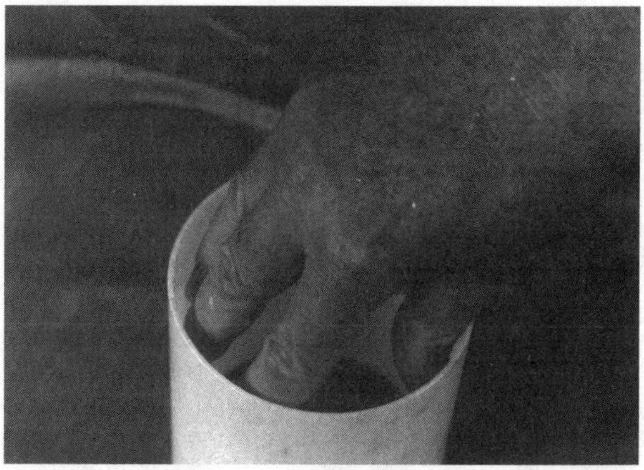 B

Figure 10-2 Nonprehensile movements do not depend upon the thumb. A—Hook movement. B—Spread movement.

To determine the type of end effector needed to do a job, a study must be made of the operation and the workpiece. The end effector may be subject to extreme temperatures or abrasive or corrosive materials. Special protective materials, as well as a shielding device may have to be used to protect the manipulator.

One automation engineer's way to determine if an operation was a candidate for robotic assembly was to try to do the work behind his back, using only one hand with the fingers taped together. This is why each operation usually requires specially made tooling.

Objects to be moved vary in shape, size, and weight. In addition, they may *change* in shape, size, or weight during a given process. The end effector must have the ability to adjust to such changes. Other conditions to be considered are the fragility of the workpiece, its surface finish, and the type of material used. For example, if an object is made of ferrous material (one that contains iron), a magnetic gripper may be used.

Sometimes, a more thorough analysis of gripper requirements must be made. Problems involving inertia, center of mass, gripping force, or friction between the part and the gripper may have to be addressed. Other concerns might involve part orientation, gripper sensing capabilities, or the robot's need to interact with other equipment.

Desirable Characteristics

The end effector should possess the necessary strength to carry out its tasks and withstand rigorous use. For certain operations, a breakaway device should be installed. Such a device will prevent damage to the robot's arm or wrist if the hand becomes stuck. End effectors that hold objects by friction are generally not bothered by this problem, since the object will slip out of the gripper when an opposing force is applied. However, such an opposing force could cause joint slippage and alter the accuracy of the robot's positioning.

Another desirable characteristic in a robotic system is ***overload sensing,*** Figure 10-3. Overload sensors will detect obstructions or overload conditions within fractions of a second. The controller then shuts down the robot before damage occurs. Unlike break-away joints, an overload sensing system has no parts that need replacing after it operates. No reprogramming of the robot is needed afterwards.

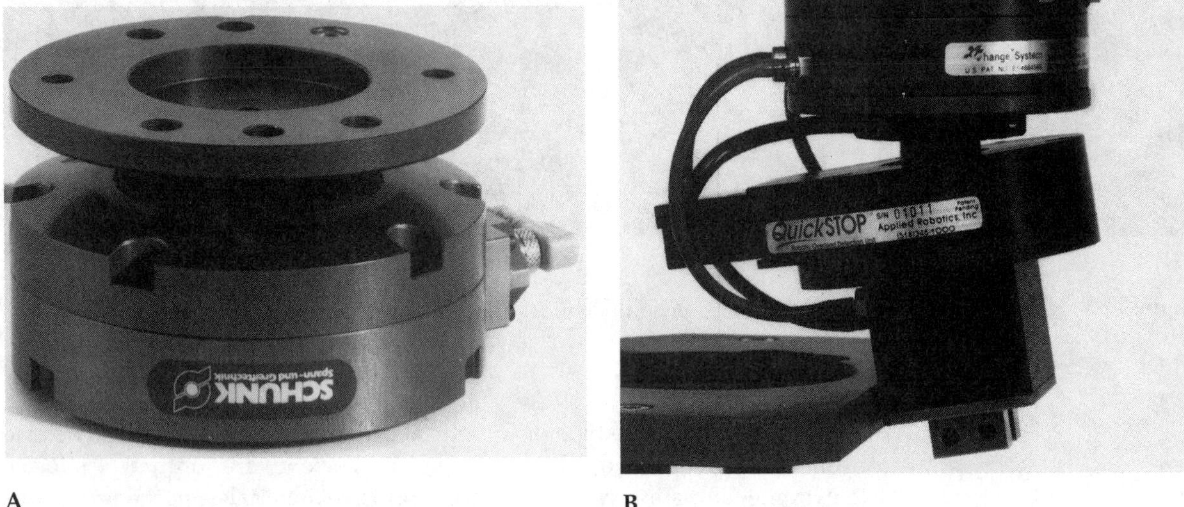

Figure 10-3 Overload protection. A—This overload protection device is used to prevent damage to the manipulator or end effector. (Schunk-USA) B—A collision sensor is used to avoid damage to the gripper, manipulator, or workpiece. (Applied Robotics, Inc.)

Another desirable robot characteristic is ***compliance*** (the ability to tolerate misalignment of mating parts). For assembly of close-fitting parts, it is essential. Compliance prevents jamming, wedging, and galling of the part. Some robots have a certain amount of compliance built into them. For robots lacking this capability, ***remote-center compliance*** (RCC) devices are available. These devices fit in the wrist of the robot. Robots equipped with them can perform precise tasks such as inserting bearings into housings with clearances of only 0.0005 in. (0.013 mm). See Figure 10-4.

Grippers

Robots use a variety of grippers to grasp, handle, and transport parts. Some common types include mechanical finger grippers, collet grippers, vacuum grippers, fragile-object grippers, electromechanical grippers, support grippers, and expandable grippers.

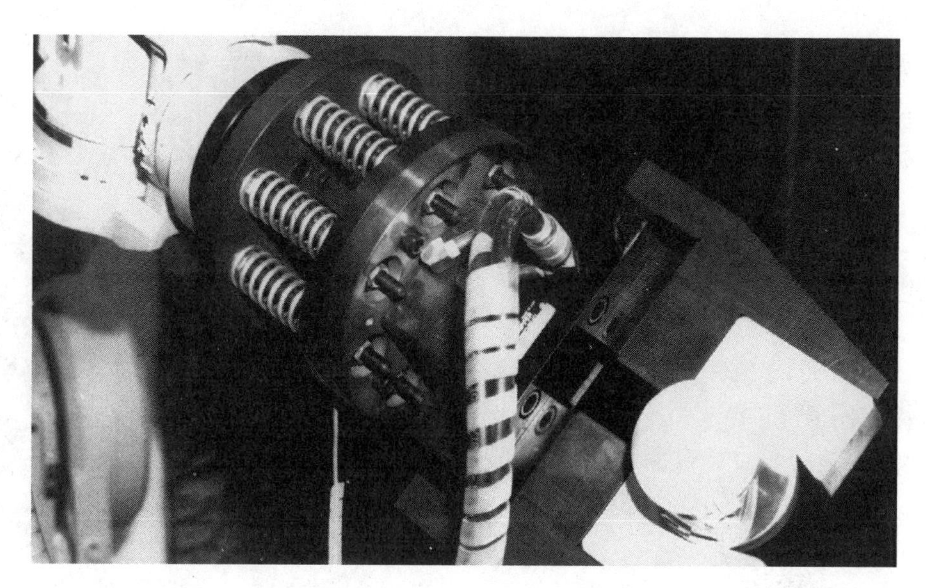

Figure 10-4
In this photo, the remote-center compliance device, which permits the robot to tolerate some misalignment of mating parts, is the assembly with coil springs that is attached to the robot's arm.

Mechanical Finger Grippers

The ***mechanical finger gripper*** is the type most commonly used for grasping objects. See Figure 10-5. Mechanical grippers are generally used for grasping parts within a confined space, reaching into channels, or picking and placing any object that has a simple shape. Mechanical linkages, gears, cables, chains, or pneumatic actuators are used to open and close the gripper's fingers. The most common mechanical gripping motions are parallel and angular, Figure 10-6.

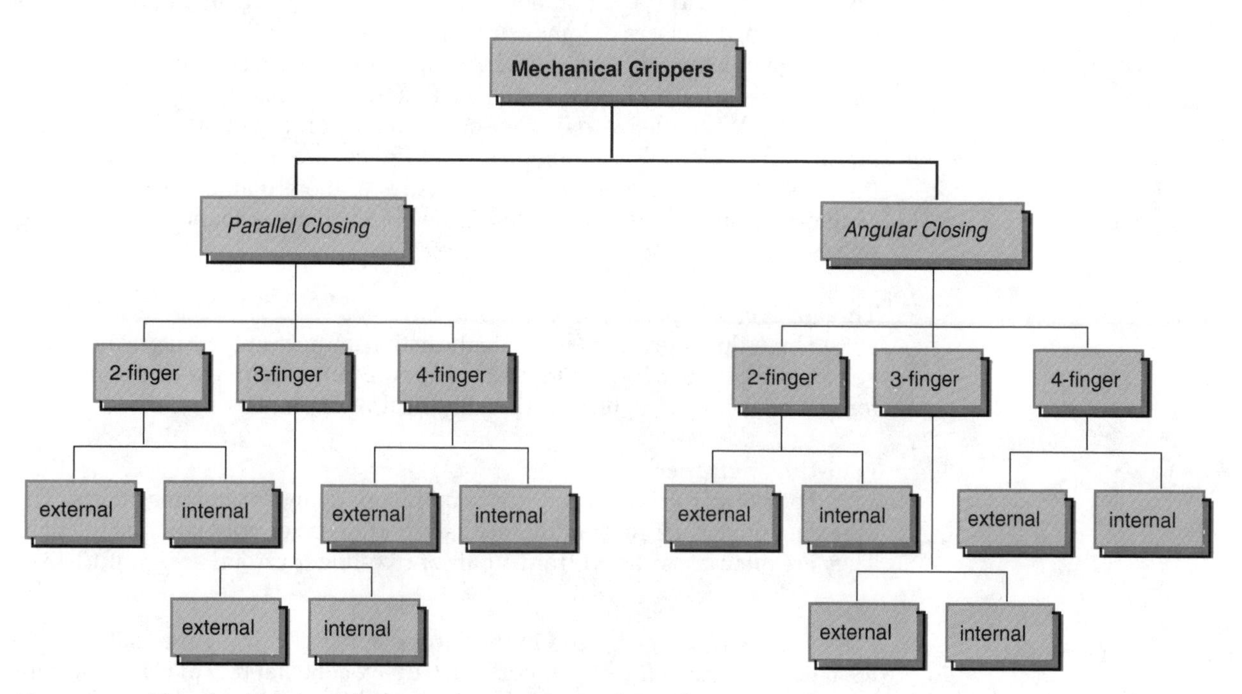

Figure 10-5 This classification scheme can be used for mechanical grippers.

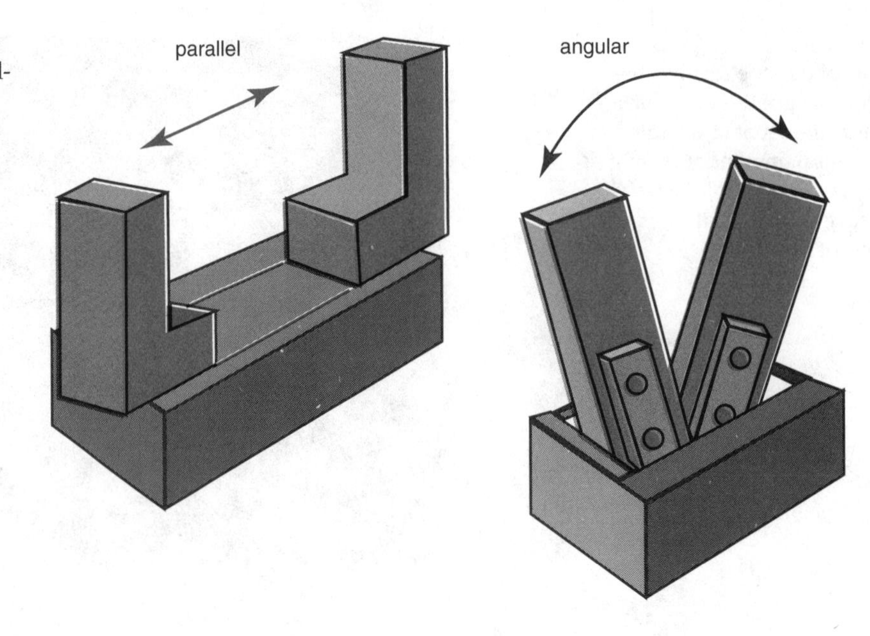

Figure 10-6
Two types of motion — parallel and angular — are made by mechanical grippers.

Two-finger grippers

Grippers with two stiff fingers, simulating motions of the human thumb and index finger, are the most common in industry. One or both fingers may move. Figure 10-7 shows external and internal gripping actions; Figure 10-8 shows a variety of two-finger grippers. In Figure 10-9, a two-finger angular gripper is equipped with a rotating joint that gives it an additional degree of freedom. The wide-jaw, parallel-motion gripper shown in Figure 10-10 is manipulating blocks as a demonstration.

To handle objects with different shapes, fingers can be interchanged. For grasping cylindrical objects, V-shaped fingers are recommended. See Figure 10-11. The "V" shape, with its two-point contact on each finger, ensures centering of the object. Self-aligning padded fingers are used to grip flat objects. It also is possible for fingers to have cavities of more than one size or shape. Multicavity fingers may be needed if the object changes shape or size during processing.

Three-finger grippers

Three-finger grippers simulate the action of the human thumb, index finger, and third finger. See Figure 10-12. They are better than two-finger grippers for grasping curved, spherical, or cylindrical workpieces.

Four-finger grippers

Four-finger grippers can grasp both square and rectangular parts. The opposing fingers, closing simultaneously, permit easier orientation of the part. They are often constructed from a pair of two-finger grippers, Figure 10-13.

Collet Grippers

Collet grippers are used to pick and place cylindrical parts that are uniform in size. Unlike finger grippers, they deliver 360 degrees of clamping contact. They have a strong clamping force for rapid part transfer and are used for

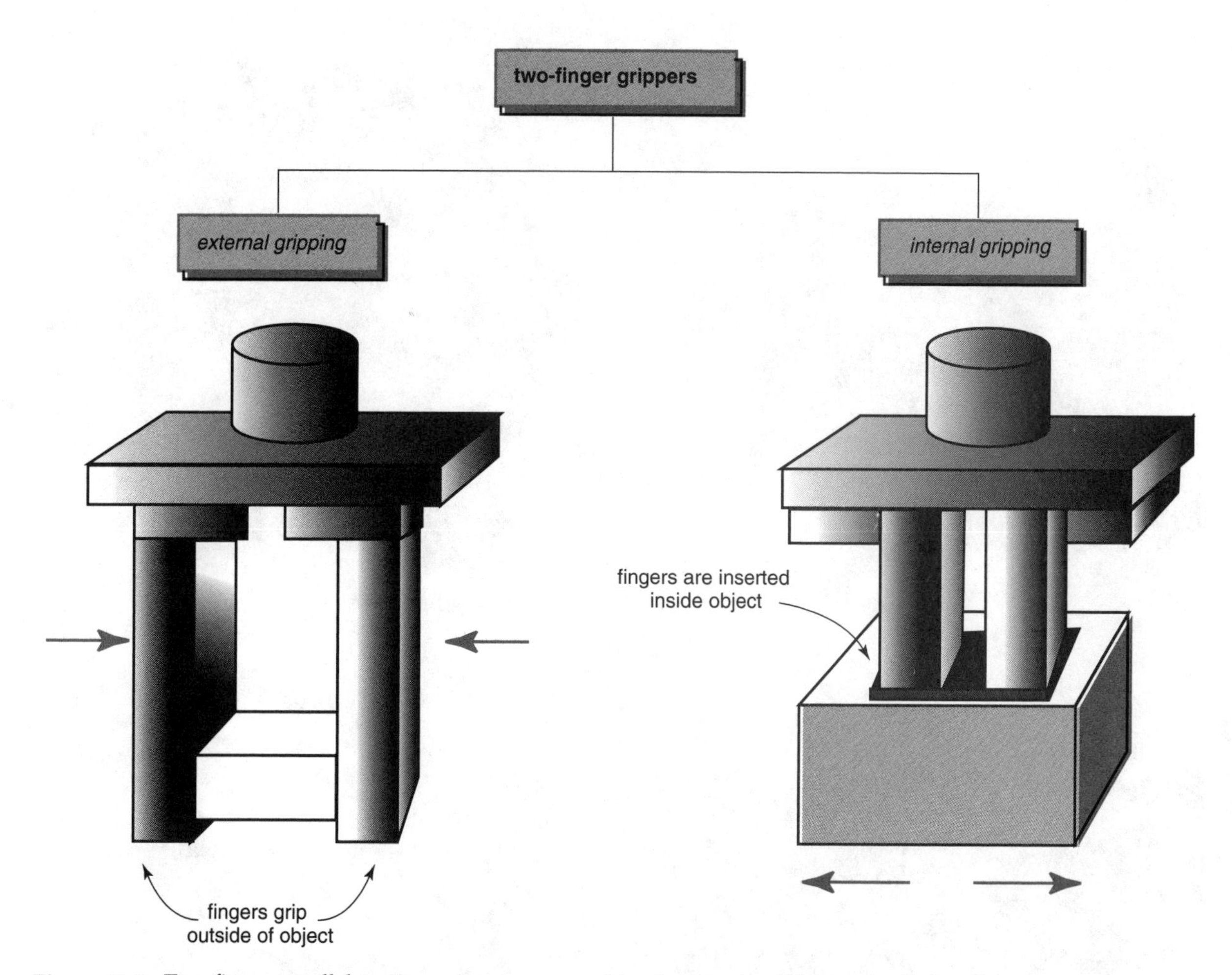

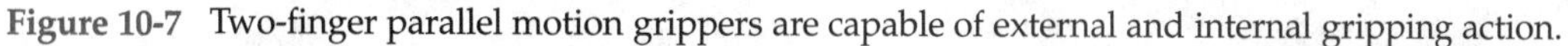

Figure 10-7 Two-finger parallel motion grippers are capable of external and internal gripping action.

grinding and deburring operations. An important characteristic is their repeatability. A solenoid valve is normally used to control them. Collet grippers are available in round, square, or hexagonal shapes.

Vacuum Grippers

The ***vacuum gripper*** consists of one or more suction cups made of natural or synthetic rubber. It is extremely lightweight and simple in construction. The number, size, and type of cups used will depend on the weight, size, shape, and type of material being handled. The cups shown in Figure 10-14 are off-the-shelf items and can be purchased from various suppliers. Multicup vacuum grippers increase the contact surface area, Figure 10-15. This permits the size and weight of the workpiece to be increased.

Vacuum grippers can be used on curved and contoured surfaces, as well as on flat surfaces. They are ideal for handling fragile parts, such as glass objects or even eggs (special soft, flexible cups have been developed for this application).

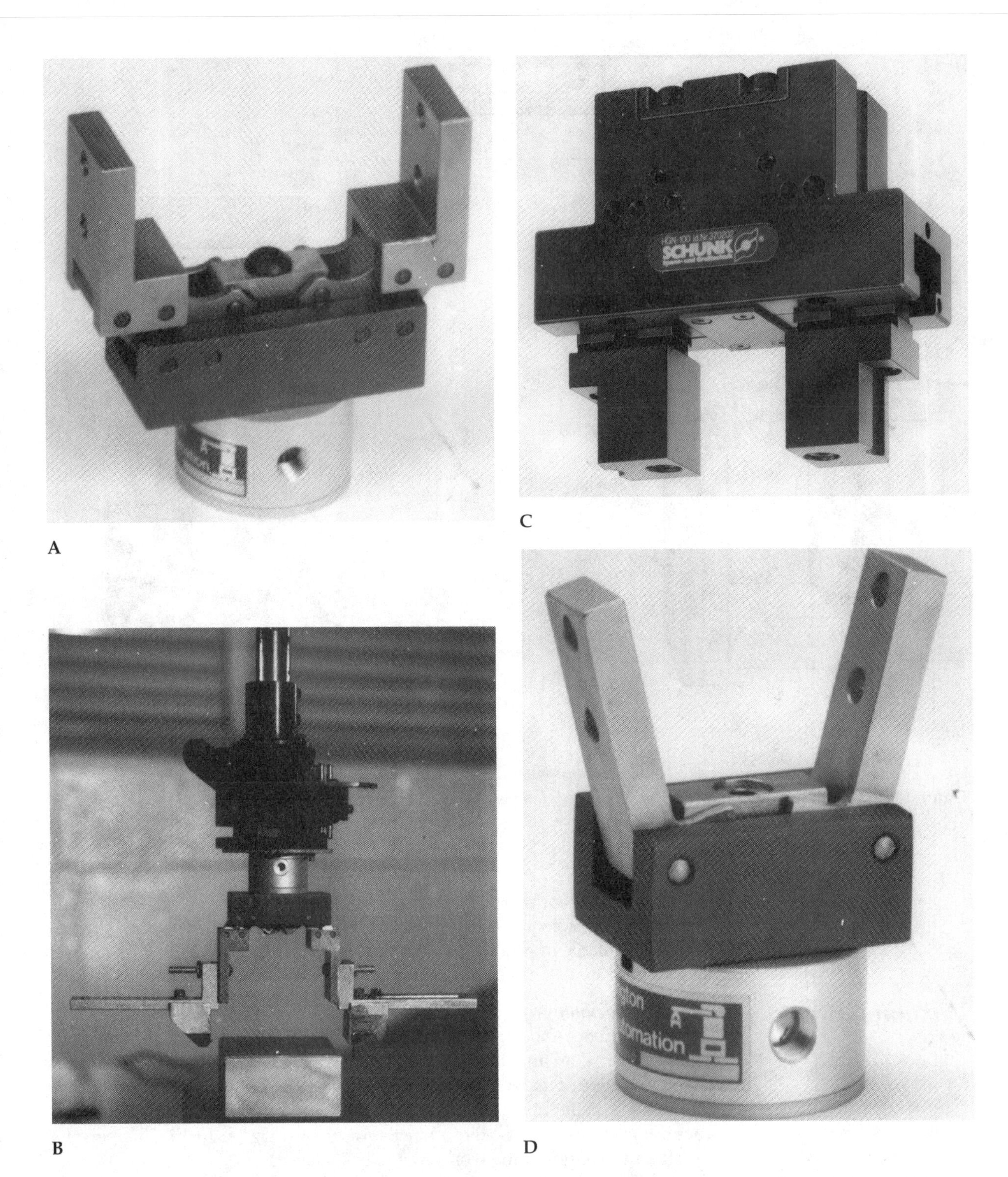

Figure 10-8 A sampling of common two-finger grippers. A—This pneumatic gripper uses a short-stroke parallel motion. (Barrington Automation) B—The jaws of this parallel motion two-finger gripper are adjustable to handle objects of various sizes. C—A hydraulically operated parallel motion internal gripper that is designed to grasp heavier loads. (Schunk-USA) D—This two-finger gripper makes an angular motion. (Barrington Automation)

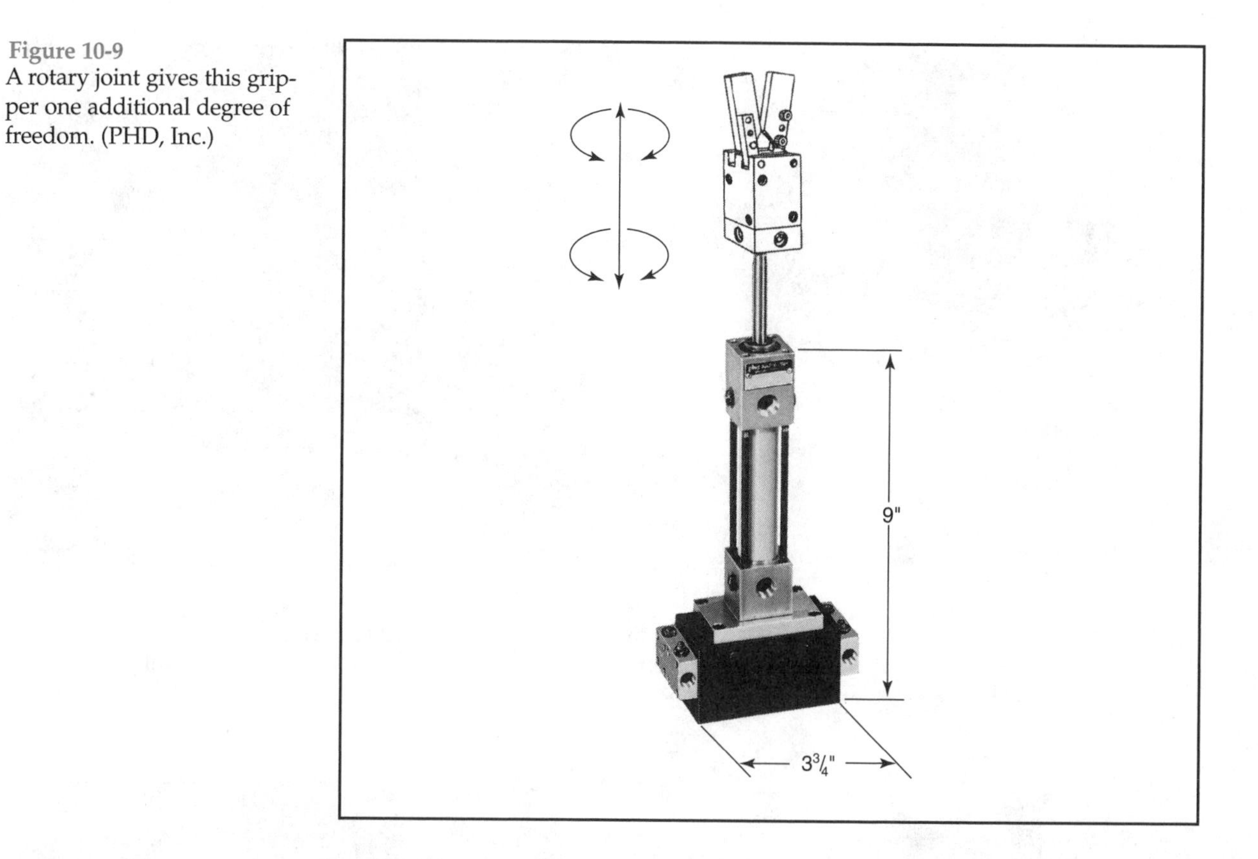

Figure 10-9
A rotary joint gives this gripper one additional degree of freedom. (PHD, Inc.)

Figure 10-10
This large parallel motion gripper is stacking blocks in a demonstration at a machinery trade show.

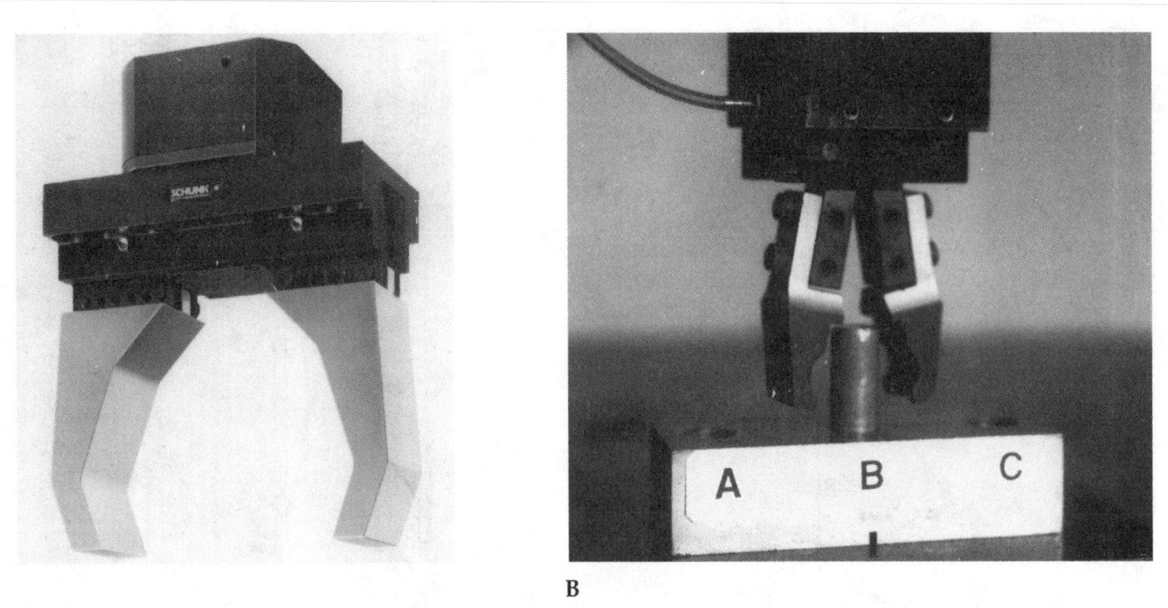

A B

Figure 10-11 Grippers for cylindrical objects. A—This two-finger gripper is capable of grasping a large cylindrical object. (Schunk-USA) B—Vertical V-grooves in the fingers allow this gripper to securely hold a cylindrical pin and insert it into a metal block.

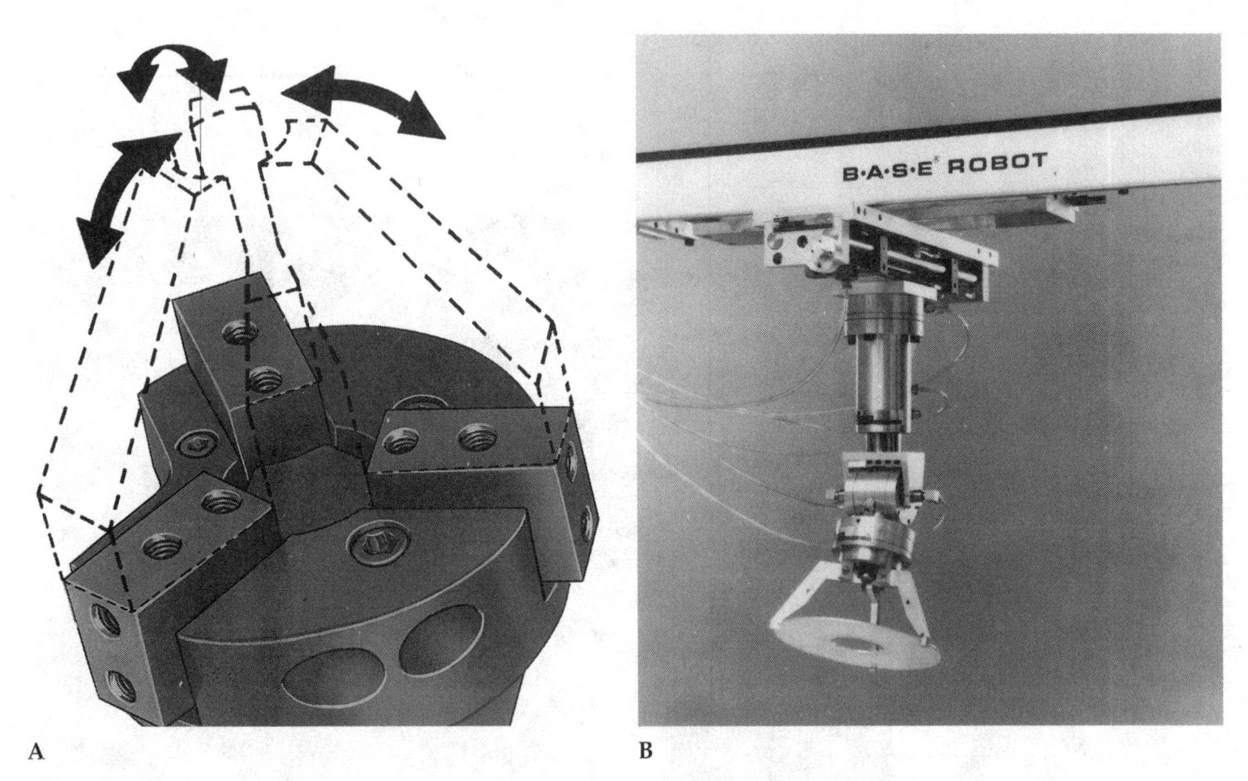

A B

Figure 10-12 Three-finger grippers. A—The action of a typical three-finger angular motion gripper is shown here. (PHD, Inc.) B—Three-finger grippers are well-suited to grasping disks, cylinders, or spheres. Note the extra degree of freedom obtained with the flipping wrist design of this end effector. (Mack Corporation)

Figure 10-13
A variety of two-, three-, and four-finger grippers stored in a holder. (Mack Corporation)

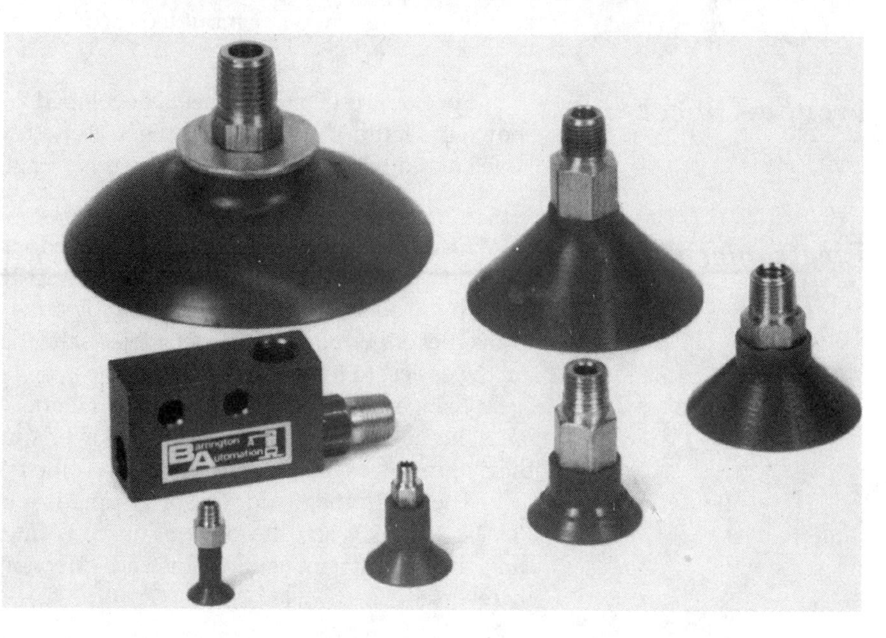

Figure 10-14
Off-the-shelf cups like these are available for a variety of vacuum gripper applications.

Figure 10-15
When a large surface area like this large cardboard box must be grasped, a multicup vacuum gripper is often used. Note that this gripper also has curved fingers used to grasp fragile objects. (Pacific Robotics, Inc.)

The flexibility of the suction cup provides the robot with a certain amount of compliance. Thus, positioning is not as critical as with some other types of grippers. To allow for unevenness in a part's surface, some vacuum cups are spring-loaded or mounted on a ball joint.

Fragile-Object Grippers

Special grippers have been developed for fragile objects. The gripper shown in Figure 10-15, for example, uses both vacuum cups and fingers. The large curved fingers are designed to grasp fragile objects.

Electromechanical Grippers

Electromechanical grippers, also called ***magnetic grippers,*** are similar in operation to vacuum grippers. See Figure 10-16. Instead of using vacuum to pick up the object, however, they employ a magnetic field created by an electromagnet or permanent magnet. Objects that have flat, smooth, clean surfaces are the easiest to handle.

Grippers made with permanent magnets work very well in explosive areas, because they do not require a power source. The part is released from the magnet by exerting force on it with some type of stripper device.

The electromagnetic type of gripper is energized by a dc power source. Release occurs when the power source is interrupted. To speed up release time, the current is not cut off. Instead, direction of current flow is momentarily reversed.

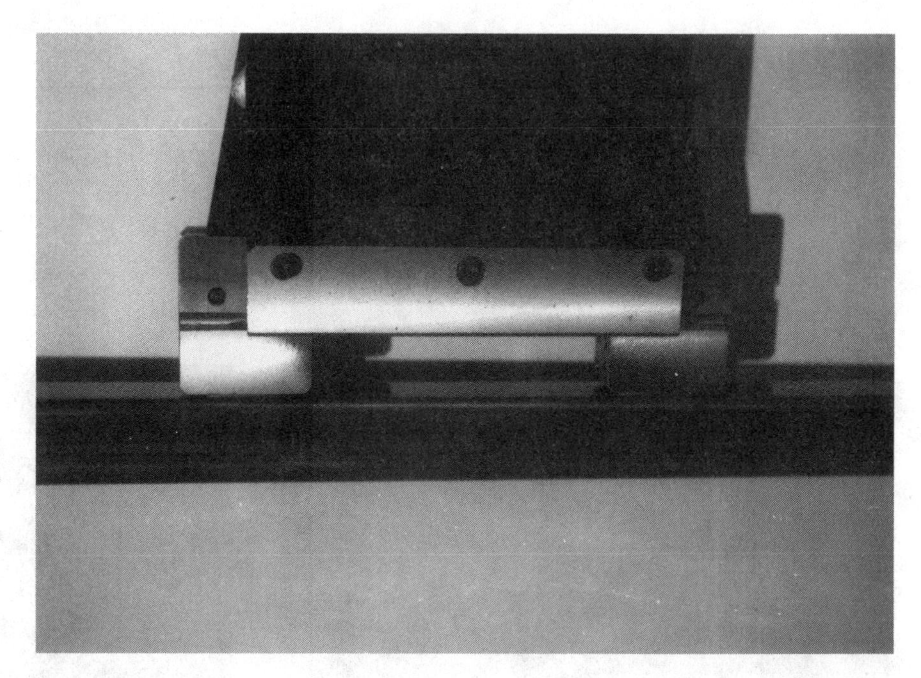

Figure 10-16
A magnetic gripper works with a permanent magnet or an electromagnet.

There are certain disadvantages associated with magnetic grippers. A major drawback is that they can be used to handle only materials containing iron. When parts are machined, metal shavings and other small metal particles are attracted to the magnet. These metal particles, if trapped between the magnet and the part, can scratch the part's surface. An accumulation of particles can cause misalignment of parts. Temperature can also be a problem. The effectiveness of the magnetic force declines when workpieces are heated to several hundred degrees, as they might be in some processing operations.

A magnetic gripper *does* have certain advantages over a vacuum gripper. It has a longer life and can handle hotter and heavier objects. Also, magnetic grippers immediately grip the part, while vacuum grippers require a certain time to build up the necessary pressure. Magnetic grippers are custom designed for specific applications. Few are available as off-the-shelf items.

Support Grippers

Support grippers are usually found on crane-type manipulators. The most common type is the hook. Support grippers often are used to support an object from beneath. A disadvantage, however, is the tendency of objects to topple over or fall with any quick movement.

Expandable Grippers

Expandable grippers are used to clamp an irregularly shaped workpiece. There are two types of expandable grippers: one surrounds objects, gripping them from the outside, the other grips hollow objects from the inside. In both cases, they make use of a hollow rubber envelope that expands when pressurized. This envelope ensures an evenly distributed surface pressure. Expandable grippers are ideal for handling fragile parts or parts that vary a great deal in size.

Tools

In addition to grippers, the robot's arm can be equipped with various types of tools. Some of the common tools used on robots today are spot welding guns, inert gas arc welders, stud welders, gluing guns, spray guns, drills, milling heads, deburrers, polishers, pneumatic screwdrivers, and nut-runners. See Figure 10-17.

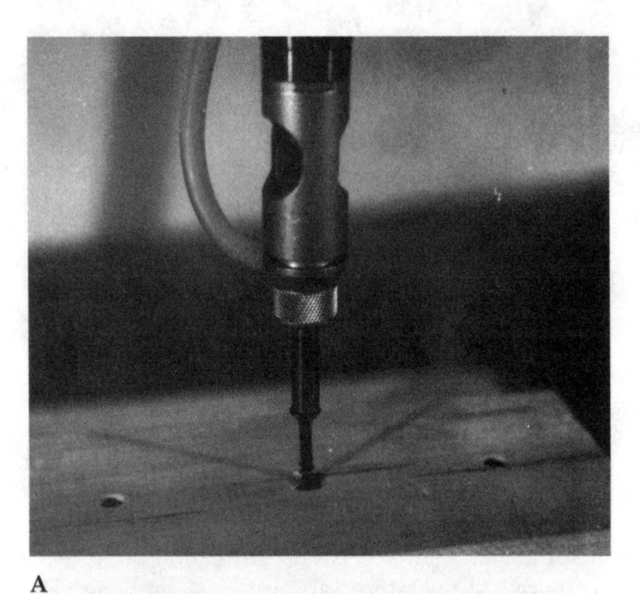

A

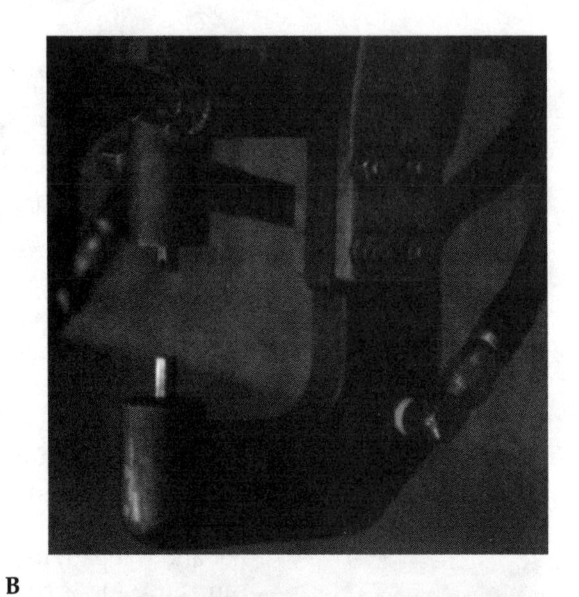

B

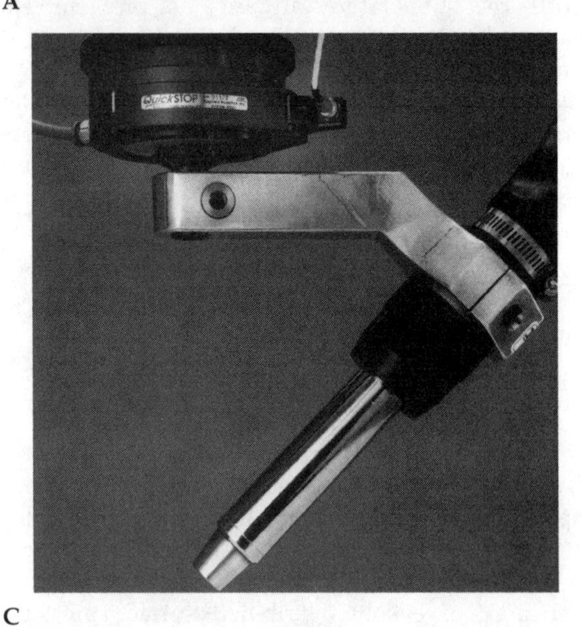

C

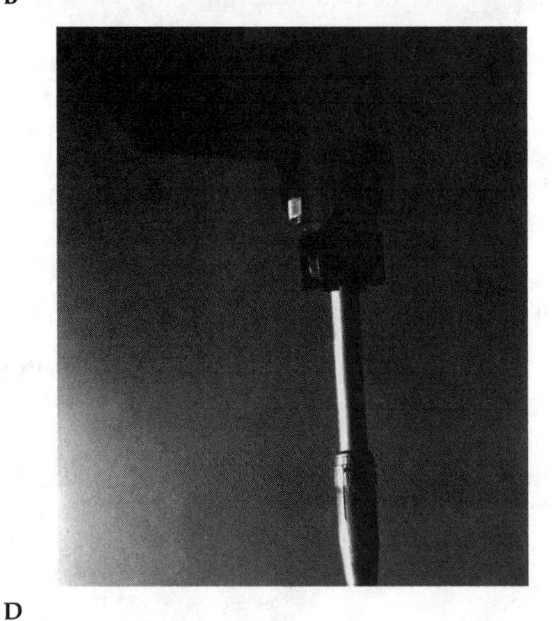

D

Figure 10-17 Various types of tools can be mounted on a robot arm. A—Screwdriver. B—Spot welding tool. C—Arc welder mounted on an overload detector. (Applied Robotics, Inc.) D—Nut driver tool.

Changeable End Effectors

One way to increase the usefulness of robots is to give them the capability to change end effectors. There are two needs that must be addressed before this can be accomplished industry wide. One is the need to standardize interface adapters — they must have compatible mounting and securing systems. All connectors for electric, hydraulic, and pneumatic parts should be provided with the adapters. Since robots come in different sizes, different sizes of standard adapters may have to be provided, as well, Figure 10-18.

Figure 10-18 Different sizes of adapters are used to permit rapid tool changing on robots of various sizes. (Applied Robotics)

Rapid tool-changing capabilities are also needed, so that a minimum of time will be lost in changing tools. Several types of ***automatic tool changers*** are shown in Figure 10-19. With the capability of readily changing end effectors, a robot can handle different part shapes and perform a wider range of assembly tasks and machining operations. By giving the robot greater flexibility, its productivity will increase.

Custom-designed End Effectors

In addition to the standard grippers and tools, custom end effectors can be designed for a particular application. Figure 10-20 shows a series of grippers in use. They illustrate the wide range of tasks robotic equipment must be capable of performing.

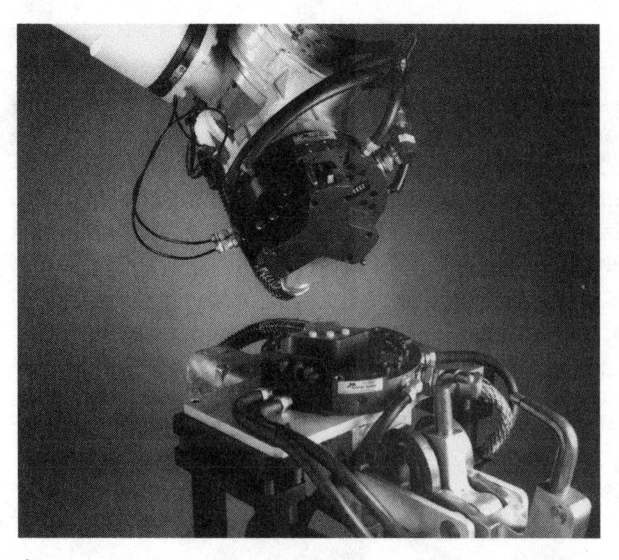

A

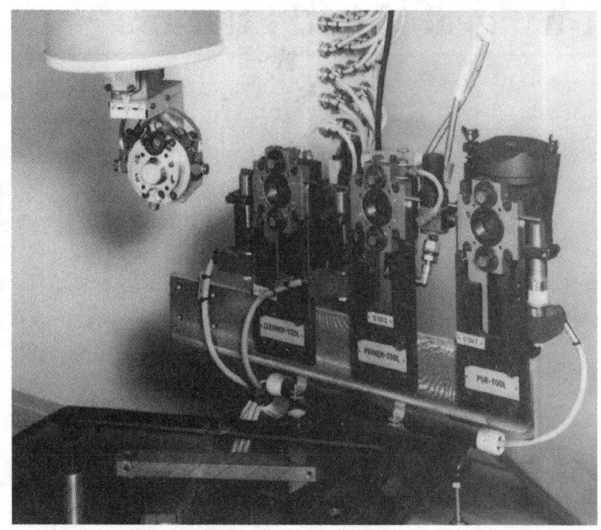

B

C

Figure 10-19
Typical automatic robot tool changer applications. A—This changer is getting ready to pick up a spot welding end effector. (Applied Robotics, Inc.) B—An automatic tool changing system permits this robot to execute three operations automatically: cleaning, primer coating, and gluing. (Schunk-USA) C—A large gantry robot is equipped with an automatic tool changing system. Note the extra-long fingers on the gripper allowing the clamping of large rings. (Schunk-USA)

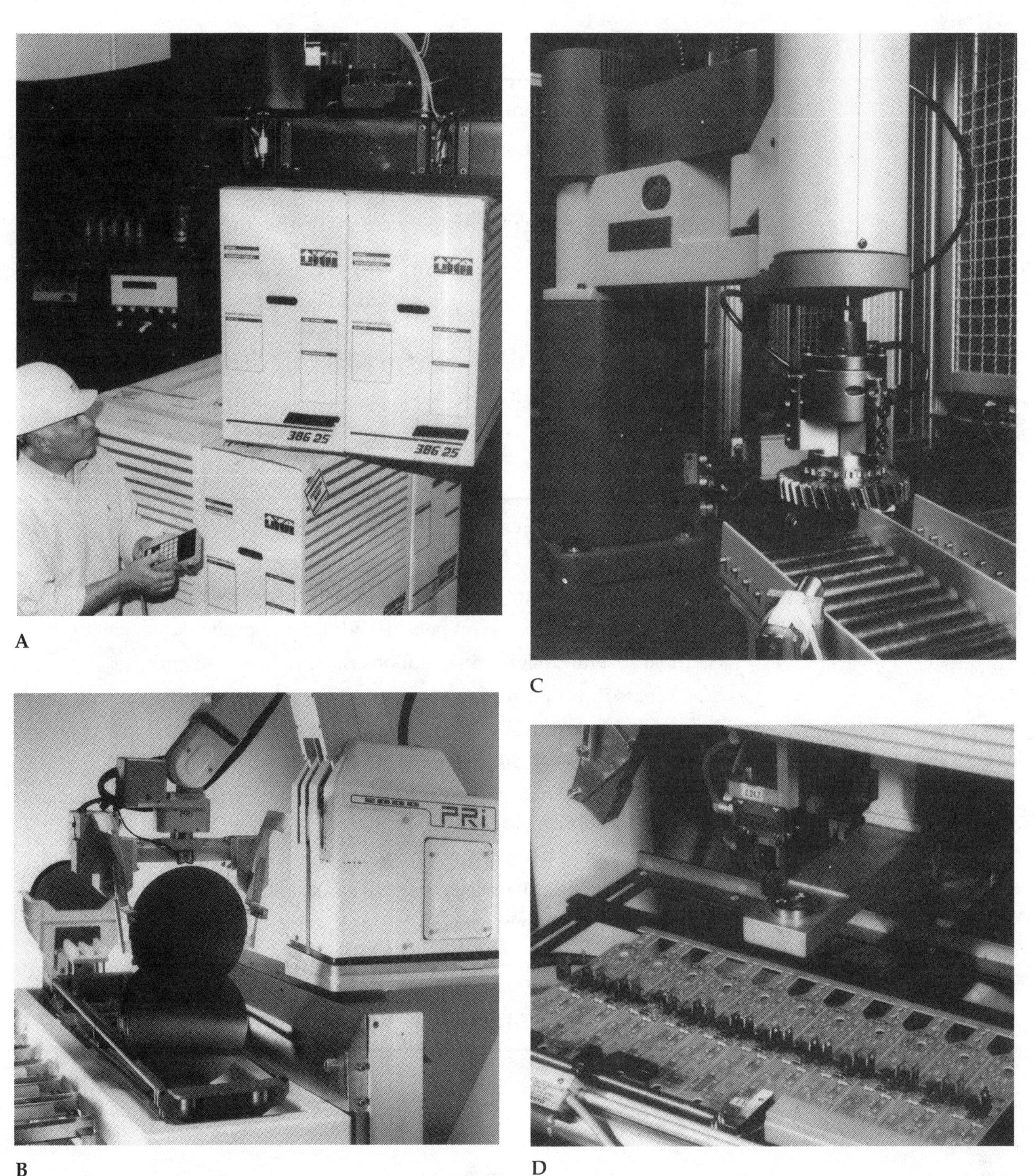

Figure 10-20
Custom-designed end effectors. A—This large parallel motion gripper is used to move loaded cartons. (Pacific Robotics, Inc.) B—The gripper on this clean-room robot transfers large silicon wafers from one container to another. (PRI-Precision Robots, Inc.) C—This internal motion three-fingered gripper is used in a production line for machining precision gears. (Schunk-USA) D —A parallel-motion gripper being used to insert components into a printed circuit board prior to soldering. (Schunk-USA)

Important Terms

automatic tool changers
collet grippers
compliance
cylindrical grip
electromechanical grippers
end-of-arm tooling
expandable grippers
grippers
hook movement
lateral grip
magnetic grippers
mechanical finger gripper
nonprehensile movements
oppositional grip
overload sensing
palmar grip
prehensile movements
remote-center compliance (RCC) devices
spherical grip
spread movement
support grippers
tools
vacuum gripper

Review Questions

Write your answers on a separate sheet of paper.

1. What are the two major classifications for end effectors?
2. In what important ways do end effectors differ from the human hand?
3. Distinguish between prehensile and nonprehensile movements.
4. List several design considerations that apply to end effectors.
5. What kinds of tasks might require the use of a three-finger gripper?
6. What kinds of tasks lend themselves to the use of a vacuum gripper?
7. What is a major disadvantage of an electromagnetic gripper?
8. What type of end effectors would you most likely use to pick up an automobile windshield? Why?
9. List six end effectors commonly found in industry.
10. What are the advantages of using automatic gripper/tool changers?
11. Why is it usually necessary for most grippers to have some amount of compliance?
12. What can be done to increase the compliance of a gripper?
13. List the advantages that robotic end effectors have over the human hand.

This end effector is used for plasma cutting. Note the use of an overload sensing device. (Motoman)

CONTROL SYSTEMS

The heart of a robotic control system is a microprocessor (computer chip) linked to input/output and monitoring devices. The control system has a series of instructions, called a program, stored in its memory. The program supplies the commands that control motors, hydraulic systems, or pneumatic systems to activate the robot's motion control mechanism. This mechanism is typically an actuator, a device that converts power into robot movement.

GE Fanuc Robotics

11 Digital Electronics

Overview

The electronic circuitry used with robots supplies commands that control motors, hydraulic systems, and pneumatic systems. These circuits also store information, count, encode, and decode. The circuits that perform these functions are called ***logic circuits.*** The technology that controls robotic and other automated systems is called ***digital electronics.***

In this chapter, the basics of digital electronics will be discussed. The chapter also covers fundamentals of computer design and function as they are applied to robots.

Electronic Information Processing

The term *computer* is used to cover a number of functions, but primarily refers to a system that will perform automatic computations. Computers range from pocket calculators to complex central units that serve an entire organization. Information is processed in two states—digital or analog.

Analog information varies continuously. An example is temperature, which is in a constant state of change. The mercury indicator in an analog thermometer may, at times, be between one degree marking and the next. ***Digital information,*** by contrast, occurs in separate full units. With a digital clock, for example, the time is always displayed as one unit (second) or the next, never between units.

Digital computers use two numbers that represent either the presence or absence of voltage. A voltage pulse is usually represented by a one (1). Absence of voltage is indicated by a zero (0). A single pulse is described as a ***bit*** of information. The term bit is a contraction of the phrase *binary digit* (the letters *bi* from the word "binary" and *t* from "digit." A group of 8 bits produces a ***byte,*** sometimes called a "binary word."

Digital Number Systems

The number system we use in everyday life is the "base 10" (decimal) system. In this system, ten digits—0, 1, 2, 3, 4, 5, 6, 7, 8, and 9—are used for counting. The number of digits used in the system is called its *base.* The decimal system, therefore, has a base of 10.

Number systems give digits a ***place value.*** This refers to the position of a digit with respect to a reference called the ***decimal point.*** The largest digit that can be used in a specific place is determined by the base of the system. In the decimal system, the first position to the *left* of the decimal point is called the *units place.* Any digit from 0 to 9 can be used in this place. When number values greater than 9 are used, they must be expressed in two or more places. The position to the left of the units place is the 10s place. The number 99 is the largest value that can be expressed by two places in the decimal system. Each place added to the left extends the number value by a power of 10.

A number value in any system can be read by adding the numbers in each place. The decimal value 2583, for example, can be expressed as: (2 x 1000) + (5 x 100) + (8 x 10) + (3 x 1). The values have increased by a power of ten for each place to the left of the decimal point. In the decimal system, this is 10^3, 10^2, 10^1, and 10^0. Mathematically, each place value is the number times a power of the system base, Figure 11-1.

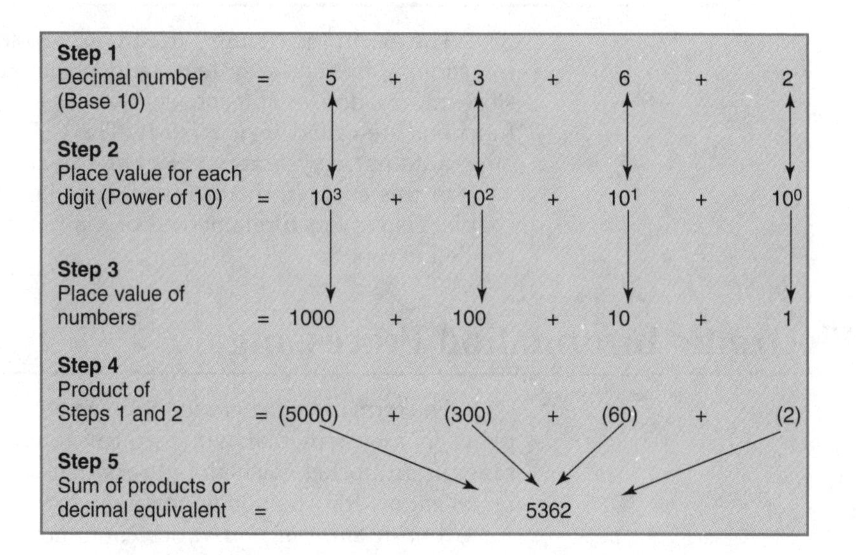

Figure 11-1
Expressing a base 10 (decimal) number.

The decimal number system is difficult to use with electronics. Instead, the binary system is typically used.

Binary number system

The ***binary number system*** uses 2 as its base. The largest value that can be expressed by a specific place is the number 1. As a result, only the numbers 0 or 1 are used. The first place to the left of the reference (in this case, called the ***binary point***) represents units, or 1s. Places to the left of the binary point are expressed in powers of 2. These include: $2^0 = 1$, $2^1 = 2$, $2^2 = 4$, $2^3 = 8$, $2^4 = 16$, $2^5 = 32$, $2^6 = 64$, and so on.

When different number systems are used, a subscript number is used to identify the base. The number 100_2 indicates that the binary ("base 2") system is being used. In the binary system, "100" is read as one-zero-zero instead of one hundred. Starting at the first digit to the left of the binary point, this number would have place values of $(0 \times 2^0) + (0 \times 2^1) + (1 \times 2^2)$ or 0 + 0 + 4. Thus, the binary number 100 is equal to 4 in the decimal number system.

The conversion of a binary number to an equivalent decimal number is shown in Figure 11-2. When converting, write down the binary number first. Starting at the binary point, indicate the decimal equivalent for each place location where a 1 is indicated. For each 0 in the binary number leave a blank space or indicate a 0. Add the place values and then record the decimal equivalent. Many electronic calculators have built-in conversions for numbers of different bases.

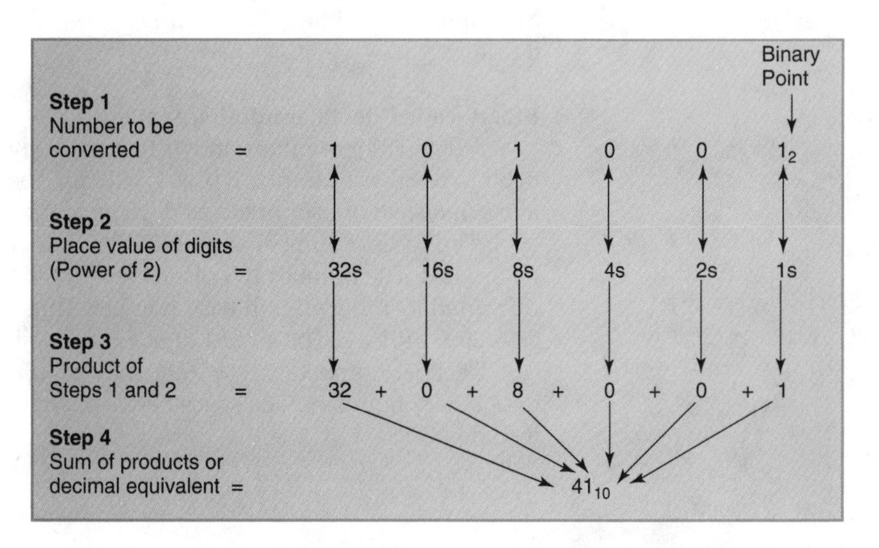

Figure 11-2
Making a binary-to-base 10 conversion.

The conversion of a decimal number to its binary equivalent is shown in Figure 11-3. As shown, the conversion is done by repeatedly dividing by 2. When the quotient is even with no remainder, the value is 0. When the quotient has a remainder, the value is 1. Division continues until the quotient is zero.

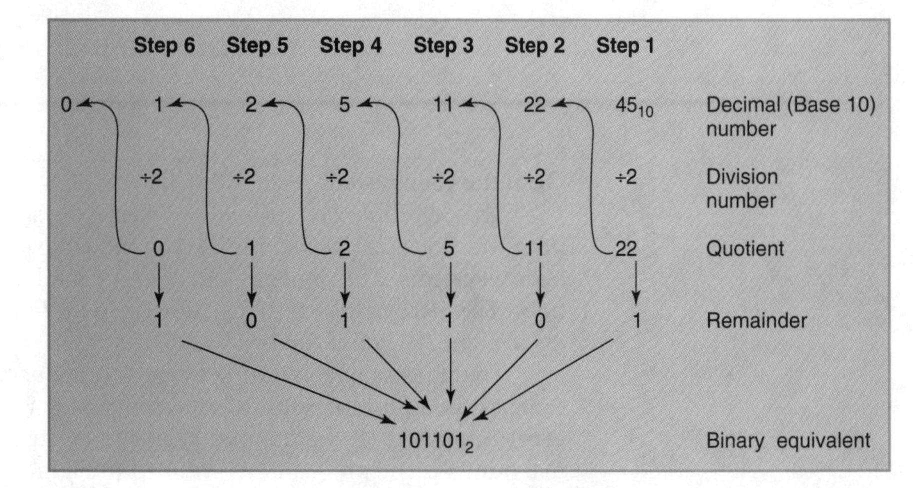

Figure 11-3
Decimal-to-binary conversion is achieved by dividing by 2.

Electronically, the value of 0 is expressed as low-voltage or no voltage. The number 1 is used to indicate a voltage larger than 0. Binary systems that use these values are said to have ***positive logic***. (Negative logic, by comparison, assigns a voltage to the 0 and no voltage to the number 1.) Positive logic will be used for all discussions and examples in this chapter.

The two operational states of a binary system, 1 and 0, are natural circuit conditions. When a circuit is turned off or has no voltage applied, it is in the "0" state. A circuit that has voltage applied is in the "1" state. This makes it possible to change states in less than a microsecond. Electronic devices thus can manipulate millions of 0s and 1s per second, and process information very quickly.

Binary-coded-decimal number system

When large numbers are written using the binary method, they are difficult to use. For example, $1101001_2 = 105_{10}$. The ***binary-coded-decimal (BCD) number system*** of counting was devised to overcome this problem.

The base 10 number is first divided into digits according to place value, Figure 11-4. The number 105_{10} is made up of the digits 1-0-5. Each digit then is converted to a four-digit binary number. Thus, 105_{10} is expressed in the form 0001 0000 0101_{BCD}. The largest digit expressed by any group of BCD numbers is 9. Decimal numbers up to 999_{10} may be converted using only three four-digit binary numbers. The space between each group of digits is important to include.

Figure 11-4
When making a BCD conversion, four binary digits represent each decimal digit.

Given decimal number		105_{10}	
Step 1 Group the digits	(1)	(0)	(5)
Step 2 Convert each digit to binary group	(0001)	(0000)	(0101)
Step 3 Combine group values	000100000101 BCD		

Octal number system

The ***octal number system***, or "base 8" system, is used to process large numbers. The octal system is based on the same principles as the decimal and binary systems. The digits 0, 1, 2, 3, 4, 5, 6, and 7 are used in the place positions. The place values of digits, moving to the left of the octal point, are powers of eight: 8^0 = units, or 1s; 8^1 = 8s; 8^2 = 64s; 8^3 = 512s; 8^4 = 4096s, and so on.

The process of converting an octal number to a decimal number is the same as used for binary-to-decimal conversion (using the powers of 8, however). Suppose that the number 372_8 is to be changed to an equivalent decimal number. The procedure is outlined in Figure 11-5.

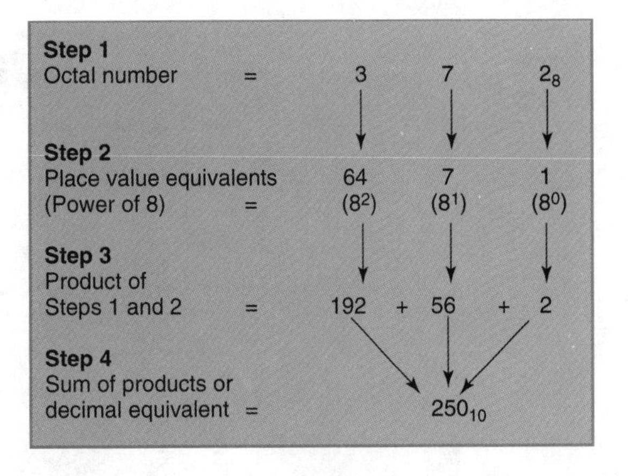

Step 1
Octal number = 3 7 2_8
Step 2
Place value equivalents
(Power of 8) = 64 (8^2) 7 (8^1) 1 (8^0)
Step 3
Product of
Steps 1 and 2 = 192 + 56 + 2
Step 4
Sum of products or
decimal equivalent = 250_{10}

Figure 11-5
Octal-to-decimal conversion requires the same basic process as binary-to-decimal conversion.

Converting an octal number to an equivalent binary number is similar to BCD conversion. The octal number is first divided into digits according to place value. Each digit is then converted into an equivalent binary number using only three digits, Figure 11-6.

Given octal number	345_8		
Step 1 Group the digits	(3)	(4)	(5)
Step 2 Convert digits to binary groups	(011)	(100)	(101)
Step 3 Combine group values to binary equivalent	11100101_2		

Figure 11-6
In octal-to-binary conversion, a maximum of 3 binary numbers is used per digit.

Converting a decimal number to an octal number means dividing by the number 8. After the quotient has been determined, the remainder is brought down as the place value. When the quotient is even with no remainder, the value is 0. The procedure for converting 4098_{10} is outlined in Figure 11-7.

If a binary number (110100100_2, for example) is to be changed to an equivalent octal number, the digits must first be divided into groups of three binary numbers starting at the binary point, Figure 11-8. Each binary group is then converted into an equivalent octal number. These numbers are then combined, while remaining in their same respective places, to represent the equivalent octal number. Converting a binary number to an octal number is often done for digital circuits. Binary numbers are first processed at a very high speed. An output circuit then accepts this signal and converts it to an octal number displayed on a readout device.

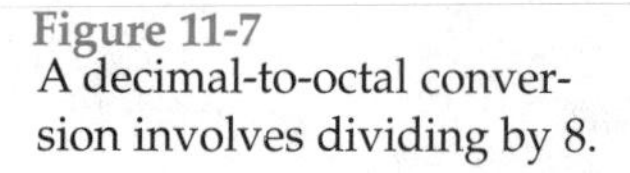

Figure 11-7
A decimal-to-octal conversion involves dividing by 8.

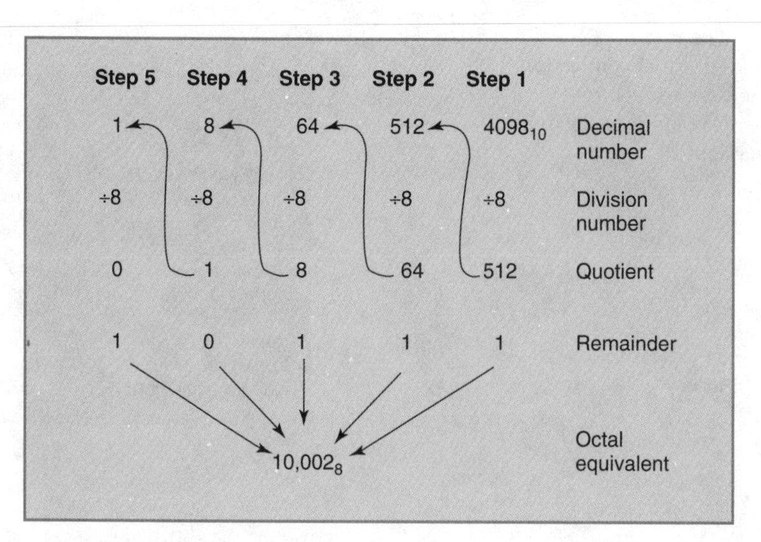

Figure 11-8
Binary-to-octal conversion is often done for digital circuits.

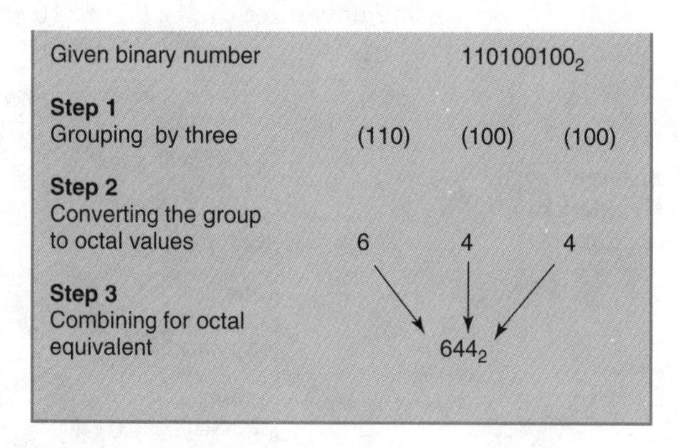

Hexadecimal number system

The ***hexadecimal number system*** is also used to process large numbers. The base of this system is 16. The largest number used in a place is 15. Digits used are the numbers 0-9 and the letters A-F. The letters A-F stand for the digits 10-15, respectively. The place value of digits to the left of the hexadecimal point are powers of 16: $16^0 = 1$, $16^1 = 16$, $16^2 = 256$, $16^3 = 4096$, $16^4 = 65{,}536$, and so on.

Changing a hexadecimal number to a decimal number uses a process similar to those used for other conversions. The hexadecimal number is written in proper digital order, Figure 11-9. The place values are then positioned under each respective digit in Step 2. In Step 3, the letters are converted to numeric values. Each of the values in Step 3 is then multiplied by its corresponding place value from Step 2. In Step 4, the products are added together. The resulting sum (Step 5) is the decimal equivalent of the hexadecimal number.

Figure 11-9
Hexadecimal-to-decimal conversion involves an extra step, converting the letters to numeric values.

		=				
Step 1:	Hexadecimal number	=	1	2	C	D_{16}
Step 2:	Place values (Power of 16)	=	4096s	256s	16s	1s
Step 3:	Conversion of letters to numbers	=	1	2	12	13
Step 4:	Product of Steps 2 and 3	=	4096 +	512 +	192 +	13
Step 5:	Sum of products or decimal equivalent	=	4813_{10}			

The process of changing a hexadecimal number to a binary equivalent is a simple grouping operation, Figure 11-10. Initially, the hexadecimal number is separated into digits. Each digit is then converted to its equivalent in binary form ("D" in hexadecimal notation equals 13 in decimal form, or 1101 in binary). Step 3 shows the groups combined to form the equivalent binary number. Note that the leading zeros from the binary number 0010 are dropped.

Figure 11-10
A hexadecimal-to-binary conversion requires grouping the binary numbers in Step 2.

Given hexadecimal number	$23CD_{16}$			
Step 1 Grouping of digits	(2)	(3)	(C)	(D)
Step 2 Converting each digit to a binary group	(0010)	(0011)	(1100)	(1101)
Step 3 Combining group values	10001111001101_2			

Decimal number-to-hexadecimal number conversion is done by dividing by 16. Remainders can be as large as 15, Figure 11-11.

Figure 11-11
A decimal-to-hexadecimal conversion uses a divisor of 16.

Step 4	Step 3	Step 2	Step 1	
1	18	298	4780_{10}	Decimal Number
÷16	÷16	÷16	÷16	Division number
0	1	18	298	Quotient
1	2	10	12	Remainder
1	2	A	C_{16}	Hexadecimal equivalent

Converting a binary number to a hexadecimal equivalent is the reverse of the hexadecimal-to-binary process, Figure 11-12. Initially, the binary number is divided in groups of four digits, starting at the binary point. Each group is then converted to the equivalent hexadecimal value and the groups are combined to form the answer.

Figure 11-12
Binary-to-hexadecimal conversion is done by first creating groups of four binary numbers, then converting them to hexadecimal values.

Given binary number		1001101101010_2		
Step 1 Grouping of fours	(0001)	(0011)	(0110)	(1010)
Step 2 Converting hexadecimal values	1 ↓ 1	3 ↓ 3	6 ↓ 6	10 ↓ A
Step 3 Combining for hexadecimal equivalent		$136A_{16}$		

Binary Logic Circuits

Binary signals are far superior to octal, decimal, or hexadecimal signals for use in logic circuits. They can be processed easily because they can be represented by two stable states of operation: on or off, 1 or 0, up or down, voltage or no voltage, right or left, and so on. There is no "in-between" state.

The symbols used to represent an operational state are very important. In positive binary logic, *voltage, on,* or *true* results from the 1 operational state. *No voltage, off,* or *false* results from the 0 condition. A circuit can be set to either state and will remain at that setting until a change occurs.

Any electronic device that can be set in one of these two operational states by an outside signal is said to be ***bistable.*** These devices include relays, lamps, switches, transistors, and diodes. A bistable device can store one binary digit or bit of information. By using many of these devices, it is possible to build a ***binary logic circuit*** that will make logical decisions based upon input signals. The basic binary logic circuits are the AND circuit, the OR circuit, and the NOT circuit. The decision made by each type of circuit is unique. These circuits are also called ***logic gates.*** The term "gate" refers to the circuit's ability to pass or block certain signals. An IF...THEN sentence is often used to describe the basic operation of a logic gate. For example, "IF the inputs applied to a gate are all 1, THEN the output will be 1."

The logic gates discussed in the following examples illustrate basic operation using simple switch and lamp input-outputs. In actual applications, logic gates are typically integrated circuits (ICs).

AND Gates

An ***AND gate,*** Figure 11-13, has two or more inputs and one output. In Figure 11-13A, when a switch is turned on, it represents a 1. Off represents a 0. The lamp displays a 1 when it is on and 0 when it is off. If both inputs are in the 1 state (closed) simultaneously, then a 1 will be output (the lamp will light. A ***truth table*** shows combinations of inputs and the resulting outputs of a logic gate. The symbol for an AND gate is shown in Figure 11-13B.

The operation of the gate is simplified by describing the input-output relationships in the truth table. The table in Figure 11-13C shows the predictable alternatives. The AND gate produces a 1 output only when switches A and B are both 1. Mathematically, this action is described as A + B = C.

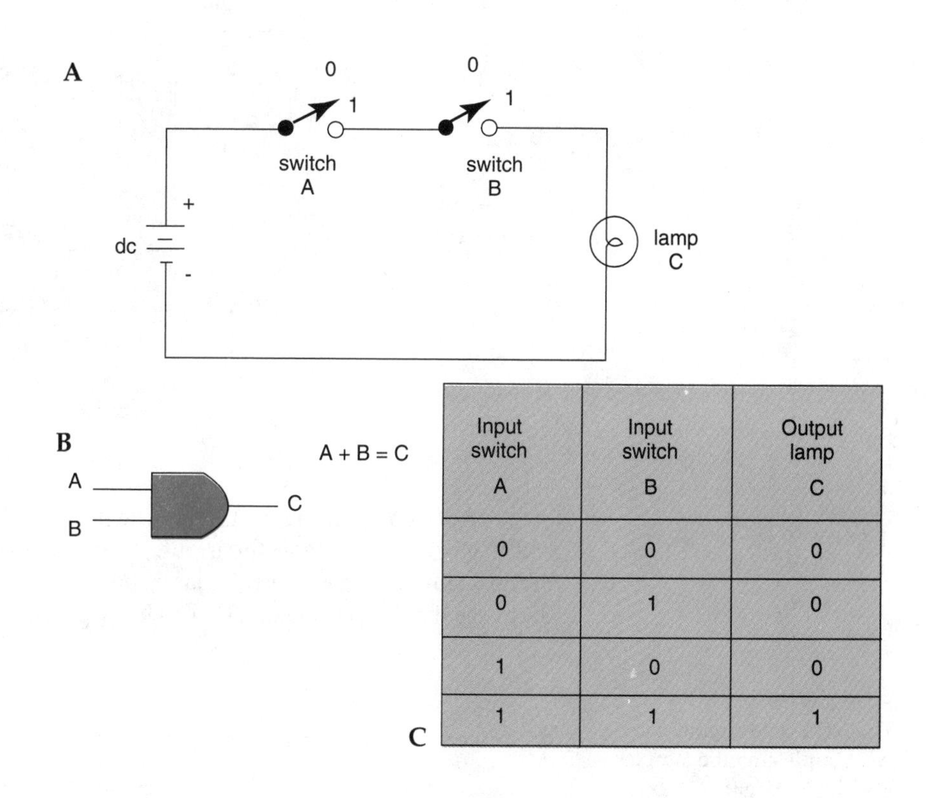

Input switch A	Input switch B	Output lamp C
0	0	0
0	1	0
1	0	0
1	1	1

Figure 11-13
The AND gate circuit. A—Simple circuit example. B—The AND gate symbol. C—The AND gate truth table.

Each input to an AND gate can create two operational states: 1 and 0. A two-input AND gate has 2^2, or 4, possible combinations that influence the output. A three-input gate has 2^3, or 8, combinations, while a four input gate has 2^4, or 16, combinations. These combinations are normally placed in the truth table in binary order. For a two-input gate this is 00, 01, 10, and 11. This represents the binary count of 0, 1, 2, and 3 in order.

OR Gate

An ***OR gate,*** Figure 11-14, has two or more inputs and one output. Like the AND gate, each input to the OR gate produces two possible states: 1 or 0. A 1 is output when both switches are 1 *or* when either switch A or B is 1. Mathematically, this action is described as A + B = C. This gate is used to make decisions as to whether or not a 1 appears at either input. The truth table, Figure 11-14C, shows that if any input is a 1, the output will be a 1.

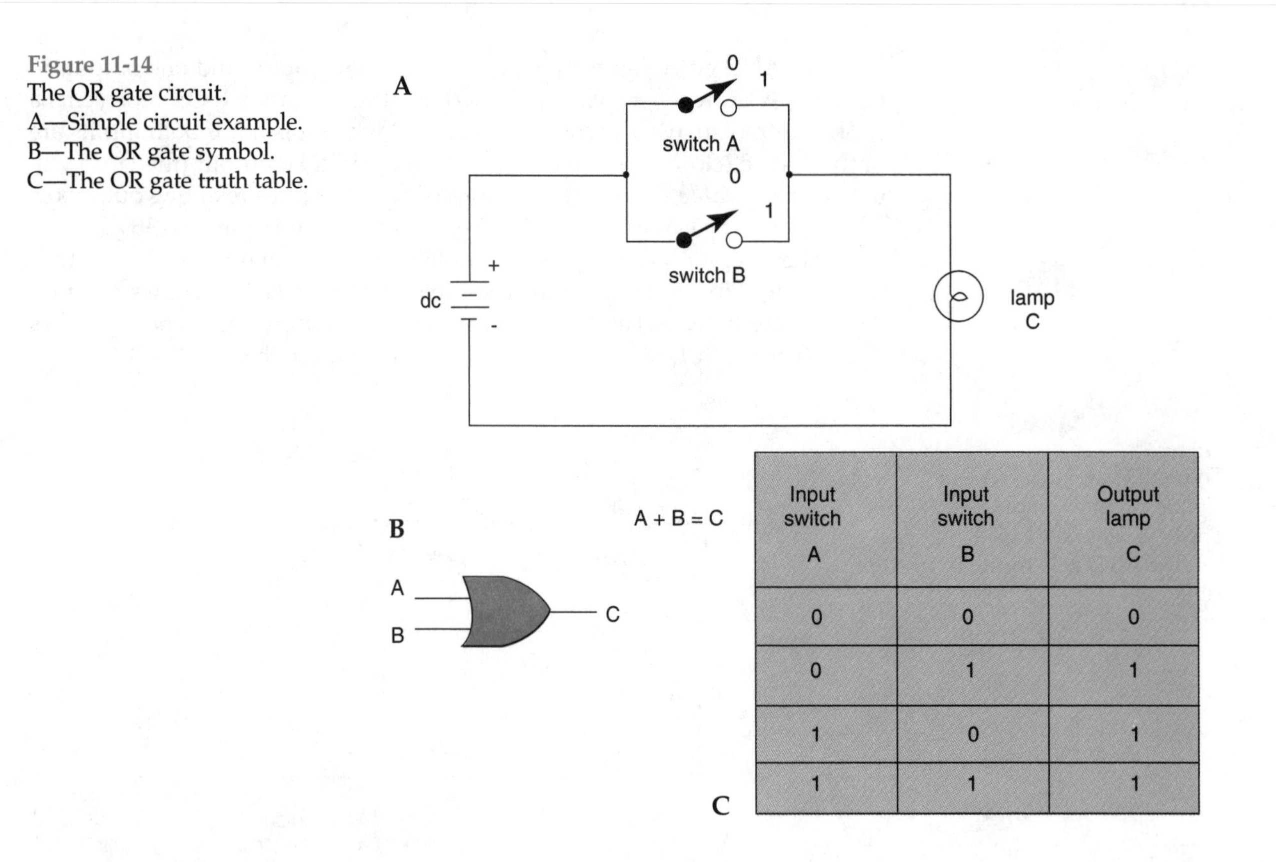

Input switch A	Input switch B	Output lamp C
0	0	0
0	1	1
1	0	1
1	1	1

Figure 11-14
The OR gate circuit.
A—Simple circuit example.
B—The OR gate symbol.
C—The OR gate truth table.

NOT Gate

A ***NOT gate,*** Figure 11-15, has one input and one output. The output of a NOT gate is opposite to the input. When the switch in Figure 11-15A is on (in the 1 state), it shorts out the lamp. Placing the switch in the off condition (0) causes the lamp to be on (1). The NOT gate is also called an ***inverter.***

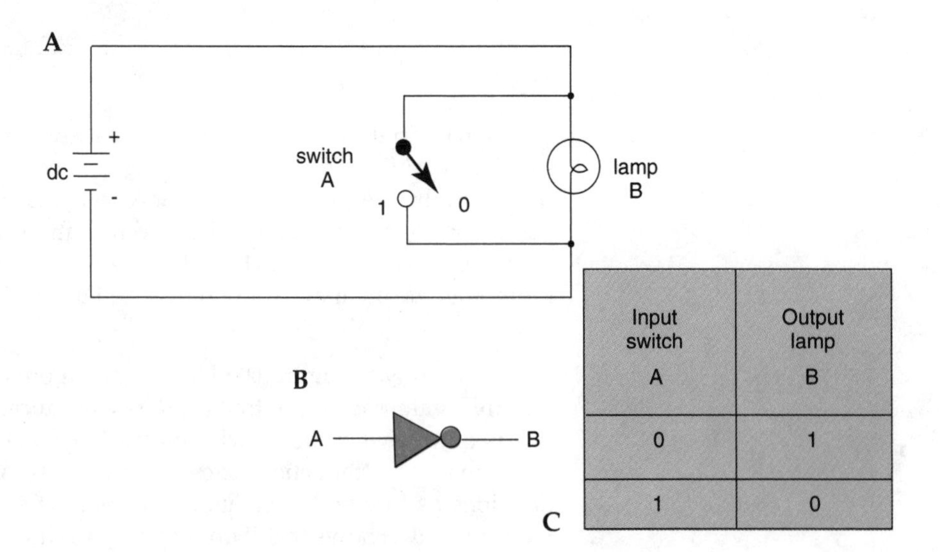

Input switch A	Output lamp B
0	1
1	0

Figure 11-15
The NOT gate circuit.
A—Simple circuit example.
B—The NOT gate symbol.
C—The NOT gate truth table.

Combination Logic Gates

When a NOT gate is combined with an AND gate, it is called a ***NAND gate.*** This is an inverted AND gate, Figure 11-16. When switches A and B are both on (in the 1 state), lamp C is off (0). When either or both switches are off, lamp C is on (in the 1 state). Mathematically, the operation of a NAND gate is $A + B = \overline{C}$. The bar over the C denotes the inversion, or negative function, of the gate.

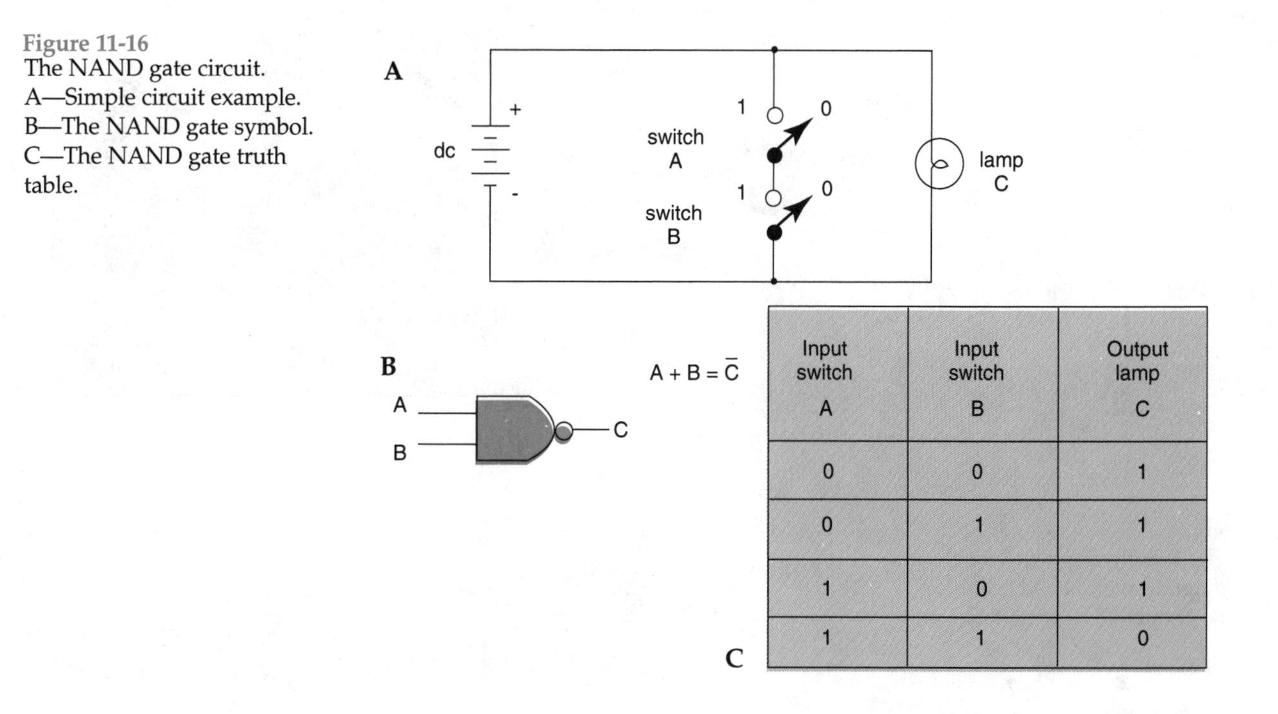

Input switch A	Input switch B	Output lamp C
0	0	1
0	1	1
1	0	1
1	1	0

Figure 11-16
The NAND gate circuit.
A—Simple circuit example.
B—The NAND gate symbol.
C—The NAND gate truth table.

The combination of a NOT gate and an OR gate results in a ***NOR gate,*** Figure 11-17. A NOR gate is the opposite of an OR gate. Mathematically, the operation of a NOR gate is stated as $A + B = \overline{C}$. A 1 will appear at the output only when A is 0 and B is 0.

Flip-Flops

Flip-flops are memory devices used in digital circuits. They are often the basic logic element for counting, temporary memory, and sequential switching operations. They can be used to hold an output state even when the input is completely removed. They can also change their output based on an appropriate input signal.

The reset-set (R-S) flip-flop is shown in Figure 11-18. Its logic diagram, symbol, and truth table are more complicated than those for a simple logic gate. They show the different states before an input occurs and how they change afterwards. Two of the operating conditions produce an unpredictable output.

Flip-flops often must be set and cleared at specific times with respect to other circuits. They operate in step with a clock pulse. The appropriate R-S inputs and clock pulse all must be present to cause a state change. This device

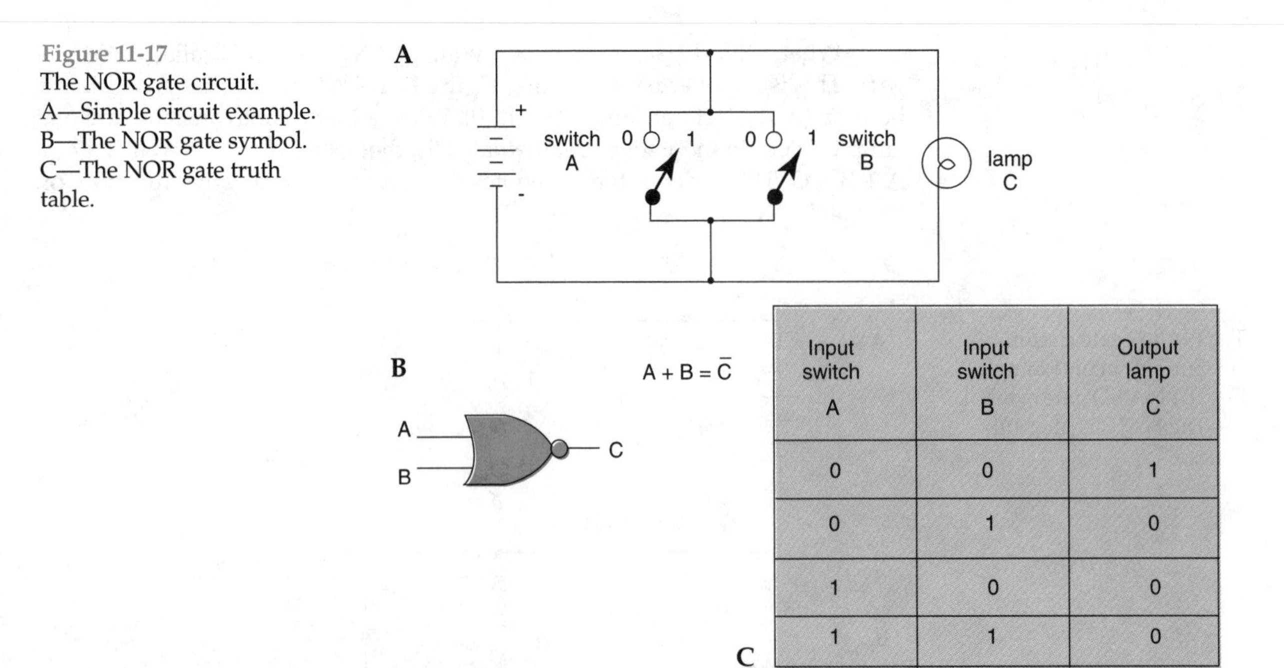

Input switch A	Input switch B	Output lamp C
0	0	1
0	1	0
1	0	0
1	1	0

Figure 11-17
The NOR gate circuit.
A—Simple circuit example.
B—The NOR gate symbol.
C—The NOR gate truth table.

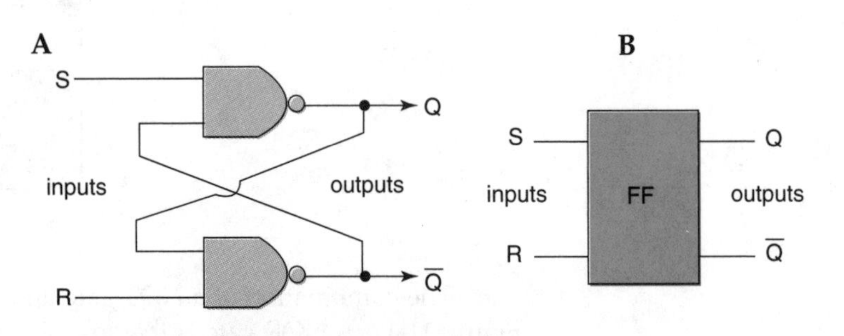

Figure 11-18
The R-S flip-flop. A—Logic diagram with NAND gates. B—Symbol. C—Truth table.

Applied inputs		Previous outputs		Resulting outputs		
S	R	Q	$\overline{Q}$	Q	$\overline{Q}$	
0	0	1	0	1/0	0/1	unpredictable
0	1	1	0	1	0	
1	0	1	0	0	1	
1	1	1	0	1	0	
0	0	0	1	0/1	1/0	unpredictable
0	1	0	1	1	0	
1	0	0	1	0	1	
1	1	0	1	0	1	

C

is called an R-S triggered flip-flop, or simply an R-S-T flip-flop, Figure 11-19. The truth table for an R-S-T flip-flop is basically the same as that for an R-S flip-flop. A state change will occur only when the clock pulse arrives at the T input. A two-input AND gate is added to the set and reset inputs for this purpose.

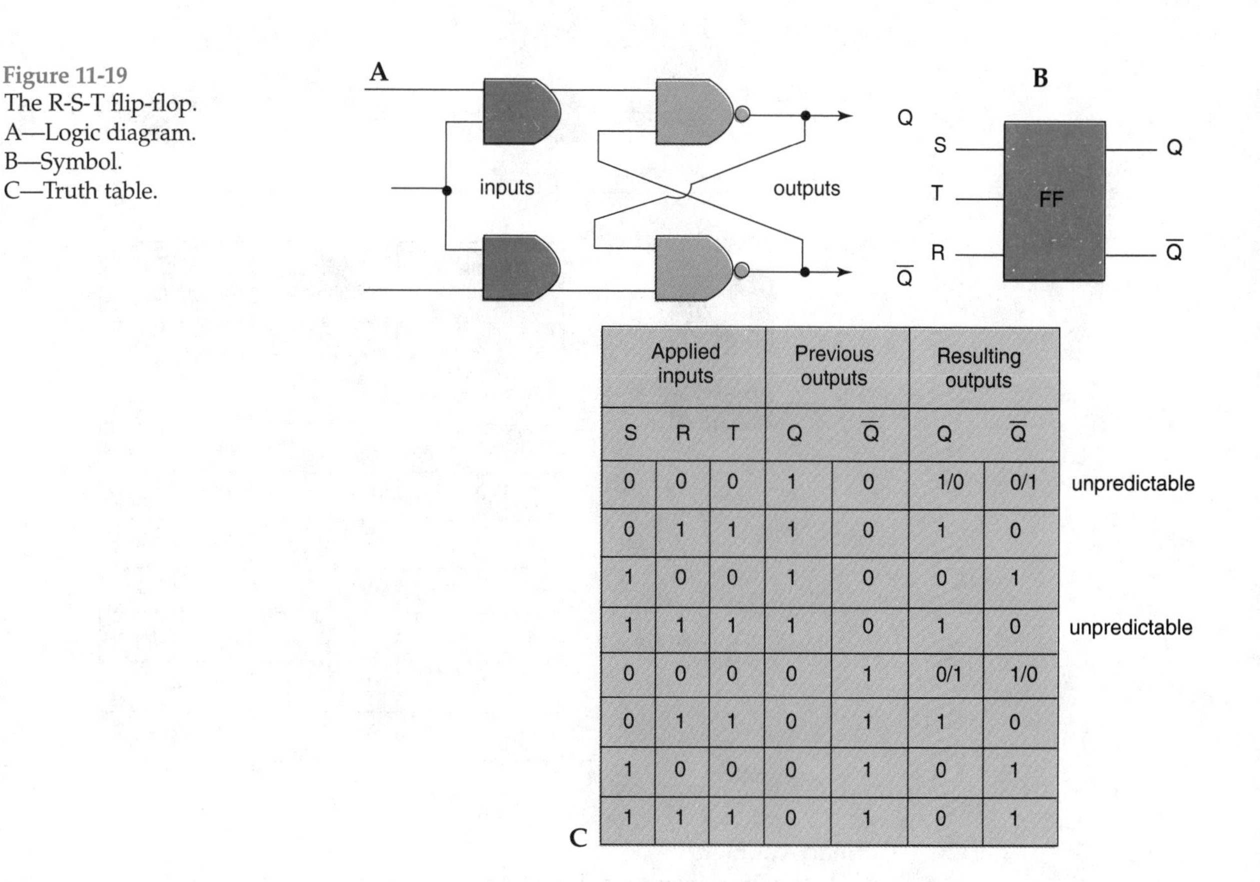

Applied inputs			Previous outputs		Resulting outputs		
S	R	T	Q	$\overline{Q}$	Q	$\overline{Q}$	
0	0	0	1	0	1/0	0/1	unpredictable
0	1	1	1	0	1	0	
1	0	0	1	0	0	1	
1	1	1	1	0	1	0	unpredictable
0	0	0	0	1	0/1	1/0	
0	1	1	0	1	1	0	
1	0	0	0	1	0	1	
1	1	1	0	1	0	1	

Figure 11-19
The R-S-T flip-flop.
A—Logic diagram.
B—Symbol.
C—Truth table.

The JK flip-flop, Figure 11-20, is unique: it has no unpredictable output states. It can be set by applying a 1 to the J input and cleared by feeding a 1 to the K input. A 1 signal applied to both J and K inputs simultaneously causes the output to change states, or "toggle." A 0 applied simultaneously to both inputs does not initiate a state change. The inputs of a JK flip-flop are controlled directly by clock pulses. Several modifications of the basic JK flip-flop are available. Some have preset and preclear inputs that are used to establish sequential operations at precise times.

Digital Counters

One of the most versatile and important logic devices is the ***counter***. Counters are used to count a wide variety of objects in various applications. However, they really count only one thing—electronic pulses. These pulses may be produced mechanically, acoustically, with a clock mechanism, or by a number of other processes. The two common forms of this device are the binary counters and the decade (BCD) counter.

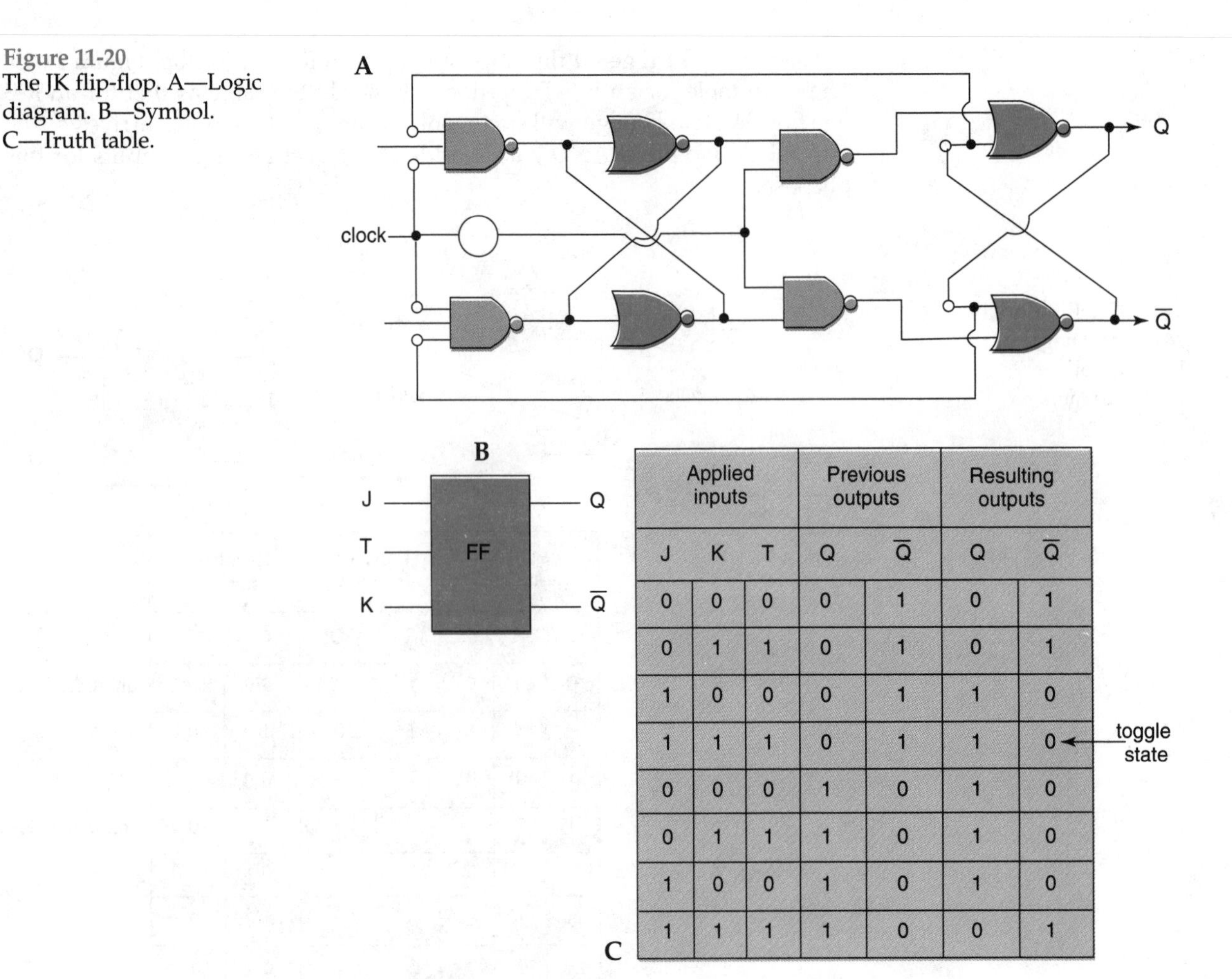

Applied inputs			Previous outputs		Resulting outputs	
J	K	T	Q	$\overline{Q}$	Q	$\overline{Q}$
0	0	0	0	1	0	1
0	1	1	0	1	0	1
1	0	0	0	1	1	0
1	1	1	0	1	1	0
0	0	0	1	0	1	0
0	1	1	1	0	1	0
1	0	0	1	0	1	0
1	1	1	1	0	0	1

Figure 11-20
The JK flip-flop. A—Logic diagram. B—Symbol. C—Truth table.

Binary counters

In ***binary counters,*** flip-flops are connected so that the Q output of the first circuit drives the trigger, or clock input, of the next circuit. See Figure 11-21. Each flip-flop, therefore, has a divide-by-two function.

The counter in Figure 11-21A is called a *binary ripple counter.* The J and K inputs for each flip-flop are held at a logic 1 level. Each clock pulse applied to the input of FF_1 will then causes a change in state. Since the flip-flops trigger only on the negative-going part of the clock pulse, the output of FF_1 will alternate between 1 and 0 with each pulse. Thus, a 1 output will appear at Q of FF_1 for every two input pulses. This means that each flip-flop has a divide-by-two function. Five flip-flops connected in this manner produce a 2^5, or 32, count. The largest count in this case is 11111_2 (31_{10}). The next applied pulse clears the counter so that 0 appears at all the Q outputs.

By grouping three flip-flops together (Figure 11-21B), a binary-coded octal (BCO) counter can be created. The number 111_2 represents the seven count, or seven units of an octal counter. Two groups of three flip-flops connected in this manner produce a maximum count of 111111_2, which represents 77_8, or 63_{10}.

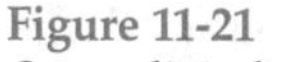

Figure 11-21
Some digital counters that use a JK flip-flop. A—Binary ripple counter (32 count). B—Octal counter (8 count). C—Hexadecimal counter (16 count).

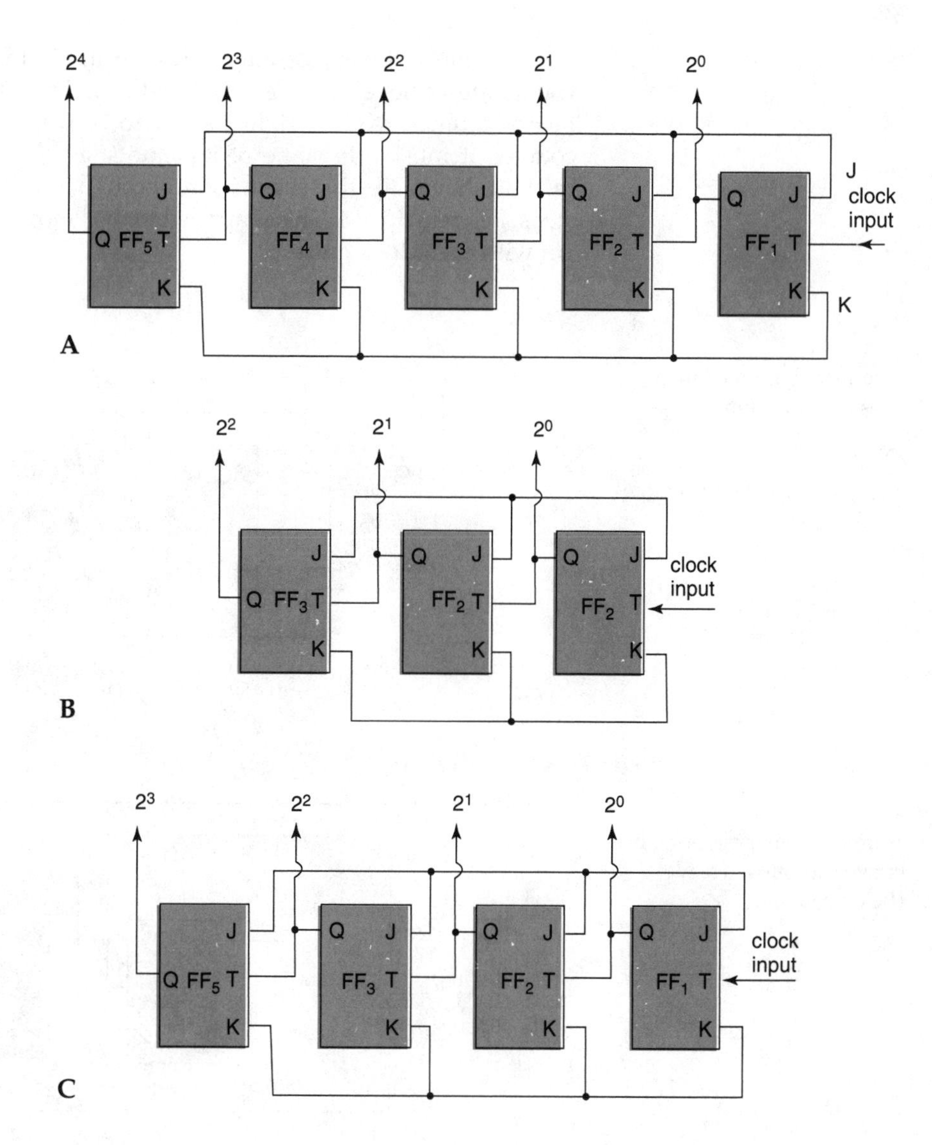

It is possible to develop the units part of a binary-coded hexadecimal (BCH) counter by placing four flip-flops together in a group. See Figure 11-21C. Thus, 1111_2 is used to represent F_{16} or 15_{10}. Two groups of four flip-flops could produce a maximum count of 11111111_2, which would represent FF_{16} or 255_{10}. Each succeeding group of four flip-flops raise the count to the next power of 16.

Decade counters

Since most of the mathematics that we use today is based upon the decimal (base 10) system, it is important to be able to count by this method. The output of a binary counter must be changed into a decimal form before it can be used in this way. The first step is to change binary signals into a binary-coded decimal (BCD) form. This is done by using a ***decade counter***.

A four-bit binary counter is shown in Figure 11-22. In this counter, 16 counts are achieved by the four flip-flops. The counts are listed below the binary counter, reading right to left. To convert this counter into a decade counter, it must skip some of its counts, as shown in Figure 11-23. In the method shown, the first seven counts occur naturally. Therefore, through these steps, FF_D remain at 1. This is applied to the J input of FF_B, permitting it to trigger with each clock pulse.

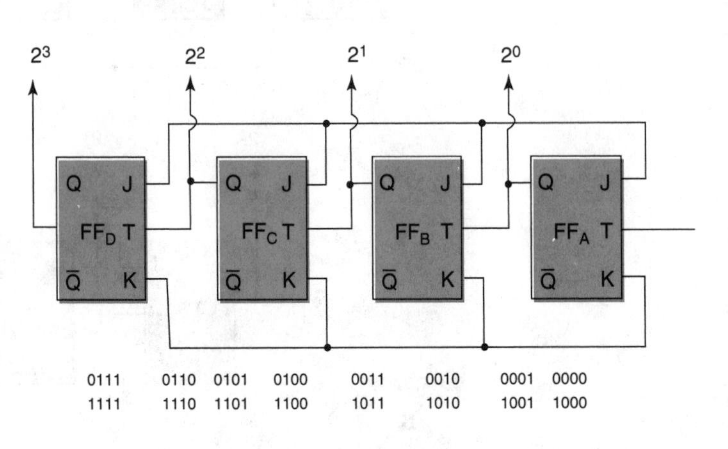

Figure 11-22
A four-bit binary counter uses four flip-flops.

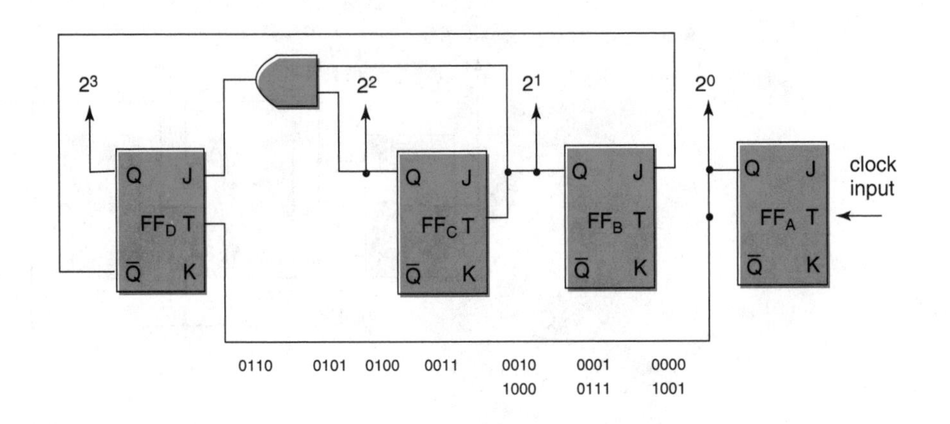

Figure 11-23
A four-bit binary counter is converted into a decade (BCD) counter.

At the seventh count, 1s appearing at the Q outputs of FF_B and FF_C are applied to the AND gate. This action produces a logic 1 and applies it to the J input of FF_D. Arrival of the next clock pulse triggers FF_A, FF_B, and FF_C into the off state and turns on FF_D. This represents the eighth count.

When FF_D is in the on state, Q is 1 and $\bar{Q}$ is 0. This causes a 0 to be fed to the J input of FF_B, which now prevents it from triggering until cleared. Arrival of the next clock pulse causes FF_A to be set to a 1. This registers a 1001_2 which is the ninth count. Arrival of the next count clears FF_A and FF_D instantly. Since FF_B and FF_C were previously cleared by the seventh count, all 0s appear at the outputs. The counter has, therefore, cycled through the ninth count and returned to zero, ready for the next input.

Computer Systems

Computer systems have certain basic parts, Figure 11-24. These parts may be arranged in a variety of different ways. The organization and design of each circuit differ considerably among manufacturers.

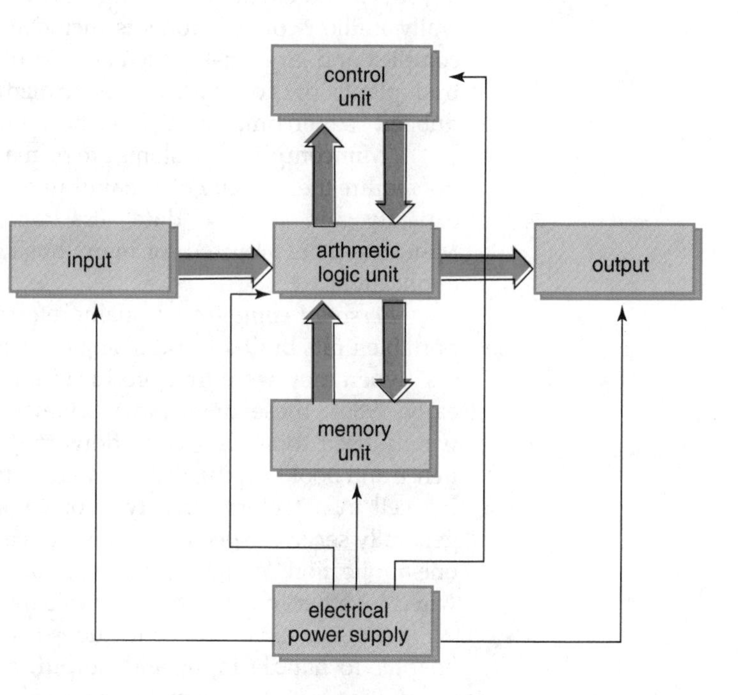

Figure 11-24
The design of a computer system may vary from one type or size to another, but all computers have the same basic parts.

A digital computer consists of input and output devices, arithmetic logic and control circuitry, and some form of memory. Before the computer is placed into operation, a program written in a language that the computer understands must be provided. The program is supplied to the input unit by magnetic disk or tape, compact disk, a keyboard, punched tape, or punched cards.

Input data is translated into some type of numeric code, which the control unit then manipulates according to programmed instructions. After all the internal operations are complete, the coding process is reversed. The machine language is then translated back into another language. This information is used to actuate the output device.

Computer memory stores the operating instructions that direct the ***central processing unit (CPU).*** Coded data in the form of 1s and 0s are "written" into memory, based on directions provided by a program. The CPU then "reads" these instructions and uses them to carry out operations. When the program is logical, processing proceeds in the correct manner with useful results.

Computers have one or more output ports that permit the CPU to communicate with other devices. Monitors (sometimes called video display terminals or VDTs), printers, magnetic disk or tape drives, and modems (interface devices allowing communication with other computers over telephone lines) are typical output devices. Operating speed is an important characteristic of the output device.

The parts of a computer are basically the same for all sizes and types. The differences among units is in physical size, amount of memory, and processing speed and capacity. The largest computers, called "mainframes" have literally millions of components included in their circuits. They are used for complex and large tasks, such as scientific computations or operating industrial plants or government departments. Several very large mainframes, dubbed "supercomputers," have been built.

"Minicomputer" systems are primarily designed for applications that do not require the capacity of a mainframe computer. They have smaller memory capacity and process data less rapidly, but are also less expensive. Minicomputers are used for many business, manufacturing, and educational applications.

Personal computers, usually referred to as "PCs," are small desktop or portable units built around a single integrated circuit. Called "microcomputers" when they were first produced for home and small business use in the early 1980s, these units have achieved considerable popularity. They are widely used today in applications ranging from handling office correspondence and bookkeeping tasks to architectural drafting to controlling a machining cell in a factory. This type of computer is often a *dedicated system*. It generally serves a specific need, solves a specific type of problem, or handles one application. While most PCs are used as "stand-alones," many others are wired together in networks to enable them to share information.

A personal computer includes a microprocessor, memory, an interface adapter to handle input and output, and several distribution paths called ***buses***, Figure 11-25. The microprocessor contains both the arithmetic logic unit and control unit of the basic computer system shown in Figure 11-24.

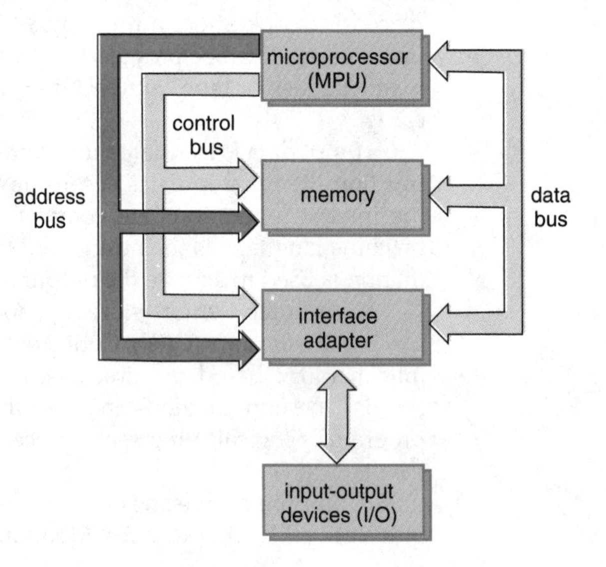

Figure 11-25
This is a simplified block diagram of a personal computer.

Information processed by a computer is of two types: instructions and data. A simple addition problem, such as 9 + 2 = 11, can be used as an analogy to illustrate the distinction. In the problem, the numbers 9, 2, and 11 are *data,* while the plus sign is an *instruction.* The data is distributed by the data bus to all parts of the system. Instructions (the *program*) are distributed by the control bus through a separate path. The address bus of the unit forms an alternate path for the distribution of address data. It is normally used to place information into memory at the correct address. The address is that spot in memory where certain information is located. With the appropriate command, data may be removed from memory and distributed as output.

The Microprocessor

A ***microprocessor unit (MPU)*** combines the arithmetic logic unit and control section of the computer. Today, a microprocessor fits into a single ***integrated circuit (IC)*** chip. A typical integrated chip, only approximately 0.5 cm (0.20 in.) square, contains thousands of transistors, resistors, and diodes. Many companies manufacture microprocessors in different designs and for different prices. Applications for them are increasing at a remarkable rate.

A microprocessor receives data in the form of 1s and 0s. It may then store this data for future processing, perform arithmetic and logic operations, and deliver the results to an output device. In a sense, a microprocessor is a "computer on a chip." The block diagram of a typical microprocessor, Figure 11-26, shows that it contains a number of components. Included are the arithmetic logic unit (ALU), an accumulator, a data register, address registers, a program counter, an instruction decoder, and a sequence controller.

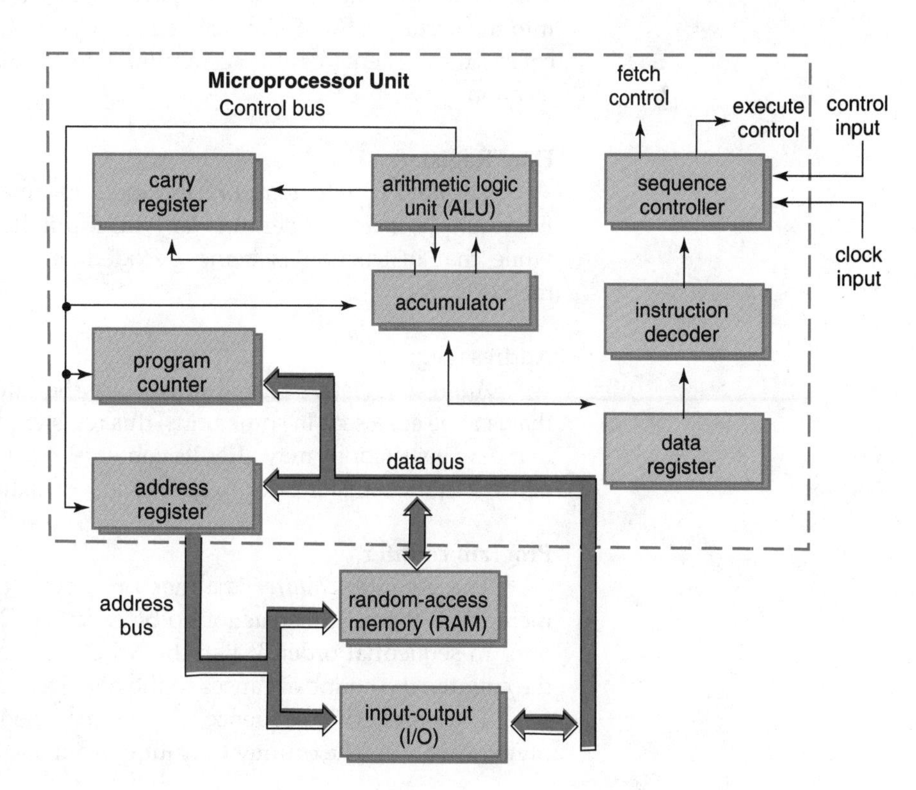

Figure 11-26
This block diagram of a microprocessor has been simplified for clarity.

Arithmetic logic unit

All microprocessors contain an ***arithmetic logic unit (ALU)***. The ALU is a calculator that performs mathematical and logic operations. It works automatically by means of signals sent from the instruction decoder. The ALU combines the input from the data register and the accumulator. Addition, subtraction, and logic comparisons using binary numbers are its primary operations.

The data supplied to the inputs of an ALU is normally in the form of 8-bit binary numbers (bytes). This data is combined by the ALU using binary arithmetic. Since a mathematical operation is performed on the two data inputs, they are often called operands.

Assume that two numbers are to be added—6_{10} and 8_{10}. Initially, the binary number 00000110_2 (6_{10}) is placed in the accumulator. The second operand, 00001000_2 (8_{10}), is then placed into the data register. When a proper control line to the ALU is activated, binary addition is performed. This produces an output of 00001110_2, or 14_{10}, which is the sum of the two operands. This value is then stored in the accumulator, replacing the operand that appeared there originally.

Accumulators

Accumulators temporarily store operands that are to be processed by the ALU. The output of the ALU also is stored in the accumulator after processing.

A typical instruction for a microprocessor is LOAD ACCUMULATOR. This means that the contents of a particular memory location are to be placed into the accumulator. A similar instruction is STORE ACCUMULATOR. This causes the contents of the accumulator to be placed in a selected memory location.

Data Registers

The ***data register*** temporarily stores information applied to the data bus. For example, it stores operands for ALU input. It may also hold an instruction while that instruction is being decoded or hold data prior to storage in memory.

Address registers

Address registers temporarily store the address of a memory location that is to be accessed. In some units, this register is programmable. This means that new instructions may alter its contents. Such a program can also build an address in the register prior to giving an instruction referring to memory.

Program counter

The ***program counter*** is a memory device that indicates the location in memory of the next instruction to be executed. This unit counts the instructions in sequential order. When the MPU receives instructions addressed by the counter, the count advances to the next location.

The numbering sequence may be modified so it does not follow strict numerical order. The counter may jump from one point to another in a routine.

Instruction decoders

Each operation that the MPU performs is identified by binary numbers known as instruction codes. Eight-bit words are commonly used. There are 2^8 (256_{10}) separate operations that can be represented by this code. After an instruction code is pulled from memory and placed in the data register, it must be decoded. The ***instruction decoder*** examines the coded word and decides which operation is to be performed by the ALU. The output of the decoder is then sent to the sequence controller.

Sequence controllers

Using clock inputs, the ***sequence controller*** maintains the proper sequence of events required to perform a task. After instructions are received and decoded, the sequence controller issues a signal that starts the proper action. In most units, the controller can respond to external signals.

Buses

The circuits of most microprocessors are connected together by a common bus network. A bus network is a series of registers connected together. See Figure 11-27.

An advantage of bus network is the ease with which data can be transferred. Each register has inputs labeled clock, enable, load, and clear. When the load and enable input lines are at 0, each register is isolated from the common bus line.

To transfer a word from one register to another, the appropriate inputs must convert to the 1 state. For example, to transfer the data of register A to register D, enable A (EA) and load D (LD) inputs are both placed in the 1 state. This causes the data of register A to appear on the bus line. When a clock pulse arrives at the common inputs, the transfer is completed.

The amount of data a bus can carry is based upon the number of conductor paths. Busses for 8, 16, and 32 bits are commonly used.

Memory

The uses to which a microcomputer can be put depends largely on the amount of memory it has available. ***Memory*** refers to the ability of the MPU to store data so that a single bit or a group of bits can be easily accessed or retrieved. Additional memory can be added by means of auxiliary chips. The two most common types of memory, ROM and RAM, are usually handled by semiconductor circuits on a single IC chip.

Read-only memory (ROM)

Most microcomputers make use of some permanently stored or rarely altered data. Examples include math tables and programs. Storage of this type is provided by ***read-only memory (ROM).*** This information is often placed in the chip when it is manufactured. ROM data is permanent: it cannot be changed, and (unlike RAM, as described in the next section) is not lost when the power source is turned off.

A variation of ROM is the read-mostly memory. This is used where read operations are needed more frequently than write operations. Programmable read-only memory (PROM) is another type. It can be erased by exposure to an

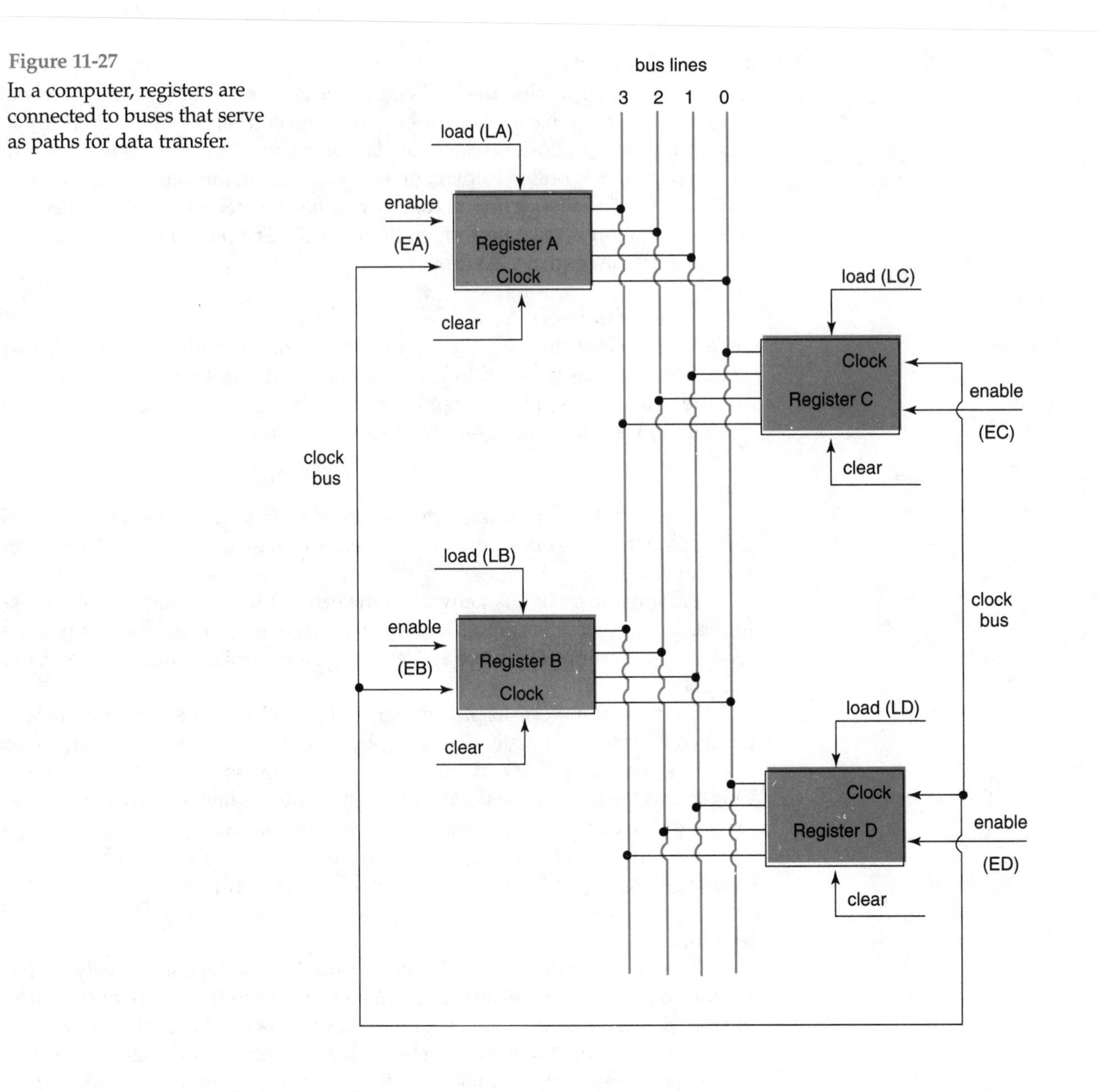

Figure 11-27
In a computer, registers are connected to buses that serve as paths for data transfer.

ultraviolet light source. After light exposure, each memory cell goes to a 0 state. New data can then be written into the chip.

Another alternative is the electrically altered read-only memory (EAROM) or the erasable programmable read-only memory (EPROM). These types of chips permit erasures of individual cells. Memory cell structure is similar to that of the PROM. The EAROM and EPROM are very useful for certain applications.

Random access memory (RAM)

Read/write memory is memory that can be written to, as well as read. In other words, it can be altered. ***Random-access memory (RAM)*** is a type of read/write memory commonly found in microcomputers because it is very fast. Large-scale integration (LSI) read/write chips can store 16,384, or 16K, bits of data in an area less than one-half of a square centimeter. The structure of this chip includes a number of separate circuits. Each can store binary data in an organized pattern. Access to each location is provided by coded information from the address bus. Data can be placed into memory or retrieved at the same rate.

Memory ICs are usually organized in a rectangular pattern of rows and columns. Figure 11-28 shows a memory *array* of eight rows that can store 8-bit words, or 64 single bits. Some read/write memory units available today have much larger capacities. To select a particular memory address, a 3-bit binary number indicates the specific row, and three additional bits indicate the column. In this example, the row address is 3_{10}, or 011_2, and the column address is 5_{10}, or 101_2. The selected memory cell address is 30. Eight-bit word storage is achieved by energizing one row and all eight columns simultaneously. The row and column decoders perform this operation.

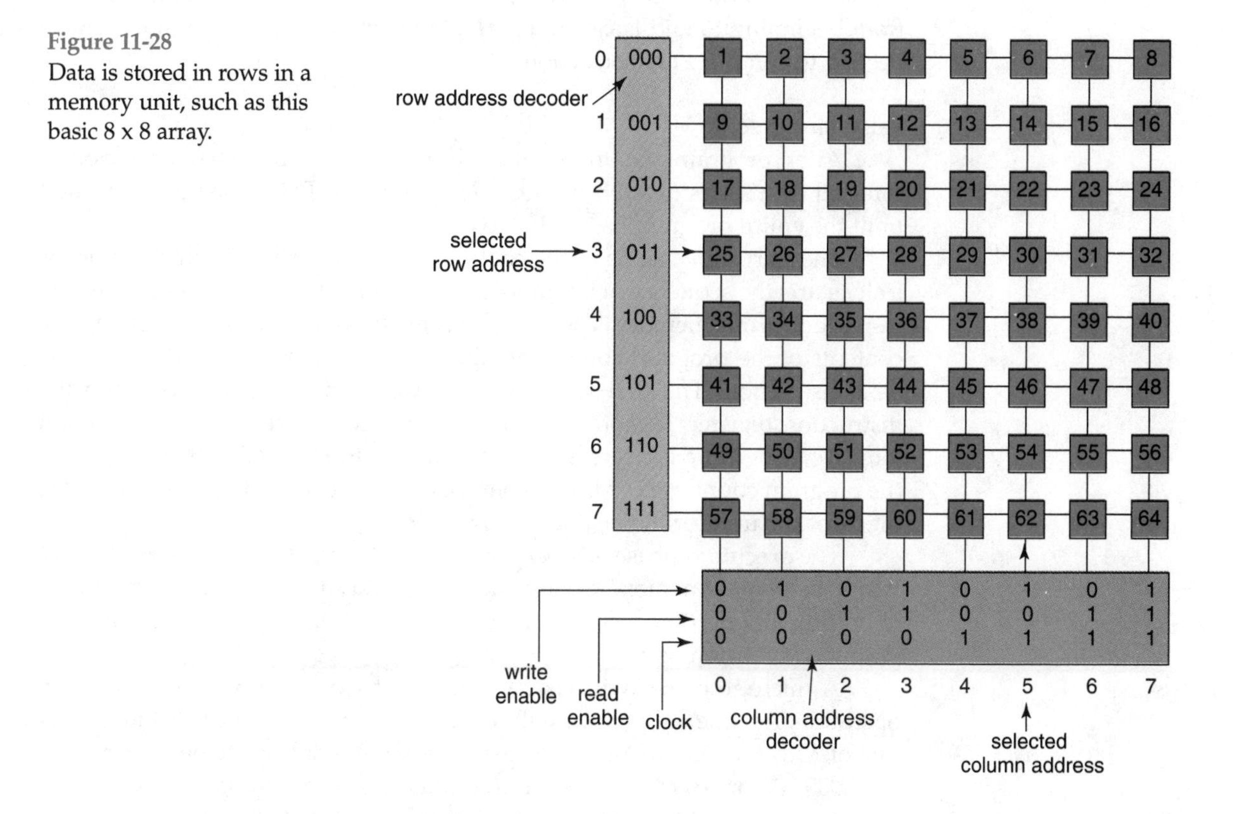

Figure 11-28
Data is stored in rows in a memory unit, such as this basic 8 x 8 array.

To write a word into memory, data sent along the address bus selects a specific address. Reading data from memory does not destroy the data. Data reading can take place as long as the unit is turned on. A loss of power, or turning the unit off, destroys the data, however. Random access memory is *volatile,* and disappears when power is interrupted.

Functions

Certain functions are basic to most computer systems. Among these are timing, fetch, execute, read memory, write memory, input-output transfer, and interrupt.

Timing

The operation of a computer involves a sequence of instructions. The sequence is controlled by a timing signal. The MPU fetches an instruction, executes the required operations, fetches the next instruction, executes it, and so on, in a cycling pattern. This means that all actions occur at or during a precise time interval. Such an orderly sequence requires a free-running electronic oscillator or clock. In some systems the clock may be built into the MPU. In others, timing units feed the system through a separate control bus.

Operations such as fetch and execute are achieved within a period called the MPU cycle. Fetch takes the same amount of time in each instance for each instruction. Execute, however, may consist of many events and sequences, depending upon the instruction. Execute timing varies a great deal.

The interval taken for a timing pulse to pass through a complete cycle from beginning to end is called a ***period***. One or more clock periods may be needed to complete an instruction.

Fetch and execute

After programmed information is placed into memory, its action is directed by a series of ***fetch*** and ***execute*** operations. The sequence is repeated until the entire program has cycled to its conclusion.

The start signal actuates the control section of the MPU, which automatically starts the sequence of operations. The first instruction is to fetch the next instruction from memory. The MPU may then issue a read instruction. The contents of the program counter are then sent to memory, which returns the next instruction. The first word of this instruction is then placed into the instruction register. If more than one word is included in an instruction, a longer cycling time is needed. After the complete instruction is in the MPU, the program counter records one count. The instruction is then decoded. This prepares the unit for the next fetch instruction.

The execution phase is based upon which instruction is to be performed. It may be to read memory, write to memory, read the input signal, or transfer to output, among others. The timing of an operation depends upon the programmed information.

A microprocessor will take a number of clock periods to perform an operation. As an example, we will use a chip that can be operated at a clock rate of 33 MHz (33 million cycles per second). A single cycle has a period of 1/33,000,000 or 0.00000003 second. Periods this small are best expressed in microseconds (*millionths* of a second) or nanoseconds (*billionths* of a second). At a clock rate of 33 MHz, the length of a single cycle would be 0.3 microseconds (μs) or 30 nanoseconds (ns). If the fastest instruction change that this chip could execute requires four clock cycles, the change would take 0.12 μs or 120 ns. If the slowest instruction would require 18 periods, it would require 5.4 μs or 540 ns. The operating time of an MPU is a good measure of its effectiveness and power.

Read memory

Read memory calls for data to be read from a specific location. The MPU issues a read operation code and sends it to the proper memory address. The read/write memory unit then sends the data into the data bus. This number is fed to the MPU, where it is placed in the accumulator after the timing pulse has been initiated.

Write memory

Write memory calls for data to be stored at a specific location. The MPU issues a write operation code and sends it to a selected read/write memory unit. Data is sent via the data bus and placed into the selected location.

Input-output transfer

Input-output (I/O) transfer operations are similar to read/write operations. The major difference is the opcode number used to call up the operation. This code actuates an I/O port, which either receives data from the input or sends it to the output device. In the system shown in Figure 11-29, data moves in either direction to the read/write memory and flows from the ROM into the data bus. The output flows from the data bus to the output device.

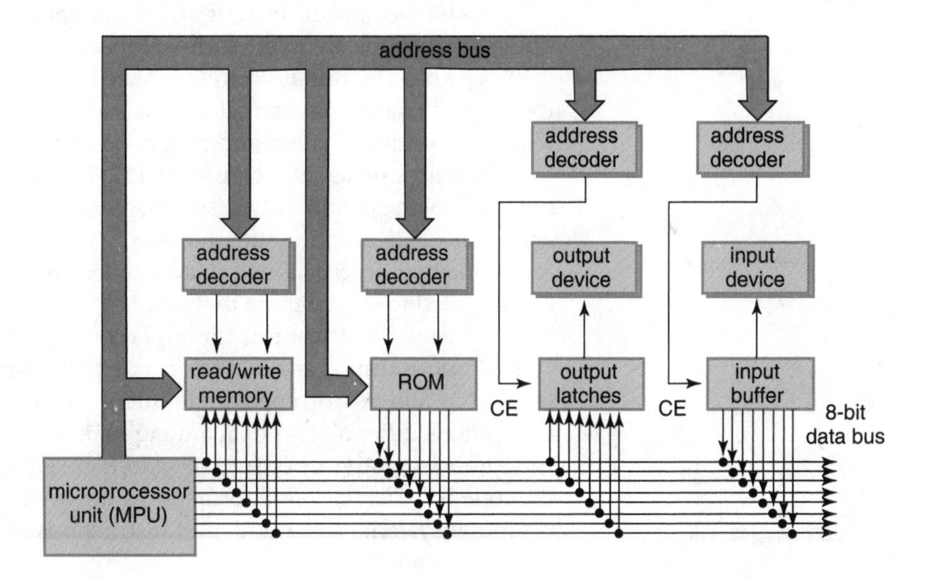

Figure 11-29
This simplified view of an MPU address bus and data bus illustrates input-output transfer. The 8-bit data bus has been superseded by 16-bit, 32-bit, and 64-bit buses in newer personal computers.

Interrupt

Interrupt operations are often used to improve efficiency. Interrupt signals come from peripheral equipment such as keyboards, displays, modems, or printers. They inform the MPU that a peripheral device needs some type of attention.

Assume that a system is designed to process a large volume of data and the output is a printer. The MPU can output data much faster than the printer can print a character. This means that the MPU has to remain idle while waiting for the printer to complete its task.

If an interrupt is used, the MPU can output a byte and then return to processing data while the printer works. When the printer is ready for the next data byte, it requests an interrupt. The MPU stops and automatically goes to a subroutine that outputs the next data byte. The MPU then continues processing information until it receives another interrupt.

Programming

In microcomputer systems, some programs are in read-only memories called firmware, and others are in the form of software. Firmware is placed on ROM chips; program changes can be made by changing the ROM. Software has the greatest flexibility. It is transferred to the system by keyboard, magnetic disk or tape, CD-ROM, or in some older systems, by punched cards or punched tape. Instructions are stored in read/write memory.

Instructions

Instructions normally appear as a set of characters or symbols that define a specific operation. These symbols are similar to those on a typewriter keyboard. Included are decimal digits 0-9, the letters A-Z, and in some cases, punctuation marks and special characters. Instructions may also appear in the form of binary numbers, hexadecimal numbers, or mnemonic codes.

Each type of MPU is designed to understand a certain set of instructions. These instructions are in the form of binary data, normally held in a read-only memory unit. The unit is address-selected and connected to the MPU through the common data bus. Instructional sets are an example of firmware—they are fixed at a specific location and cannot be changed.

Instructions usually consist of 1, 2, or 3 bytes of data. This type of data must follow the commands in successive memory locations. These commands are usually called addressing modes.

One-byte instructions are also called inherent-mode instructions. They are designed to send data to the accumulator registers of the ALU. No address code is needed because it is an implied machine instruction. The instruction "CLA," for example, is a 1-byte opcode that clears the contents of accumulator A. No address or specific definition of data is needed. Inherent mode instructions differ a great deal among manufacturers.

Representative opcode instructions are shown in Figure 11-30. Note that the code is in hexadecimal form, and a mnemonic is given for each one. Each computer system has a number of instructions of this 1-byte type that contain only an opcode.

Immediate addressing

Immediate addressing is done with a 2-byte instruction that contains an opcode and an operand. The opcode appears in the first byte, followed by the 8-byte operand. A common practice is to place intermediate addressing instructions in the first 256 memory locations. Since this is the fastest mode of operation, these instructions can be retrieved very quickly.

Relative addressing

Relative addressing instructions transfer program control to a location other than the next consecutive memory address. Transfer is often limited to a

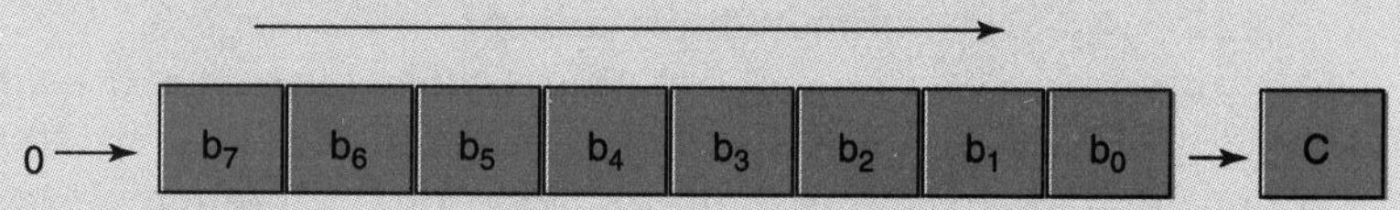

Mnemonic	Opcode	Meaning
ABA	1B	Add the contents of accumulators A and B. The result is stored in accumulator A. The contents of B are not altered.
CLA	4F	Clear accumulator A to all zeros.
CLB	5F	Clear accumulator B.
CBA	11	Compare accumulators: Subtract the contents of ACCB from ACCA. The ALU is involved but the contents of the accumulators are not altered. The comparison is reflected in the condition register.
COMA	43	Find the ones complement of the data in accumulator A, and replace its contents with its ones complement. (The ones complement is simple inversion of all bits.)
COMB	53	Replace the contents of ACCB with its ones complement.
DAA	19	Adjust the two hexadecimal digits in accumulator A to valid BCD digits. Set the carry bit in the condition register when appropriate. The correction is accomplished by adding 06,60, or 66 to the contents of ACCA.
DECA	4A	Decrement accumulator A. Subtract 1 from the contents of accumulator A. Store result ACCA.
DECB	5A	Decrement accumulator B. Store in accumulator B.
LSRA	44	Logic shift right, accumulator A or B. 0 → b_7 b_6 b_5 b_4 b_3 b_2 b_1 b_0 → C
SBA	10	Subtract the contents of accumulator B from the contents of accumulator A. Store results in accumulator A.
TAB	16	Transfer the contents of ACCA to accumulator B. The contents of register A are unchanged.
TBA	17	Transfer the contents of ACCB to accumulator A. The contents of ACCA are unchanged.
NEGA	40	Replace the contents of ACCA with its twos complements. This operation generates a negative number.
NEGB	50	Replace the contents of ACCB with its twos complements. This operation generates a negative number.
INCA	4C	Increment accumulator A. Add 1 to the contents of ACCA and store in ACCA.
INCB	5C	Increment accumulator B. Store results in AACB.
ROLA	49	Rotate left, accumulator A.

Figure 11-30
An example of opcode instructions. The code is in hexadecimal form.

number of locations in front or in back of the present location. The 2-byte instruction contains an opcode in the first byte and a memory location in the second byte. The second byte points to the location of the next instruction which is to be executed.

Indexed addressing

Indexed addressing is similar to relative addressing. The second byte of the 2-byte instruction is added to the contents of the index register to form a new, or "effective," address. This address is obtained during execution rather than being held at a predetermined location. It is held in a temporary memory address register to ensure that it will not be altered or destroyed during processing.

Direct addressing

In direct addressing, the address is located in the byte of memory following the opcode. This permits addressing the first 256 bytes of memory, from 0000_{16} to $00FF_{16}$.

Extended addressing

Extended addressing increases the ability of direct addressing to accommodate more data. It is used for memory locations above $00FF_{16}$ and requires 3 bytes of data for the instruction. The first byte is a standard opcode. The second byte is an address location for the most significant 8 bits of data. The third byte holds the address of the least significant 8 bits of data being processed.

Program planning

A computer cannot solve even the simplest problem without the help of a well-defined program. The system follows this program to accomplish a task. This is why programming is an essential part of all computer applications.

The programmer must be fully aware of the instructions used by the system. The instruction set of a computer is the basis for all programs.

A programmer should be able to decide what specific instructions are needed to perform a given task. A limited number of operations can usually be developed without the aid of a diagrammed plan of procedure. Complex problems, however, require a specific plan in order to avoid confusion or the loss of an important step. Flowcharts are commonly used to aid in this type of planning. See Figure 11-31.

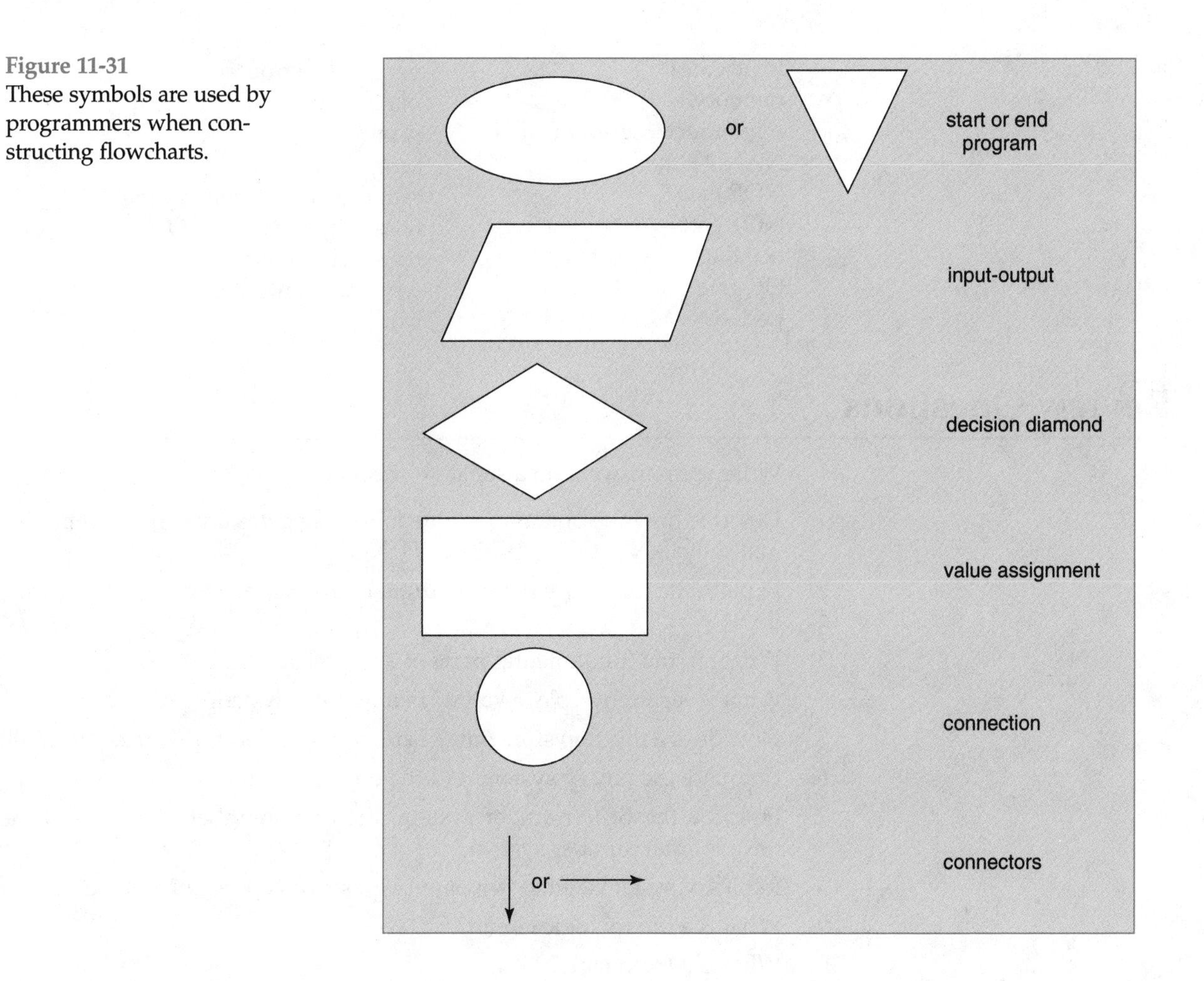

Figure 11-31
These symbols are used by programmers when constructing flowcharts.

Important Terms

accumulators
address registers
analog information
AND gate
arithmetic logic unit (ALU)
binary-coded-decimal (BCD) number system
binary counters
binary logic circuit
binary number system
binary point
bistable
bit
buses
byte
central processing unit (CPU)
counter
data register
decade counter
decimal point
digital electronics
digital information
execute
fetch
flip-flops
hexadecimal number system
input-output (I/O) transfer
instruction decoder
integrated circuit (IC)
interrupt
inverter
logic circuits

logic gates
memory
microprocessor unit (MPU)
NAND gate
NOR gate
NOT gate
octal number system
OR gate
period
personal computers
place value
positive logic
program counter
random-access memory (RAM)
read-only memory (ROM)
read/write memory
sequence controller
truth table

Review Questions

Write your answers on a separate sheet of paper.

1. Describe how the operation of the following gates are expressed mathematically: AND, OR, NOT, NAND, NOR.
2. Explain the difference between digital information and analog information.
3. What are the fundamental parts of a digital system?
4. What is meant by "place value" in a number system?
5. How does a flip-flop store data? Explain how a flip-flop is used to count.
6. Compare the binary system and the decimal system.
7. Describe the BCD number system, the octal number system, and the hexadecimal number system.
8. In binary systems, what is meant by positive and negative logic?
9. What is a binary logic circuit?
10. What is a truth table?
11. Describe the fundamental parts of a computer.
12. Explain the differences between software and hardware.
13. What does encoding and decoding mean?
14. What are a "bit" and a "word" in binary terms?
15. What is meant by the term "byte"?

12 Programmable Logic Controllers

Overview

For many years, machines were controlled by electromechanical devices, such as relays, solenoid valves, motors, linear actuators, and timers. Although they sometimes performed some basic logic functions, most controllers simply turned the machine on or off. Production line sequences were controlled by motors with timers. Nearly all controllers were permanently hard-wired into the system, making changes difficult and expensive. In industries where frequent changes were needed, this type of control was costly. However, it was in many cases not simply the best way, but the *only* way that control could be achieved with any success.

Things changed when digital electronics came on the scene. It made possible the devices known as *programmable logic controllers.* PLCs have numerous applications in controlling robots used as part of computer-integrated and flexible manufacturing systems. This chapter discusses the development of programmable logic controllers, their components and operation, and how they are programmed.

Development of PLCs

In the late 1960s, digital electronic devices began to replace the electromechanical equipment used to control machines. Further refinement of these devices led to the development of the ***programmable logic controller (PLC),*** Figure 12-1. Early PLCs could perform only AND and OR functions. Today, they can make a variety of calculations. In fact, their ability to do simple arithmetic is the main thing that distinguishes them from electromagnetic controllers. Today, they perform logical operations compatible with the traditional relay logic formerly used. They are very versatile and can be easily changed to meet the demands of complex situations. They have reduced downtime when making changeovers, occupy very little space, and have improved efficiency.

A PLC can respond to numerical data, such as size, weight, temperature, or pressure. For example, it can accept a temperature signal, multiply it by a constant, and print out the results in degrees Fahrenheit. The resulting action

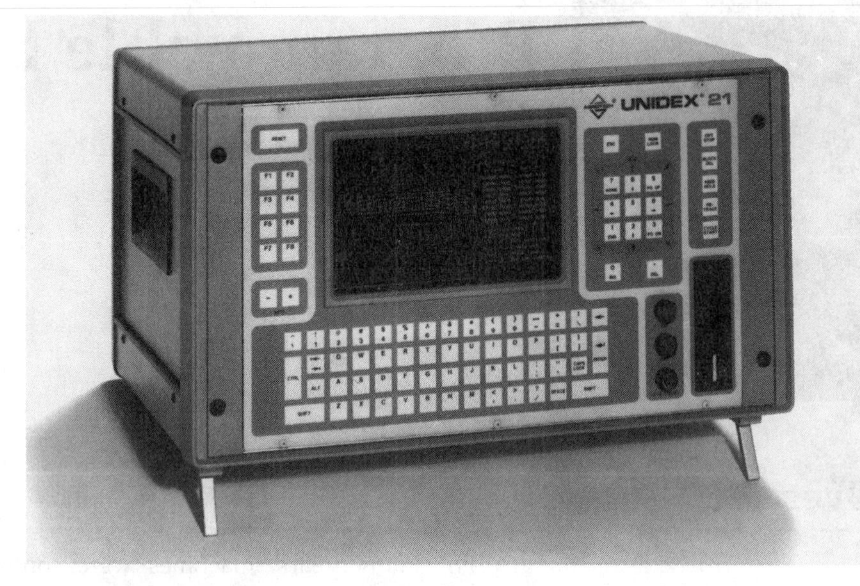

Figure 12-1
Programmable logic controllers (PLCs) are used to control machine tools. There are many different sizes and types available. (Aerotech, Inc.)

can be retained in memory, recorded, displayed on a monitor, or used to set off an alarm. PLCs can also handle trigonometric functions, square roots, and logarithms. These calculations are important in process control, process modeling, real-time error correction, and production line control.

In a programmable logic controller, instructions can be entered or changed without using complex languages and routines. The operator needs only a sound understanding of relay logic and its design. Functions can be altered by simply depressing a few buttons or keys to change the logic sequence. In many cases, an entire system change can be made. The new procedure may be completely different from the original. Once the system has been modified, the program is retained in memory.

Controller Systems

A block diagram for a simplified PLC system is shown in Figure 12-2. Its parts include the input/output structure, the processor, the memory unit, the monitor, and the programmer. In general, the devices making up the PLC are more complex than a single block can indicate. Most programmable logic controllers, in fact, are considered dedicated computers. Their degree of sophistication or power depends on the application in which they are used. Many PLCs interface with a mainframe computer. Others are independent and respond only to those things needed to control a specific machine.

Components

PLCs typically have both a standard power supply and a backup in case of power failure. The ***central processing unit (CPU)*** is its nerve center. A microprocessor is the basic logic element. It has address outputs and address buses, stores and handles data, and monitors the status of input and output signals. The output is based on information being processed through the system.

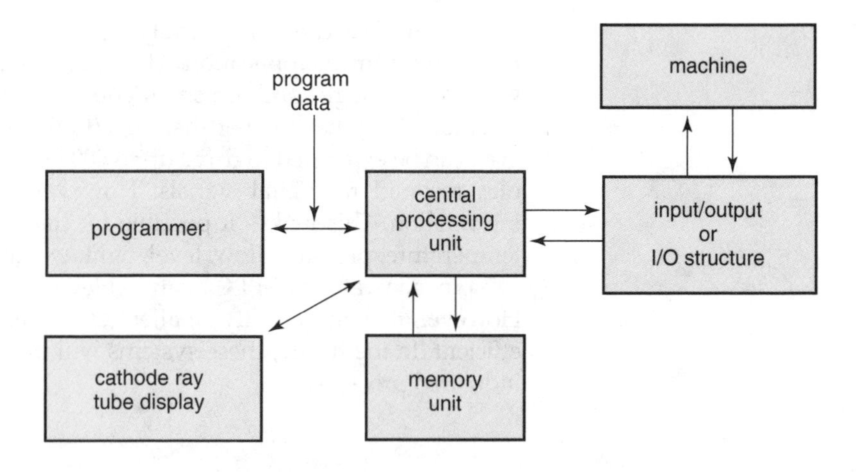

Figure 12-2
This block diagram represents the parts of a PLC.

The arithmetic processing unit (APU) is part of the microprocessor. It computes basic arithmetic, logarithmic, and trigonometric functions. The APU generally functions independently once the initialization (starting sequence) is complete.

The PLC typically has memory modules that are expandable. Possible choices are an ultraviolet-erasable programmable read-only memory (UV-PROM) or semiconductor random-access memory (RAM). Both types can be intermixed in the PLC assembly.

A display monitor and keyboard (buttons) are also components. Control buttons are also generally found on the front panel and are used to select such functions as timer control, parts count, or temperature.

Interface modules may be installed to connect the PLC with a network, or with such peripheral devices as printers, monitors, modems, or data storage devices. As a rule, each type of connection requires its own interface module.

The plug-in input-output (I/O) modules are housed in the assembly. For output, the module may be a solid-state relay, electromechanical relay, timer, or power transistor. It makes a direct connection between the PLC and the load device. This type of construction eliminates hard-wired circuits and reduces interfacing problems between the system and the load. Input modules are designed to interface signals. Their circuitry accommodates such things as thermocouples, light sensors, and practically any type of analog input signal.

Communication with a programmable logic controller may be done using an external monitor (CRT or cathode ray tube). Designed for operation in environments where electromagnetic noise, high temperature, humidity, and mechanical shock occur, such a monitor is typically housed in a rugged case. The CRT can be easily moved to the work site where the controller is located. When the environment is not harsh, a standard CRT may be used.

Mini Systems

Continuing improvements in integrated circuit and power transistor technology are responsible for the development of ***mini-PLCs.*** These systems can control simple machine operations and many types of manufacturing processes. A number of companies are now producing them.

Mini-PLCs can economically replace as many as four electromechanical relays in a control application. They provide timer and counter functions, as well as relay logic, and are small enough to fit into a standard 19-inch rack assembly. Most have fewer than 32 I/O ports or modules. Some units, however, can be expanded to drive up to 400 I/O devices. As a rule, the I/O modules respond to digital signals. However, some units can handle analog information. This makes it possible for the system to respond to changes in temperature, pressure, flow, level, light, weight, and other analog data.

In general, mini-PLCs can achieve the same control as larger PLCs. However, they are usually smaller, less expensive, simpler to use, and more efficient. In the future, these systems will play a greater role in the control of industrial processes.

Programming

A PLC is programmed (given instructions to store and execute) through the use of a keyboard, pushbuttons, magnetic disks or tape, or a network connection to a computer. How the PLC performs is based on its programming. In general, PLCs are programmed using assembly language or relay logic.

Assembly Language

Assembly language is a machine-oriented instructional set for a specific microprocessor. In assembly language, the operation code is in the form of symbols or words called ***mnemonics***. A mnemonic is easier to recognize and remember than binary code. An advantage of assembly language is that it makes keeping track of loops and variables easier than other programming methods. It also makes it easier for someone else to look at the program and see how it works. However, assembly language programs are long and complicated. Many errors can occur. Also, the mnemonics must ultimately be translated into binary data before they can be used by the machine. This can be done by writing down each one, line by line, and then entering the translation into memory. An easier (and more usual) method is to use a program called an ***assembler*** that does the translation automatically. The data is then usually input directly into memory.

For some microprocessors, assembly language uses bit and byte I/O modes and provides a number of special control functions. For example, a common microprocessor has 244 instructions with 1-3 bytes of code per instruction (1 byte of operand and 0-2 bytes of data). Sample instructions include:

LAL—Load the accumulator with the content of the L register.

AD—Add 1 to the contents of the accumulator and store the result in the accumulator.

LLE—Load the L register with the contents of the E register.

Assembly language is generally considered a low-level language. It is efficient with respect to the amount of memory and execution time required. It is somewhat difficult to use and debug, however, unless the programmer works with the language frequently. Other languages used by PLCs are typically variations of assembly language.

Relay Logic

Some PLCs are programmed using ***relay logic***. A ladder diagram is entered into the system by means of the keyboard. Figure 12-3 shows the layout of a PLC keyboard for relay logic. It has network control, relay control, and a numerical entry section.

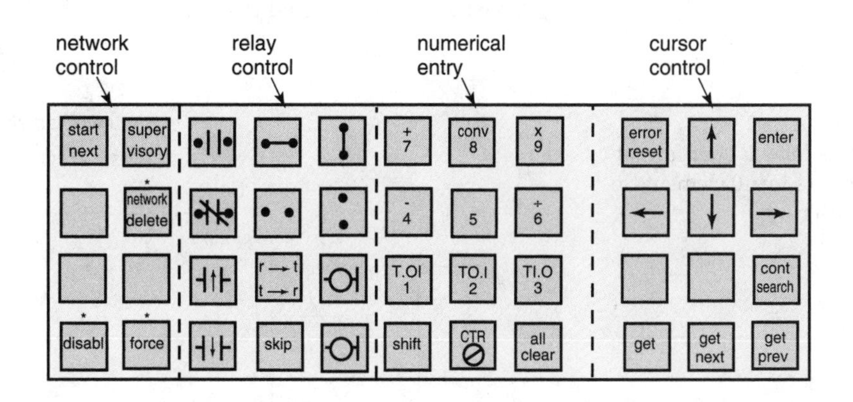

Figure 12-3
This diagram shows the layout of a typical relay logic PLC keyboard.

When using relay logic, the basic programming element is the ***relay contact***. This contact may be normally open (NO) or normally closed (NC). In Figure 12-4, the normally open contacts are on the left and the normally closed contacts are on the right. The line to the right of the two lower contacts is for connection to optional branch circuits. Below each contact is a four-digit reference number. This number identifies specific contacts used in the system. Contacts are connected, in either series or parallel, to form the horizontal rungs of a ***relay ladder***.

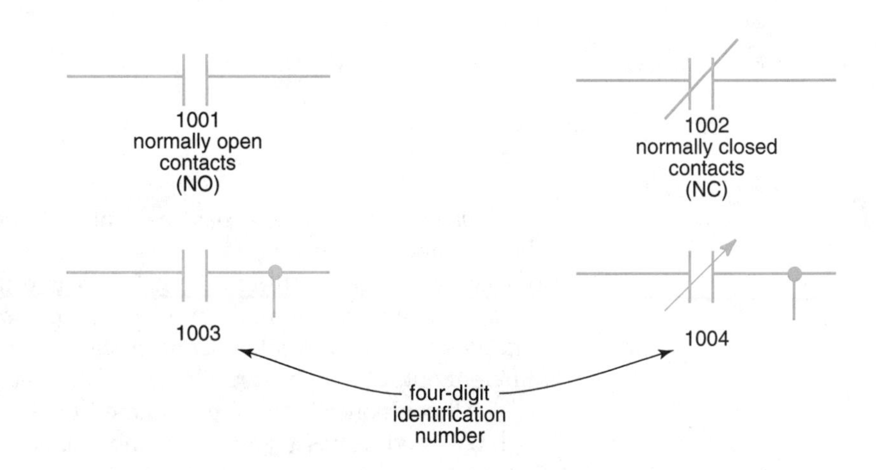

Figure 12-4
Relay contacts may be of the normally open or normally closed type.

Figure 12-5 shows a typical relay ladder diagram. This format allows for up to ten elements in each horizontal rung. The rungs are connected into networks. A network can be a single rung or a combination of several, as long as the elements in each rung are connected. Up to seven rungs may be contained in a network. The left rail of the ladder is the common connecting element. On the right rail, up to seven coils are connected in random order. These coils are given an identification number. The number can be used only once during the operational sequence. The number of devices and registers available for use depends on the capacity of the system. Power flow is only from left to right. The operating sequence occurs vertically from top to bottom. In loop operations, it can be recycled from the bottom back to the top.

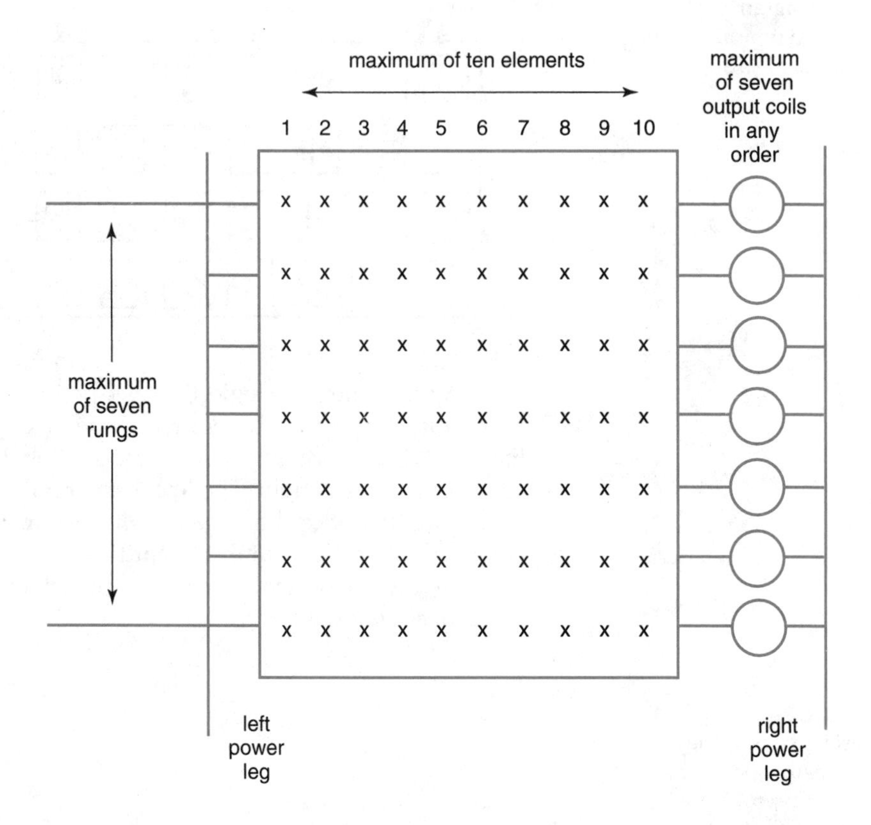

Figure 12-5
This relay ladder format allows for 10 elements in each rung.

Once a relay program has been entered into the PLC, it is monitored on the CRT and may be modified using the keyboard. On the monitor, the path of electrical current through the network is brighter than the other components. Changes can be made by placing a cursor on the device to be altered and pressing the correct key. Contact status can be changed or bypassed and different outputs can be turned on or off by this process.

All components and registers are identified by numbers. For example, each I/O module has a reference number or address that identifies its location in the system. Each manufacturer has its own numbering system. An example is shown in Figure 12-6. Notice that the outputs are identified by a number beginning with zero, while input numbers begin with one. The identity of each I/O device is extremely important during programming. For this reason, the numbers cannot be altered or used more than once.

Once a relay diagram has been entered, the PLC examines the network and "solves" the logic in numerical order. Network 1 is solved first, then net-

Figure 12-6
Reference numbers are used to identify input/output modules. In this example, all output numbers begin with 0, and all input numbers with 1.

I/O Reverence Configuration

module number (top to bottom)	circuit number	channel one housing number							
		one		two		three		four	
		output	input	output	input	output	input	output	input
1	1	0001	1001	0033	1033	0065	1065	0097	1097
	2	0002	1002	0034	1034	0066	1066	0098	1098
	3	0003	1003	0035	1035	0067	1067	0099	1099
	4	0004	1004	0036	1036	0068	1068	0100	1100
2	1	0005	1005	0037	1037	0069	1069	0101	1101
	2	0006	1006	0038	1038	0070	1070	0102	1102
	3	0007	1007	0039	1039	0071	1071	0103	1103
	4	0008	1008	0040	1040	0072	1072	0104	1104
3	1	0009	1009	0041	1041	0073	1073	0105	1105
	2	0010	1010	0042	1042	0074	1074	0106	1106
	3	0011	1011	0043	1043	0075	1075	0107	1107
	4	0012	1012	0044	1044	0076	1076	0108	1108
4	1	0013	1013	0045	1045	0077	1077	0109	1109
	2	0014	1014	0046	1046	0078	1078	0110	1110
	3	0015	1015	0047	1047	0079	1079	0111	1111
	4	0016	1016	0048	1048	0080	1080	0112	1112
5	1	0017	1017	0049	1049	0081	1081	0113	1113
	2	0018	1018	0050	1050	0082	1082	0114	1114
	3	0019	1019	0051	1051	0083	1083	0115	1115
	4	0020	1020	0052	1052	0084	1084	0116	1116
6	1	0021	1021	0053	1053	0085	1085	0117	1117
	2	0022	1022	0054	1054	0086	1086	0118	1118
	3	0023	1023	0055	1055	0087	1087	0119	1119
	4	0024	1024	0056	1056	0088	1088	0120	1120
7	1	0025	1025	0057	1057	0089	1089	0121	1121
	2	0026	1026	0058	1058	0090	1090	0122	1122
	3	0027	1027	0059	1059	0091	1091	0123	1123
	4	0028	1028	0060	1060	0092	1092	0124	1124
8	1	0029	1029	0061	1061	0093	1093	0125	1125
	2	0030	1030	0062	1062	0094	1094	0126	1126
	3	0031	1031	0063	1063	0095	1095	0127	1127
	4	0032	1032	0064	1064	0096	1096	0128	1128

works 2, 3, 4, and so on. Each network is solved by a scanning process. Scanning consists of a series of high speed pulses. Each pulse must pass through the network, following the proper sequence. Scanning moves from the left rail to the right rail and from the top to the bottom. It occurs the instant power is applied to the processor and continues for as long as it remains energized.

Scanning usually occurs so rapidly that the PLC appears to solve the logic steps instantly. The result is that each network is immediately available again to all following networks, regardless of any change in its status or in numerical value. This assures that each network is solved in its correct numerical order and not by the numerical value assigned to a specific coil or contact.

Programming methods

During programming, the appropriate component number and function are entered. The procedure is then placed in memory. For the example that follows, programming is done using a relay ladder that can accommodate up to ten elements in each rung. As many as seven of these rungs can be connected to form a network (see Figure 12-5).

When programming a relay ladder diagram into a PLC, the separate devices and registers are numbered, as shown in Figure 12-6. The numbering system used depends on the memory size of the system. In a low-capacity system, number assignments might be 0001 to 0064 for output coils and 0258 to 0320 for internal coils. A larger system might use number assignments of 0001 to 0256 for output coils and 0256 to 0512 for internal coils. Any coil can be used only once. However, references to contacts controlled by a specific coil can be used as many times as needed. Output coils that are not used to drive a specific load can be used in the programming procedure.

When programming the response of a particular input, the input may be identified as a normally closed (NC) or normally open (NO) relay contact. The coil for the selected contact takes on the same number. While the coil is identified as a circle on the diagram, the contacts are identified by the standard contact symbol. See Figure 12-7. The contacts can be programmed for the NO or NC condition, depending on their intended functions.

Any external input that is normally closed, such as a safety switch, overload switch, or stop push button, must be treated differently. These are pro-

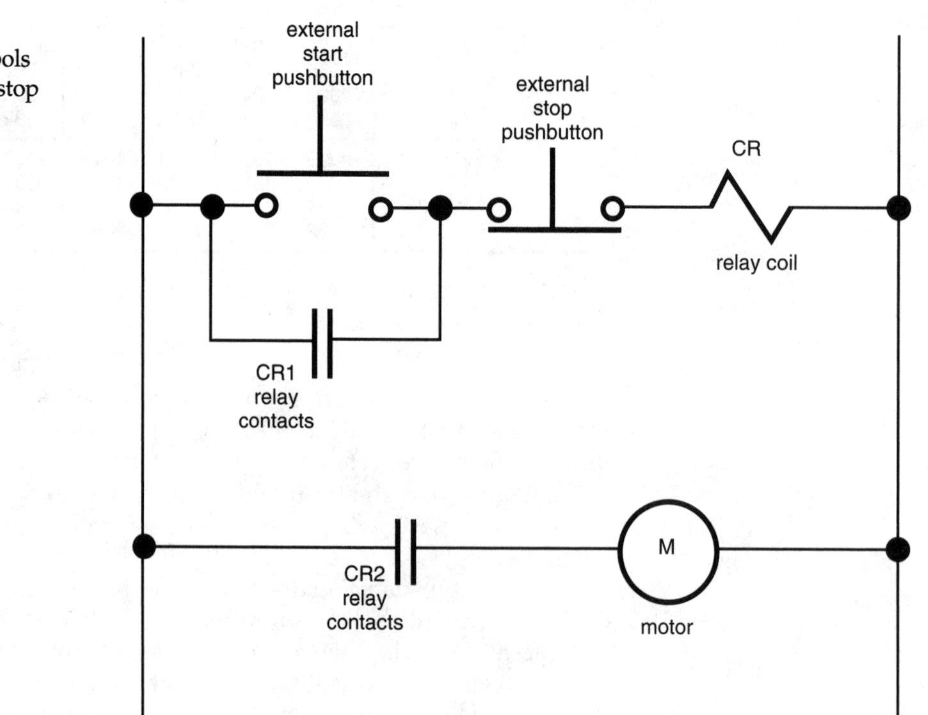

Figure 12-7
These relay ladder symbols are used for a start-and-stop controller.

grammed on the CRT as normally open (NO). An external NC push button, for example, would be entered on the CRT as an open contact. By switching the external contact function and its input signal, a double inversion occurs, causing the desired condition.

Assume that the logic shown in Figure 12-7 is used for a simple start-stop motor controller. The start and stop buttons are located externally. Pushing the start button closes contacts CR1 and energizes the relay coil. Contacts CR2

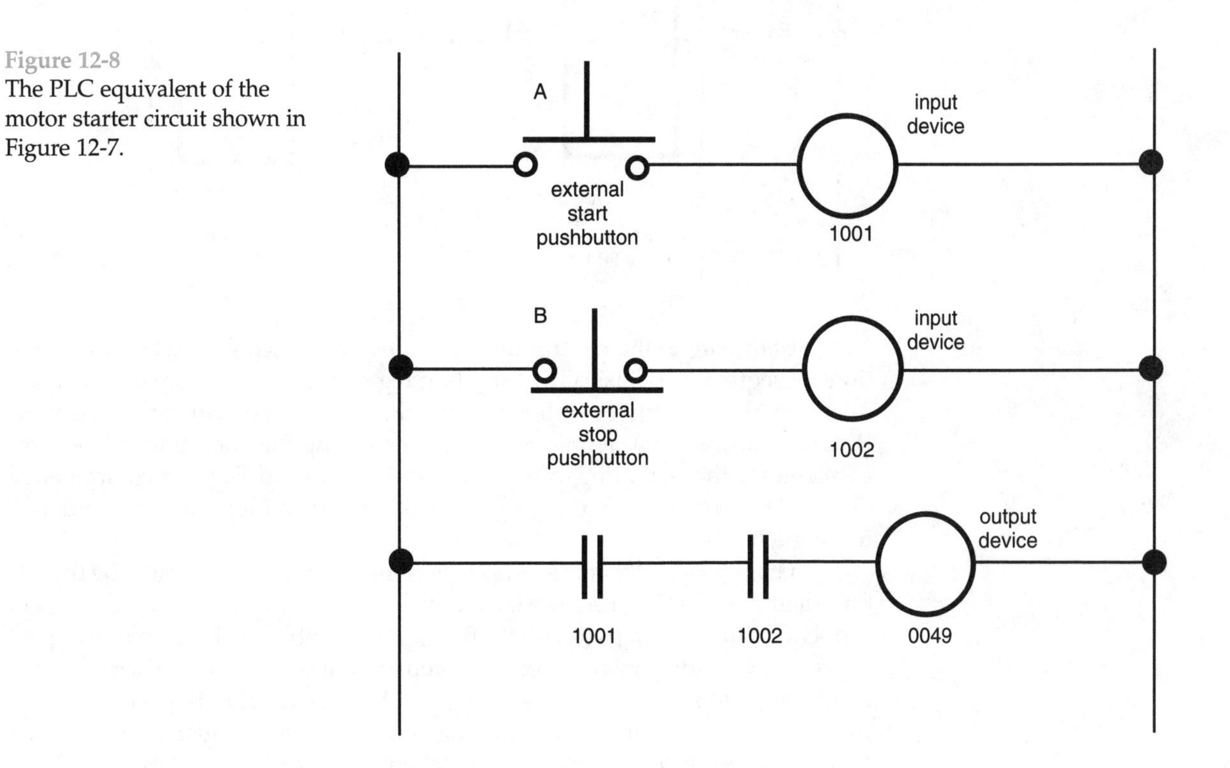

Figure 12-8
The PLC equivalent of the motor starter circuit shown in Figure 12-7.

close at the same time, completing the circuit to the motor, and turning it on. The motor will continue to run as long as power is supplied. Pushing the stop button turns off the motor and removes the latch from the start button.

The PLC equivalent of this motor controller is shown in Figure 12-8. Numbers refer to the specific components of the PLC. Input devices are numbered 1001 and 1002. This includes the input module and its resulting switching operation. The output device is numbered 0049. The start and stop buttons are connected externally, so they do not have a number.

An actual PLC circuit for starting and stopping a motor is shown in Figure 12-9. The I/O modules interface with the microprocessor. The input and output modules are independent parts of the system, but both are controlled by the microprocessor. This circuit would be displayed on the CRT and could be modified with a few simple keystrokes. It is somewhat more complex than the ladder diagram equivalent. However, it is more versatile and can be quickly changed.

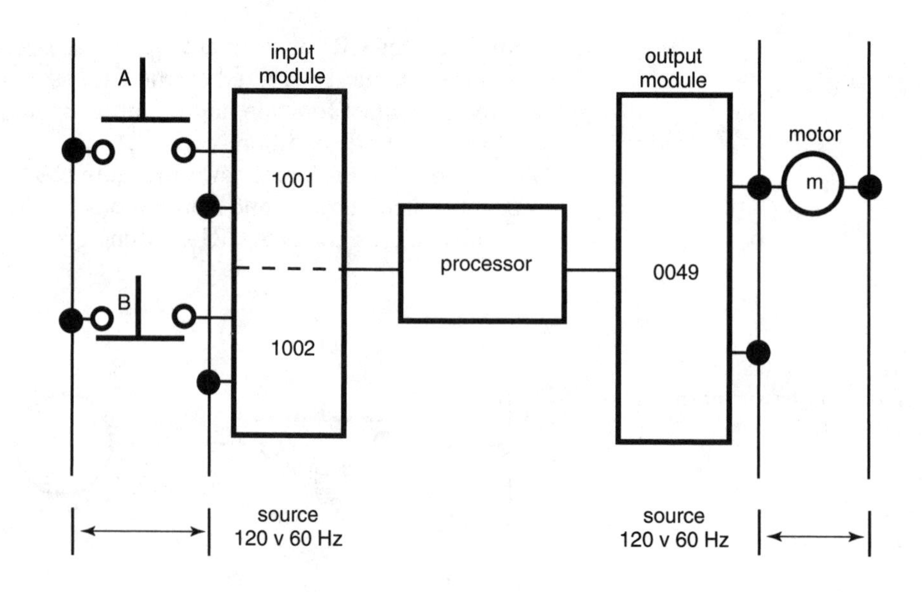

Figure 12-9
This diagram shows the actual PLC circuit of a motor controller.

Sequencing Operations

Sequencing is the control of a load device in each of its stepping positions. A sequencer is basically a counting register. Sequencers are widely used in automatic assembly operations and in the control of production lines. Each PLC manufacturer has a unique way of identifying the operation, its location in memory, the actuating procedure, and the numbering of components. Generally, a sequencer is represented as a block in ladder diagrams and as a rectangle on a CRT.

In Figure 12-10, the number of steps being controlled is indicated by the top four-digit number. The number 0006 indicates six steps. In general, sequencers can achieve up to 32 steps (shown as 0032). This number can be altered using the keyboard. Counting information is stored in a specific register. Registers are identified by four-digit numbers. In Figure 12-10, the number is 4053.

Control of a sequencer is similar to that of other registers. It has two inputs, *set* and *reset*. The output is attached to six independent devices. Each time the set input is energized, the number value of the register increases 1. This is the equivalent of moving a stepper relay one step. When three steps are achieved, the third output device is actuated. In this case, a four-digit numbering system is used to identify the output device. Output device 2304 would be actuated by the third input.

When the reset input is energized, the stepping count goes to zero and the sequence returns to the starting position. This is not affected by the location of its present count or step position. The sequencer will not energize any of the outputs during the transition period.

Important Terms

assembler
assembly language
central processing unit (CPU)
mini-PLCs
mnemonics
programmable logic controller (PLC)
relay contact
relay ladder
relay logic
sequencing

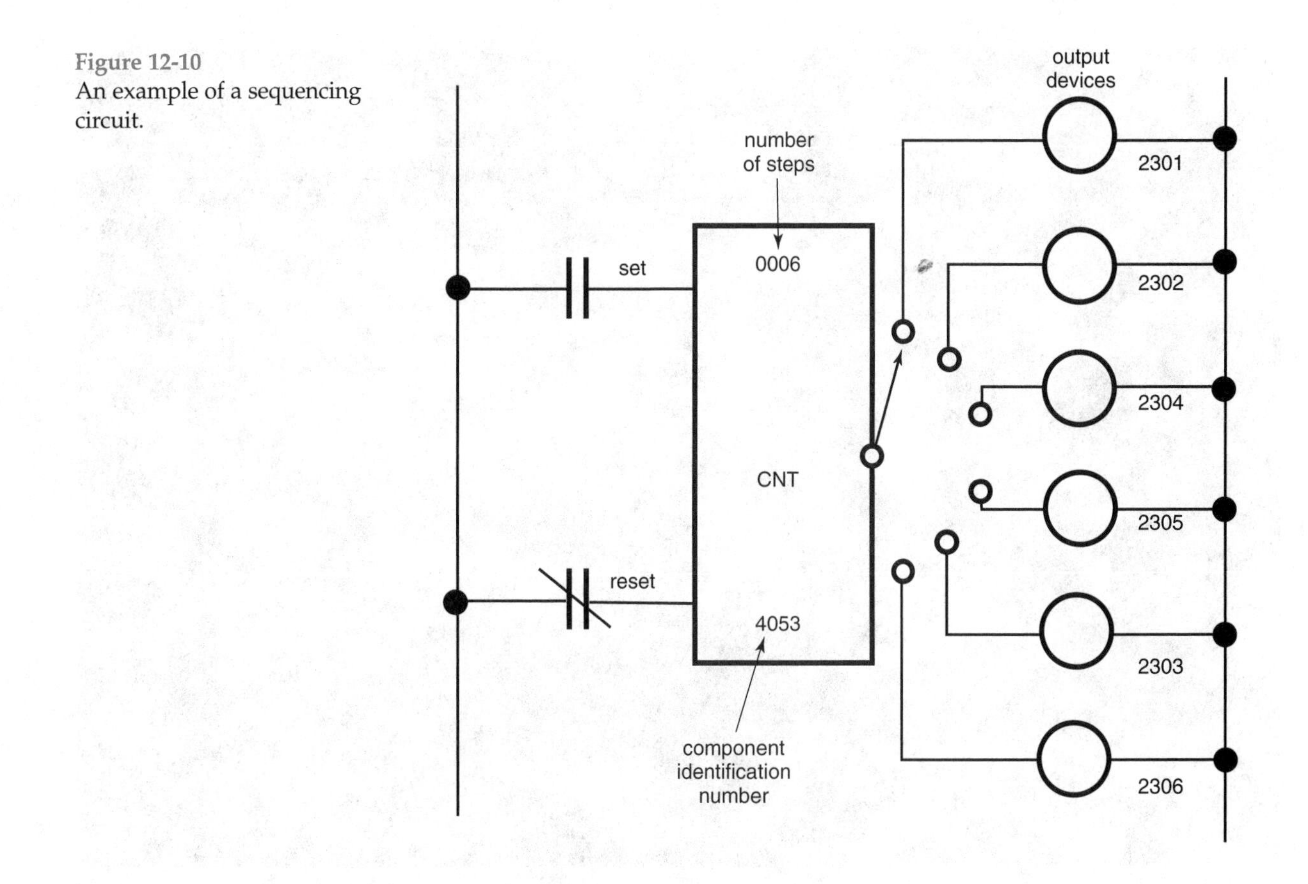

Figure 12-10
An example of a sequencing circuit.

Review Questions

Write your answers on a separate sheet of paper.

1. What are the fundamental parts of a programmable logic controller (PLC) and explain their functions?
2. What is a mini-PLC? How does it differ from a full-scale PLC?
3. What is meant by the term "assembly language"? Briefly describe its function.
4. How does the relay language of a PLC differ from other general programming languages?
5. What are some of the programming basics that must be taken into account when using relay language to program a PLC?
6. How is the timing function of a programmable logic controller achieved?
7. What is sequencing? How is it achieved in a programmable logic controller?
8. Arithmetic functions achieved by a PLC have a number of common features. What are they?
9. Describe the scanning process done by a PLC.
10. What is a relay ladder diagram?

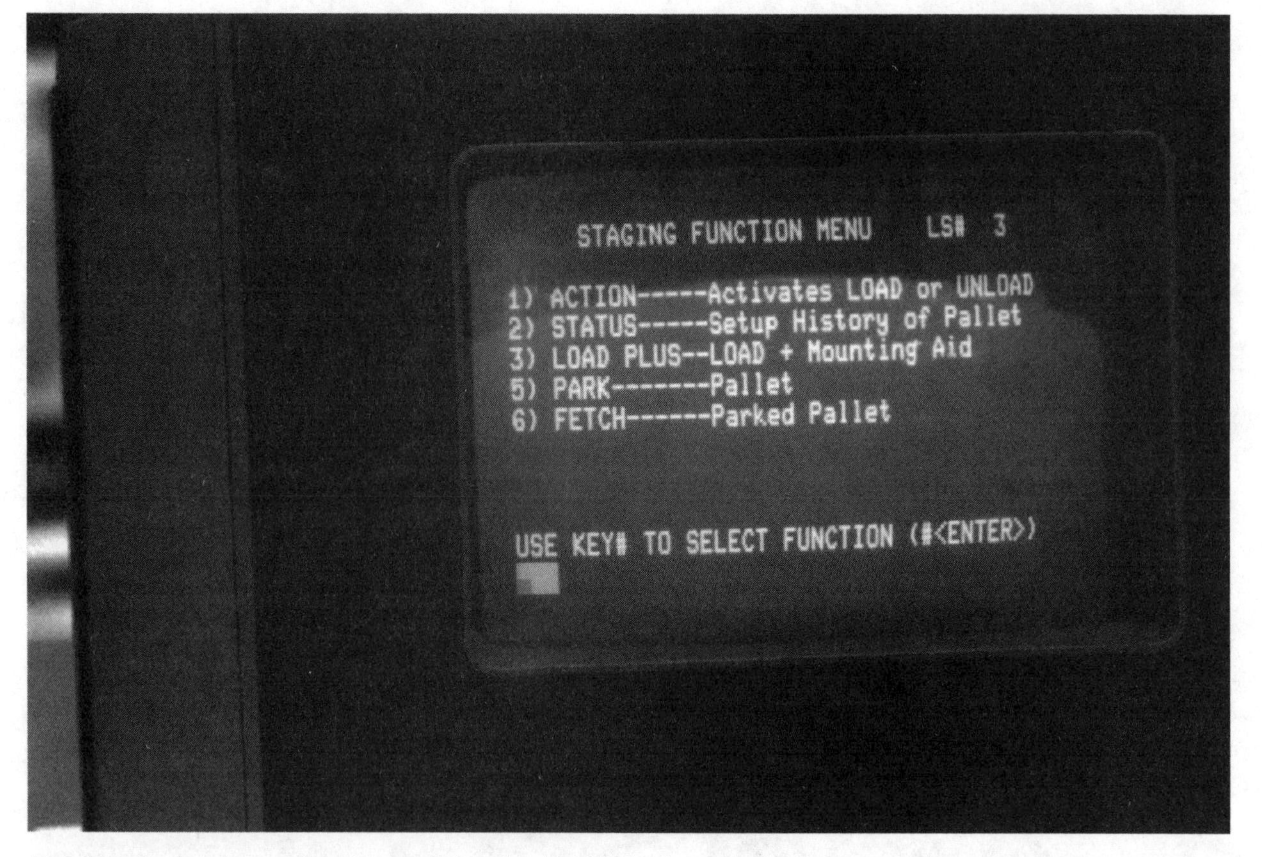

This PLC display shows a function menu for a load and unload operation. The menu includes commands to control the operation of a pick-and-place robot. (Cincinnati Milacron)

13 Robot Interfacing and Vision Systems

Overview

Robots do not work in isolation. They interact continuously with other production equipment, Figure 13-1. This is called *interfacing*.

Figure 13-1
In a robotic workcell, robots interface with other equipment.

Vision systems give the robot "eyes" and allow it to perform a wider variety of tasks. Machine vision systems are being built into automated workcells to determine part orientation, and to handle measurement and inspect parts, among other tasks.

In this chapter, the principles and components of interfacing are discussed. Machine vision systems and their applications are also covered.

Interfacing

An ***interface*** is the common point at which two or more systems communicate with each other. External end effector sensors, workcell limit switches, external relays, operator alarms, safety equipment, bar code readers, conveyors, transfer lines, programmable logic controllers (PLCs), and machine vision systems all interface with robots.

For a robot to communicate, it must be provided with ***input ports*** and ***output ports***. The digital signals used for communication travel through these ports. There are three basic types: dedicated digital input and output ports, serial ports, and parallel ports.

Dedicated Digital Input Ports

Through a digital input port, the robot receives a signal conveying the state (on or off) of an external switch. Limit switches, such as the one shown in Figure 13-2, communicate information about such things as the presence or absence of a part at a given location. This information is used by the robot's computer program to initiate some action. For example, Figure 13-3 shows a limit switch connected to a typical digital input port. When no part is present, the limit switch is open. The base of the input transistor Q1 is held at ground potential by resistor R1. In this condition, transistor Q1 is in a nonconducting state. When a part is present, the limit switch is closed, supplying the base of input transistor Q1 with 24v direct current. In this condition, the transistor is in a fully conducting state. The controller is programmed to detect these two conditions (presence of a part and a change of transistor state). This information is then passed on to the robot's computer program, and the appropriate action is carried out.

Another use of digital input ports is signals from safety devices. Such safety devices as carpet switches, light curtains, motion detectors, or door switches on safety fences, can be connected to a digital input port. The controller shuts down the robot when an operator or other person or object enters the work envelope. Although Figure 13-3 shows only one digital input port, most robots come equipped with 20 or more.

Dedicated Digital Output Ports

One way that a robot interacts with external equipment, such as a conveyor belt drive motor, is through digital output ports. Output ports permit digital signals to be sent from the robot controller to equipment controls. See Figure 13-4.

Whenever the control signal from the robot's computer program is at a low voltage level, the base of transistor Q1 is nonconducting. The output current flow between the emitter and collector of Q1 is zero. The circuit is open, or in a state of high resistance. In this condition, the motor control relay CR1 will *not* be energized; the motor will be off. When the computer sends a high-level signal (12v or 24v) to the base of Q1, however, the current flow will be at maximum. The resistance between the emitter and collector of Q1 will be very low and the motor control relay will be grounded. Under these conditions, relay CR1 is energized, and the motor is switched on. Most robot controllers come equipped with 20 or more digital output ports to match the number of input ports.

Figure 13-2
Limit switches like this one communicate information via input ports. (Euchner-USA, Inc.)

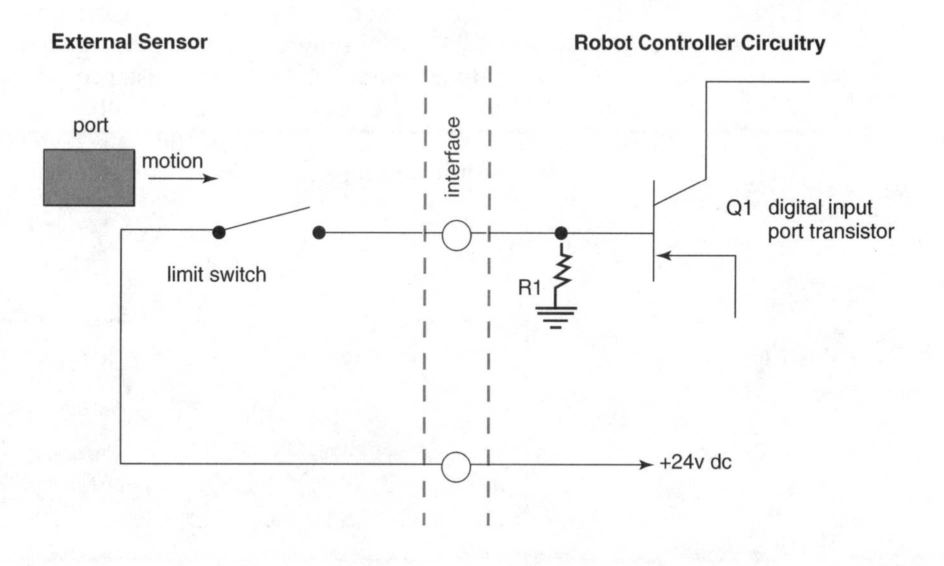

Figure 13-3
When a part moves past the limit switch, the contacts close and 24 volts dc is applied to the base of digital input port transistor (Q1). The controller circuitry detects this change in condition and initiates the appropriate action.

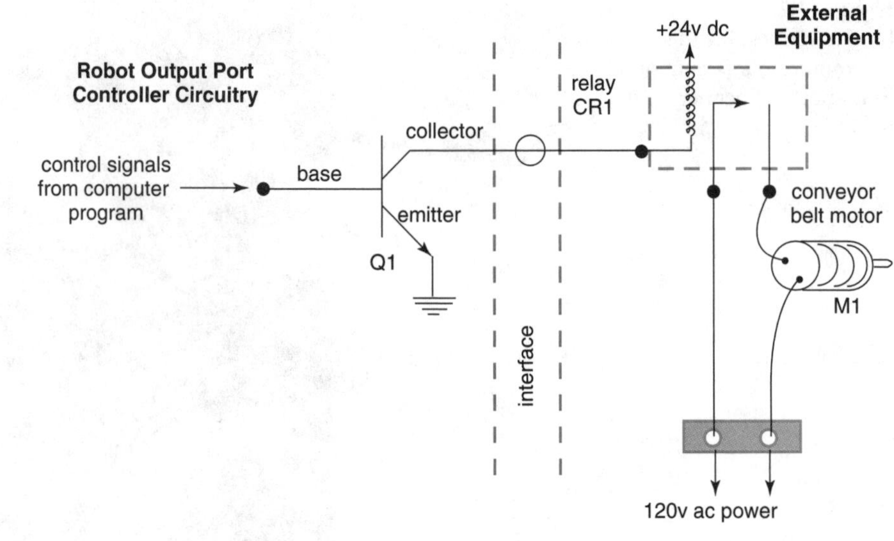

Figure 13-4
When enough voltage is applied to digital output port transistor Q1, the lower end of relay CR1 is grounded. Grounding the lower end of the relay coil causes the relay contacts to close. This supplies ac voltage to the conveyor belt motor (M1).

Serial Ports

Serial transmission delivers digital data over long distances, but at a fairly slow rate. The rate is slow because the data must be transmitted one bit at a time. Two computers can communicate with each other over telephone circuits via ***serial ports*** connected to modems. Existing transmission lines can be used, which is an advantage.

The most common serial interface used in robotics is the standard RS232C, Figure 13-5. It was developed to standardize the interface between data terminal equipment and data communications equipment. The RS232C interface is most commonly used when binary data is to be transmitted over short distances, such as between two computers or between a computer and a piece of peripheral equipment. The circuit performs three functions: data transfer, timing, and control. Figure 13-6A defines its connections. Applications that do not involve long-distance communications require only a few RS232C connections. These are the signal ground (AB), transmitted data (BA), and received data (BB). Long distance communications (those involving modems) require use of many of the other connections listed. Figure 13-6B shows how AB, BA, and BB connections communicate between the microprocessor in a robot's controller and an external video terminal.

Figure 13-5
Typical serial interface standards.

Current Loops	Teletype to computer
RS232C	Computer to terminal
RS422 & RS423	Peripheral to computer (distances greater than 20m)

A

Interchange Circuit	Pin Number	Description	Ground	Data		Control		Timing	
				From DCE	From DCE	From DCE	To DCE	To DCE	To DCE
AA	1	Protective ground	X						
AB	7	Signal ground/common return	X						
BA	2	Transmitted data			X				
BB	3	Received data		X					
CA	4	Request to send					X		
CB	5	Clear to send				X			
CC	6	Data set ready				X			
CD	20	Data terminal ready					X		
CE	22	Ring indicator				X			
CF	8	Received line signal detector				X			
CG	21	Signal quality detector				X			
CH	23	Data signal rate selector (DTE)					X		
CI	23	Data signal rate selector (DCE)				X			
DA	24	Transmitter signal element timing (DTE)							X
DB	15	Transmitter signal element timing (DCE)						X	
DD	17	Receiver signal element timing (DCE)						X	
SBA	14	Secondary transmitted data			X				
SBB	16	Secondary received data		X					
SCA	19	Secondary request to send					X		
SCB	13	Secondary clear to send				X			
SCF	12	Secondary received line signal detector				X			

B

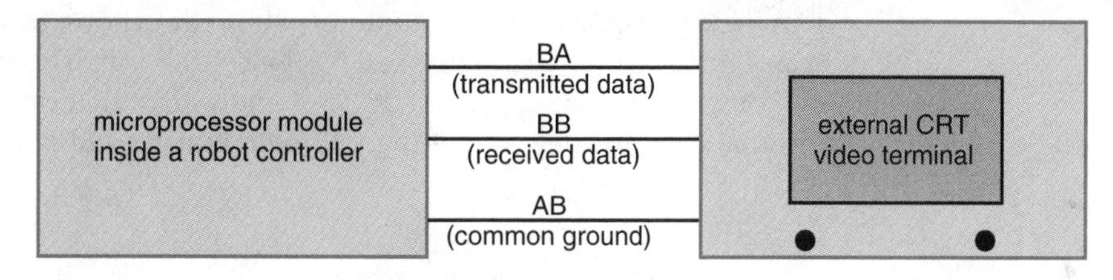

Figure 13-6 RS232C interfacing. A—This chart lists pin assignments and interchange circuit functions for the RS232C serial interface standard. B—The external CRT video terminal is interfaced to a robot controller's microprocessor module via RS232C connections.

Another common use of the RS232C serial port is to connect a robot to a PC. The PC can be used off-line to write and debug the initial program. Then the PC can be connected to the robot, via the RS232C port, and the robot taught the exact moves needed to run the application. Once the program is complete, it can be downloaded into the robot's memory.

The most common connector used with an RS232C cable is a 25-pin DB25. When preparing cables to be used with the RS232C port, do not exceed a total cable capacitance of 2500 picofarads (pf). The capacitance limits the distance between the two pieces of equipment. Another limitation is a 20-kilobit maximum data transfer rate. Although the RS232C standard does not specify the data format, the most common is 8-bit ASCII.

Standards have now been developed to allow serial transmission over distances up to 1000 meters. These newer standards are the RS422 for unbalanced transmission lines and the RS423 for balanced transmission lines.

Parallel Ports

Another way that digital equipment can communicate with peripheral equipment is by means of ***parallel transmission***. Multiple bits of data are sent at the same time, following side-by-side paths like the lanes of a highway. Parallel transmission is done through ***parallel ports***. Although parallel transmission is faster, it is generally considered too costly to use over long distances.

One commonly used parallel port standard is the IEEE 488. It was developed in 1975 by the Institute of Electrical and Electronic Engineers to reduce the amount of time needed to set up test equipment. (At that time, test equipment had to have its own independent interface. Interfacing more than two devices to a computer could get complicated and expensive.) The IEEE 488 standard allows computer-controlled test equipment to be set up in a matter of hours. These advantages make it useful for any industry that uses automated testing.

The IEEE 488 standard is used when a computer is connected to measuring equipment by means of a general-purpose interface bus (GPIB). The GPIB is a cable that interfaces system controllers with programmable instruments. Message length over the 8-bit serial bus is variable, but the rate should not exceed 1 megabyte per second. A maximum of 15 devices can be supported by the bus. Total cable length is 20 meters, or 2 meters times the number of connected devices (whichever is less). Cable between any two devices must not exceed 4 m.

Using the GPIB to connect test instruments to a system controller or PC is shown in Figure 13-7. The bus carries data in both directions, 8 bits at a time. The data and the direction are controlled by the control bus. The bus can connect to the controller various kinds of devices: equipment that can only "talk," equipment that can only "listen," and equipment that can either talk or listen.

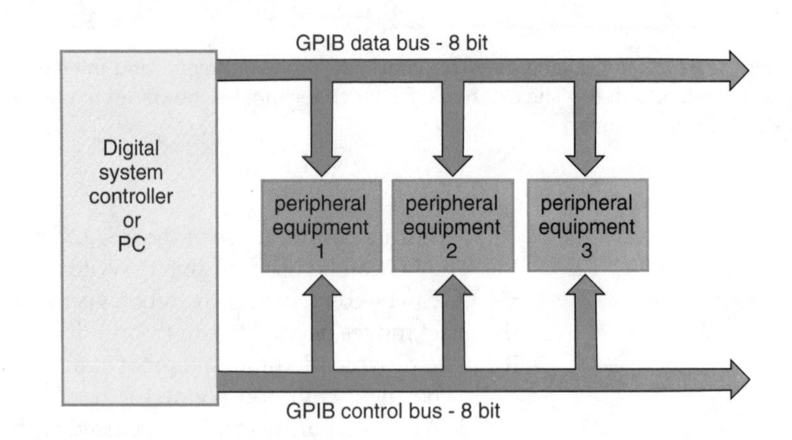

Figure 13-7
The GPIB parallel interface bus is used here to connect external test equipment to a digital system controller or to a PC.

The controller sets the direction of data transfer. Many controllers can be attached to the GPIB, but only one controller at a time may have command. If a second controller needs to communicate with the devices on the bus, the first must surrender its control. The controller is generally a computer with a GPIB interface card. Microprocessor-based computers and minicomputers both work well as controllers.

Machine Vision

Machine vision systems give the robot "eyes" and allow it to perform in a more intelligent manner. They can be used for such applications as robot guidance, determining part orientation, measuring and inspecting parts, and identifying images. See Figure 13-8. Machine vision is becoming an essential requirement in modern automated workcells. When robotic workcells are not equipped with machine vision, all parts must be pre-oriented before the robot is able to grasp them and perform its task.

Figure 13-8
The video camera "eyes" of this system are suspended over the conveyor. (Automatix)

Machine vision can help eliminate the need for many machine operators. Machines equipped with vision can perform many complex tasks once done only by humans. Used for assembly and inspection, they have had a major impact in the automotive and electronics industries. For example, today's printed circuit boards often contain thousands of individual parts. If even a tiny fraction of these parts is incorrectly placed on the board, the cost of finding and repairing the problem is enormous. When human workers perform circuit board assembly, fatigue can cause as much as a 20 percent error rate. Using robots with machine vision is much more successful.

In the automotive industry, such simple tasks as determining the presence or absence of parts in subsystems coming off the assembly line can be done with machine vision. General Motors has stated that it would like to install tens of thousands of vision systems in the future.

Fundamentals of Machine Vision

Machine vision systems use video cameras and computers to translate light energy into an image. These images are then used to determine part orientation or some other task. Even the most effective machine vision systems cannot come close to achieving the human eye's level of perception and sophistication.

Machine vision technology is relatively new. Some of the earliest systems were introduced in the late 1970s. These early systems were rather expensive, due to the high cost of computer memory needed to handle the images. They were also inefficient, because all but mainframe computers contained fairly slow processors. Today, however, faster computers and less expensive memory chips are making vision systems practical. The data obtained can be used to make important decisions regarding action taken by either human operators or a robot. Vision systems can also be used in conjunction with material inventory and material flow within a factory.

Four functions take place during image processing. These functions are acquisition, preprocessing, analysis, and interpretation, Figure 13-9. Figure 13-10 shows the relationship among the subsystems that carry out these functions.

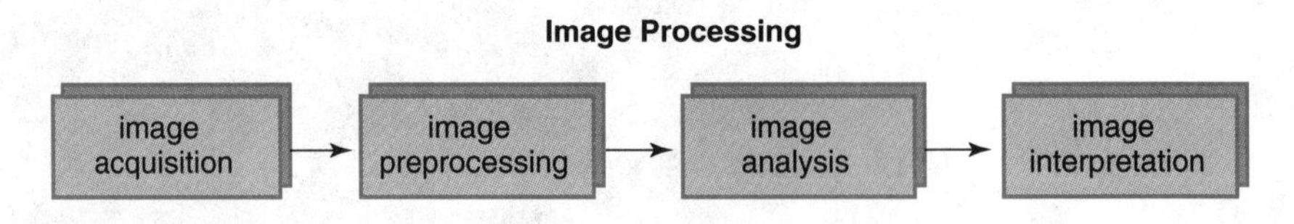

Figure 13-9
A typical machine vision system carries out four image processing functions.

Image acquisition

Image acquisition consists of illuminating the workpiece and digitally scanning its image. Fluorescent lamps, incandescent bulbs, strobe lights, or arc lamps usually provide the illumination. The type and amount of light needed depends on the application. Front lighting is used to enhance surface features, such as bar codes or labels. Side lighting is used when three-dimensional (3D) images are desired. Back lighting provides a silhouette of the object.

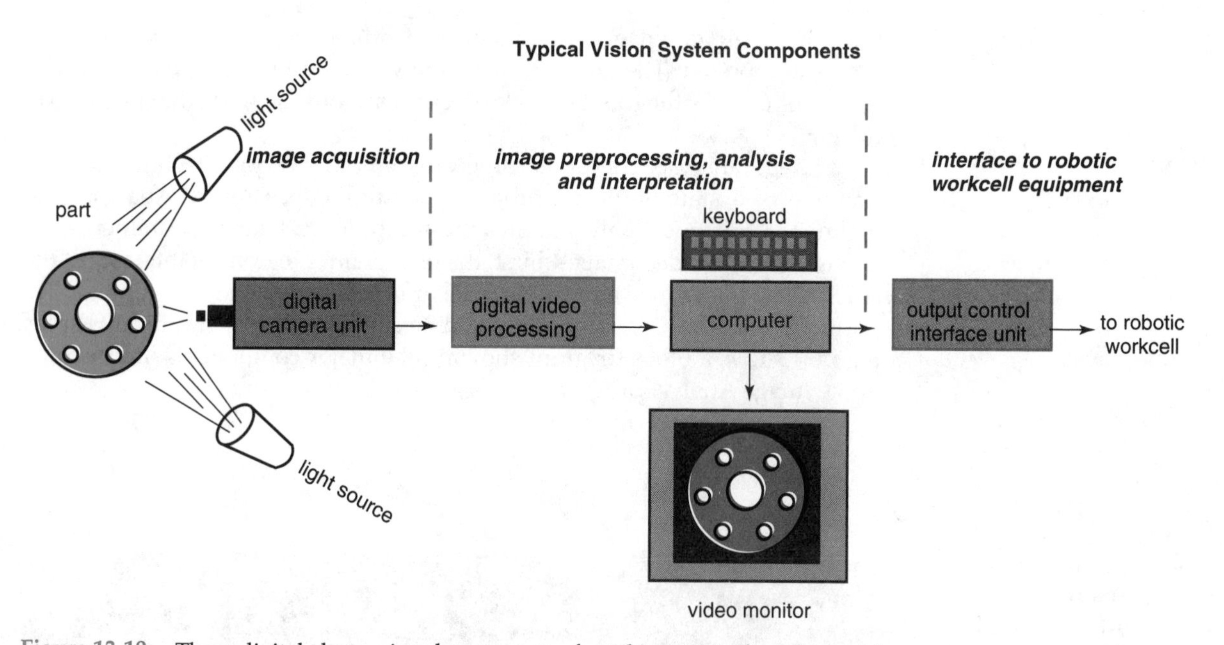

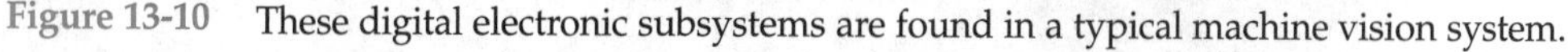
Figure 13-10 These digital electronic subsystems are found in a typical machine vision system.

Scanning is done by a video camera. The solid-state digital cameras commonly used with vision systems have either a charge-coupled device (CCD) or a charge-injected device (CID). At the heart of the camera is a silicon chip used with an array of photosensitive elements. The light reflected into the camera lens from the workpiece falls onto this photosensitive array and is converted into an analog electrical signal. The image is broken down into dots of light called picture elements, or ***pixels***. Pixels are the means used to form the image on a TV screen. A typical digital camera can produce an image consisting of 256 by 256 pixels.

Older machine vision systems often used vidicon tube cameras. However, the newer solid-state cameras have several advantages over vidicon cameras. They are more rugged, smaller in size and weight, and more sensitive to low levels of light. Because they are solid-state, no tubes have to be replaced. They require fewer adjustments and the image is less distorted. They also operate at a higher speed, which creates less image "bloom" (glare from light-colored objects).

Image preprocessing

During ***image preprocessing,*** an analog-to-digital (AD) converter is used to change the analog signal into an equivalent digital signal. The digital signal represents light intensity values over the entire image. These values are stored in memory, permitting the digital image to be analyzed and interpreted.

Image analysis and interpretation

Image analysis involves gathering information from the image that will enable the robot to do a task. Analysis is performed by computer software. Using algorithms (special codes), the software identifies and measures features of the digital image. After all of the features are analyzed, the information is interpreted. The goal of a machine vision system is ***image interpretation***. This enables the robot to make decisions about the work that needs to be done.

Image analysis computer programs employ various techniques. One example of a state-of-the-art program is Automatix's Image Analyst.® See Figure 13-11. Image Analyst allows the computer to take measurements and other data from video images. First, the user defines regions of interest (ROIs). An ROI can be as small as a single pixel or it can include the entire image. The positions and sizes of the ROIs are determined using a graphical display. Figure 13-12 is a block diagram showing the major components and features of the Image Analyst program.

Figure 13-11
Regions of interest can be selected using this image analysis software. (Automatix)

Another example of analysis is shown in Figure 13-13. This vision system is providing 100 percent inspection of similar parts as they are manufactured. The system uses a state-of-the-art computer and software. Any defects found are automatically reported by type, location, and size on a high-resolution color display.

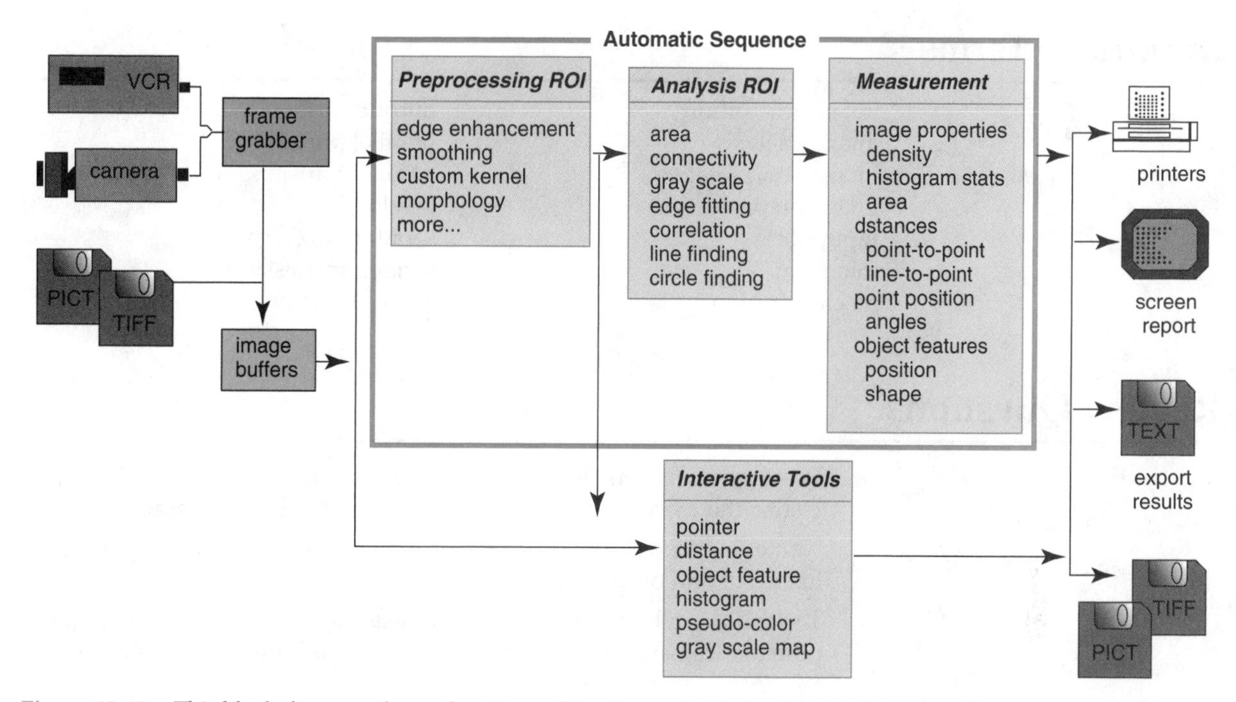

Figure 13-12 This block diagram shows functions of the Image Analyst software. (Automatix)

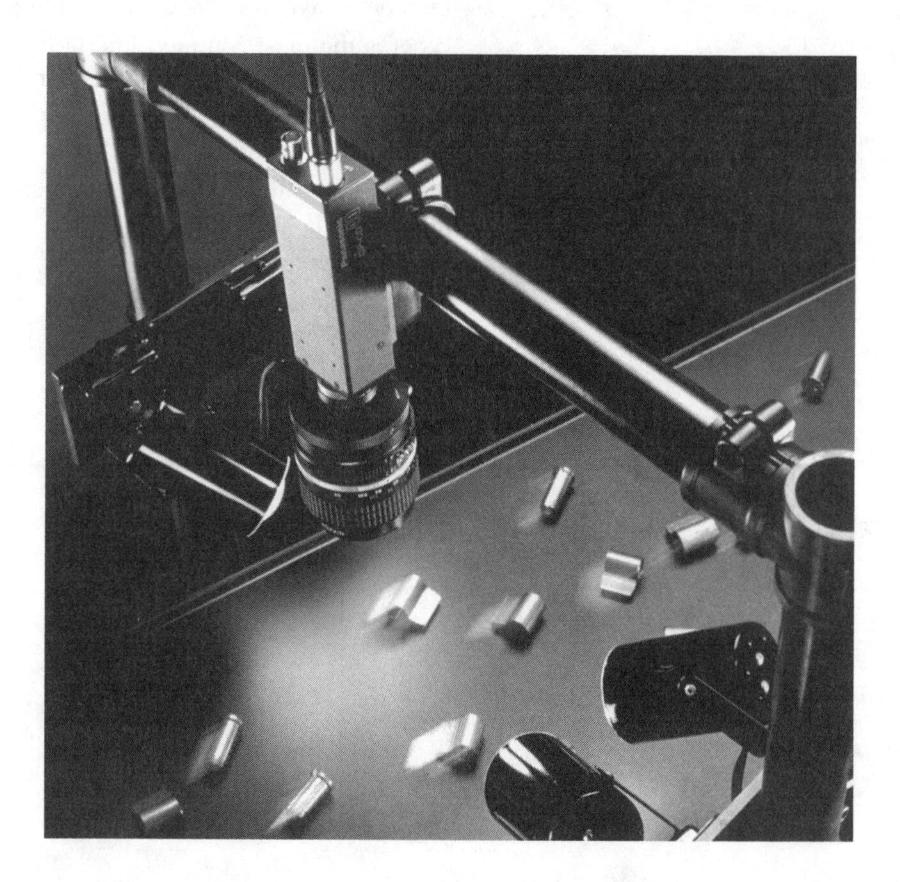

Figure 13-13
This vision system provides 100 percent inspection of similar but different parts as they are being manufactured. (Adept Technology)

Important Terms

image acquisition
image analysis
image interpretation
image preprocessing
input ports
interface
machine vision systems
output ports
parallel ports
parallel transmission
pixels
serial ports
serial transmission

Review Questions

Write your answers on a separate sheet of paper.

1. Define the term "interfacing" and discuss its relation to automation.
2. Name several external devices which can be connected to a robot's digital input and output ports.
3. To send digital data from a robot to external equipment at the highest possible transmission rate, would you use a parallel or a serial port? Why?
4. Describe the advantages and disadvantages of a serial port.
5. List some advantages and disadvantages of a parallel port.
6. What is the most common interface used to transmit serial 8-bit data over short distances, such as between two computers?
7. What is a general-purpose interface bus (GPIB)? Describe its applications and limitations.
8. List three uses for a robotic vision system.
9. Name and describe the four functions employed by a vision system during image processing.
10. Two standards have been developed for long-distance serial data transmission. Name these standards and discuss the primary difference between them.

14 The Future of Robotics

Overview

The future of any technology is difficult to predict. However, the many changes that have occurred as a result of advances in computer applications have affected robotics, as well. Computer-controlled manufacturing, also referred to as ***computer-integrated manufacturing*** or ***CIM,*** is sometimes called the Second Industrial Revolution. Uses for robots outside of industry will also continue to expand. Robot applications are limited only by the imagination.

In this chapter, you will be introduced to the factory of the future, including a close look at a flexible manufacturing lab. Non-factory robots, artificial intelligence, and training programs are also covered.

The Factory of the Future

For decades, scientists, engineers, and plant managers have dreamed of a totally automated factory without human workers. Just how far are we from realizing this dream? Not very far, according to *Business Week* magazine. In an article on General Motor's Saginaw-Vanguard plant, the magazine noted,

"Technology is king here. Lasers inspect parts and check for wear on machine tools. Experimental robots piece together components shaped by automated equipment. Driverless, automatic-guided vehicles whiz about, picking up and delivering parts . . .

"At one spot, for example, pairs of robot arms assemble components the way a human would—holding them in midair. This method eliminates the complicated fixtures and parts feeders normally required. Hence, the system can be retooled for a new part simply by switching the software programs that guide the robots. . . .

"Hardly a human works among the multitude of robots. Only 42 hourly workers are spread over two shifts, and eventually the plant will add an overnight shift with no human presence of any kind."

The Japanese are also hard at work on the ideal factory. The *Wall Street Journal* reported on a new Nissan Motor Co. manufacturing facility:

"Nissan managers call it the 'Dream Factory.' Inside, instead of a conveyor, is a convoy of 'intelligent motor-driven dollies,' little yellow platforms that tote cars at variable speeds down the production line, sending out a stream of computer-controlled signals to coach both robots and workers along the way."

The technology for fully-automated workerless factories exists today. Implementing such technology is not easy, however. In the 1980s, General Motors found that simply increasing the number of robots on an assembly line did not necessarily translate into increased sales and profits. Debugging the robots was a major problem. Some robots even sprayed one another instead of the cars. The time and cost invested in getting a multitude of robots programmed and running error-free has been greatly underestimated.

The hesitation of many companies to fully automate production lines has been due, in great part, to a misunderstanding about how to use current technology. In some cases, applying robots to poorly understood production problems made a bad situation even worse. Today, American manufacturers lag behind their Japanese and European counterparts in the knowledge and expertise needed to successfully apply robotics technology. George Devol, who patented the first industrial robot in 1954, criticized the "lack of interest in our manufacturing industry" in automated systems, and noted that "European and Japanese companies are embracing this concept." Relatively few totally automated systems can be found in U.S. industry at this time. As robots become smarter through the use of vision systems, smart sensing systems, and artificial intelligence, however, a greater degree of automation can be expected.

How should a successful, fully automated factory work? Eastern Kentucky University has developed a fully automated flexible manufacturing laboratory to teach the principles of robotics. Its operation is described in the following section.

A Robotic Flexible Manufacturing System (FMS)

Even a single robotic workcell requires many external pieces of equipment, each coordinated with the robot. In many cases, this external equipment can cost several times as much as the robot itself. In more complex, fully automated workcells, several robots must work in coordination with one another and with external equipment.

The automated flexible manufacturing system at Eastern Kentucky University has four industrial robots and their associated equipment. A ***flexible manufacturing system (FMS)*** is capable of making many different products without retooling or other changes. The workcell in this example shows how many of the concepts discussed in this book come together, Figure 14-1.

Each of the four robots is set up as a workstation and performs a specific task on the workpiece, Figure 14-2. The four tasks are: Station 1, loading and milling; Station 2, edge routing; Station 3, chip removal and cleaning; and Station 4, offloading of finished parts. In addition to the robots, other equipment includes:

Δ A computer numerically controlled (CNC) milling machine used to machine the logo into the workpiece. See Figure 14-3.

Δ A flexible material handling conveyor system used to move the workpiece from one station to the next.

Figure 14-1
A view of the flexible machining workcell used at Eastern Kentucky University's FMS laboratory. Note the four robots that are used with the parts conveyor system.

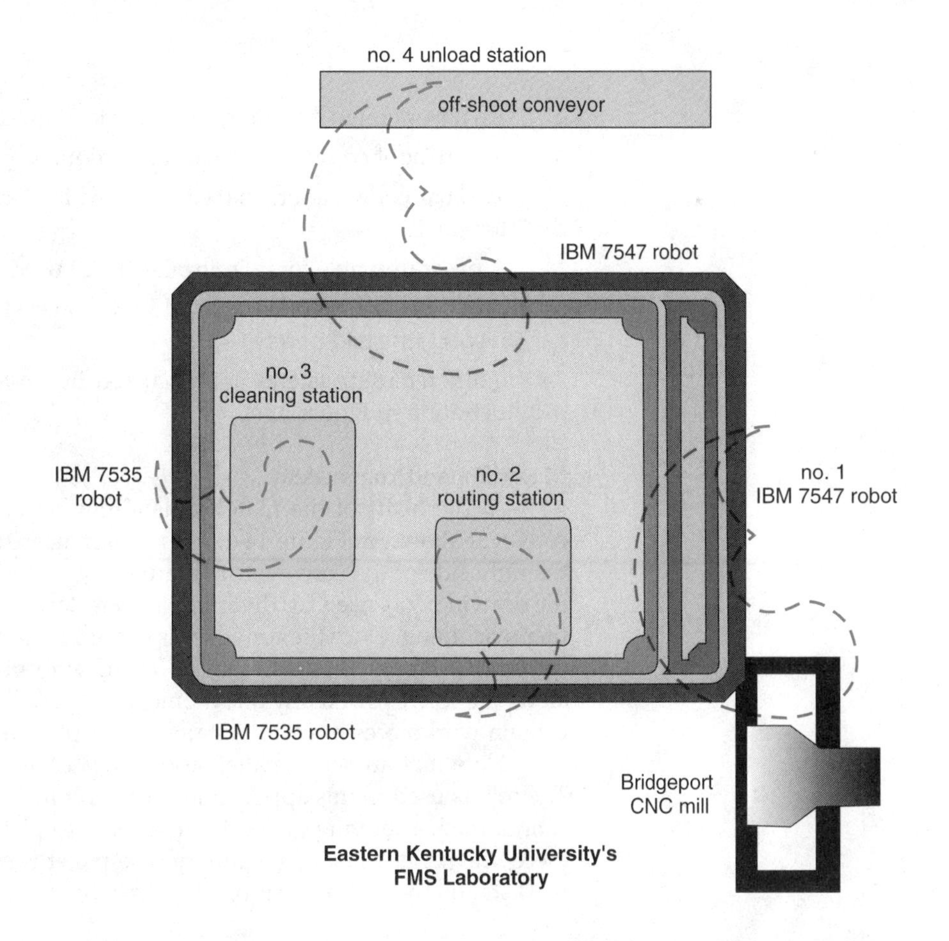

Figure 14-2
The work envelope for each robot and its relationship to peripheral equipment is shown in this plan.

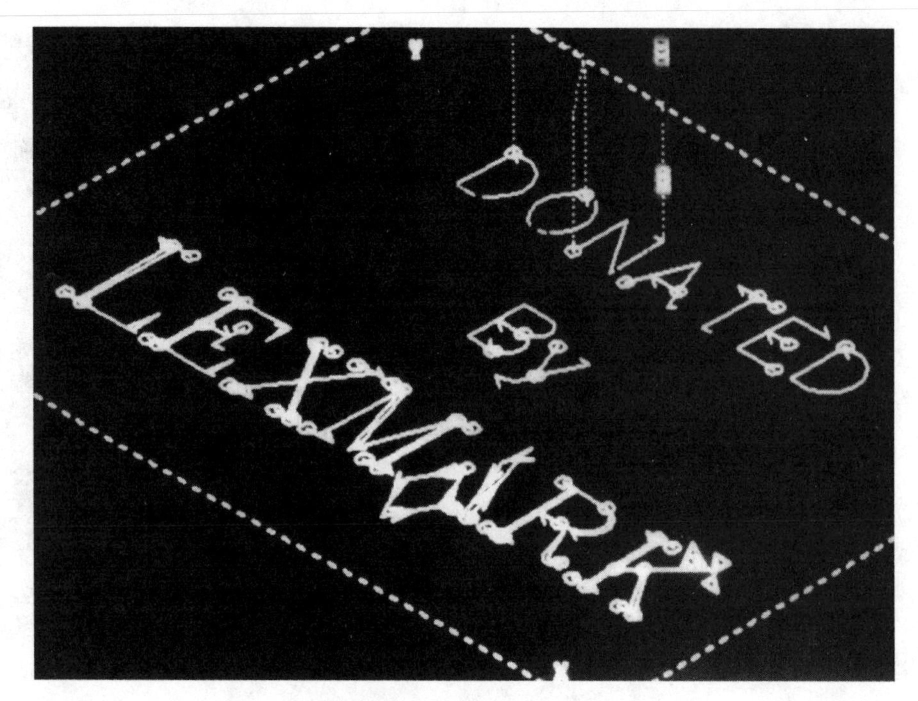

Figure 14-3
This CAD drawing shows the tool path for the logo of the product to be manufactured. A milling machine at Station 1 machines this design into the workpiece.

- Δ An air-operated chip-removal and cleaning station.
- Δ An offshoot conveyor unloading station.
- Δ Two bar code readers that can be used for parts tracking and inventory activities.
- Δ A programmable logic controller (PLC) used for system coordination.
- Δ Provision has been made to add a vision system that will be used to inspect finished parts.

As the different subsystems are discussed in the following sections, refer to the illustration in Figure 14-4.

Material handling system

At the heart of the material handling system is the rectangular chain conveyor shown in Figure 14-1. This system has an auxiliary track parallel to the right side of the conveyor. This auxiliary (outer) track is not being used in the drawing. It is used for diverting and stacking workpieces that require further machining. Once the conveyor is energized, it runs continuously until the manufacturing cycle is terminated. The chain pulls a series of small pallets along the conveyor. At any one moment, some pallets are empty, while others contain workpieces in various states of completion.

Note that an active pallet stop is located near each robot workstation. The robots used in this application do not yet incorporate vision systems and cannot track a moving pallet. For the robot to pick up or put down a workpiece, the pallet must be stationary. The pallet is stopped at the appropriate workstation by the pallet stop, which is activated by a microswitch. Since the

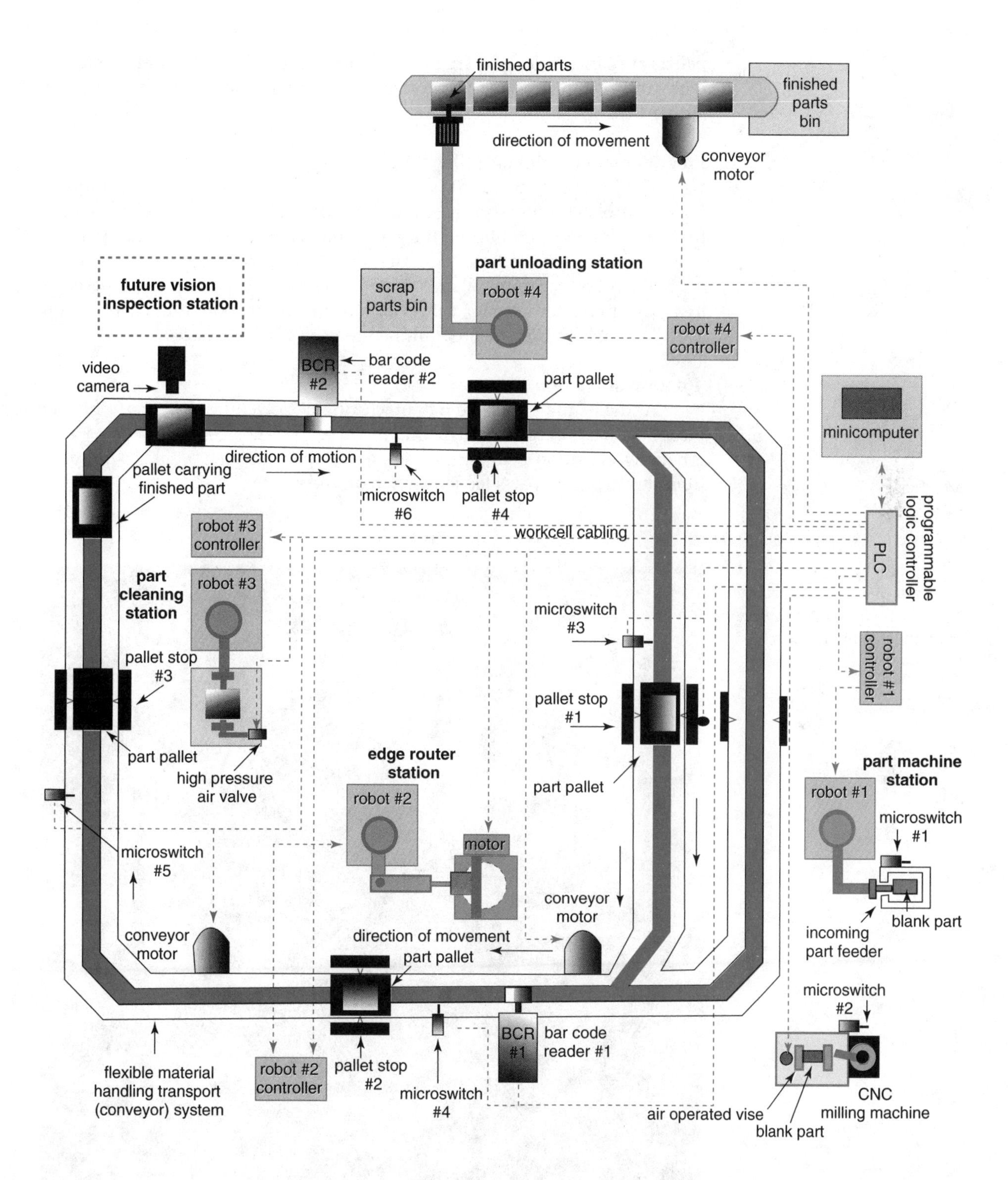

Figure 14-4
This detailed drawing describes the major components of the flexible manufacturing system (FMS). Note the four robots which are used as workstations, and how they relate to the operations. Station 1 handles loading and milling; Station 2, edge routing; Station 3, chip removal and cleaning; Station 4, parts offloading.

pallets are carried along by friction, the capturing of a pallet does not interfere with conveyor operation. The conveyor and pallet stops are also visible in Figure 14-1.

Programmable logic controller (PLC)

The brain behind the operation of the FMS system is a programmable logic controller, shown on the right in Figure 14-4. The PLC serves several functions. It communicates with the four individual robots and monitors the sensor switches located throughout the cell. It also issues commands that begin the robot programs, start conveyors, stop pallets, determine if parts are in the incoming parts feeder, collect data from the bar code readers, and so on. The PLC must be custom-programmed to perform these tasks.

Robot workstation 1

At the beginning of the manufacturing cycle, a human operator fills the incoming parts feeder with blank workpieces. The feeder is located near robotic workstation 1. In Figure 14-5, robot 1 is retrieving a blank workpiece from the incoming parts feeder, shown on the right.

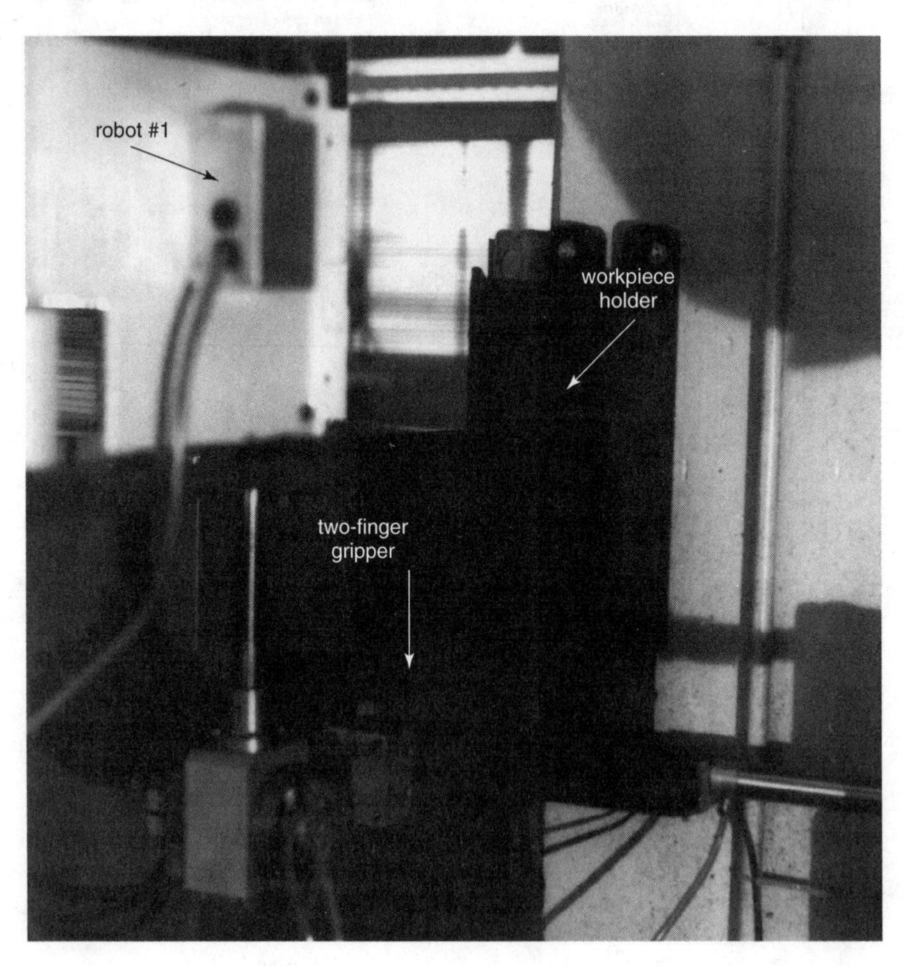

Figure 14-5
To begin the operation, robot 1 retrieves a blank workpiece from the incoming parts feeder.

Next, the operator starts the PLC. The PLC then checks the closure of microswitch 1 (refer to Figure 14-4). This switch is located at the incoming parts feeder and signals the PLC that a part is available for machining. If the parts feeder is empty or jammed, the microswitch remains open and the PLC triggers a warning alarm. A human operator then must check the problem.

If a workpiece is available, the PLC issues a command to the robot's controller to run its "get blank workpiece from parts feeder" subroutine. Robot 1 moves to the parts feeder and grasps a workpiece. It then makes a 90-degree turn and places the workpiece into an air-operated vise at the milling station shown in Figure 14-6. Microswitch 2 is located at the vise and signals the PLC that the part has been clamped into the vise. The PLC now issues a command to turn on the milling machine and mill the logo into the blank workpiece.

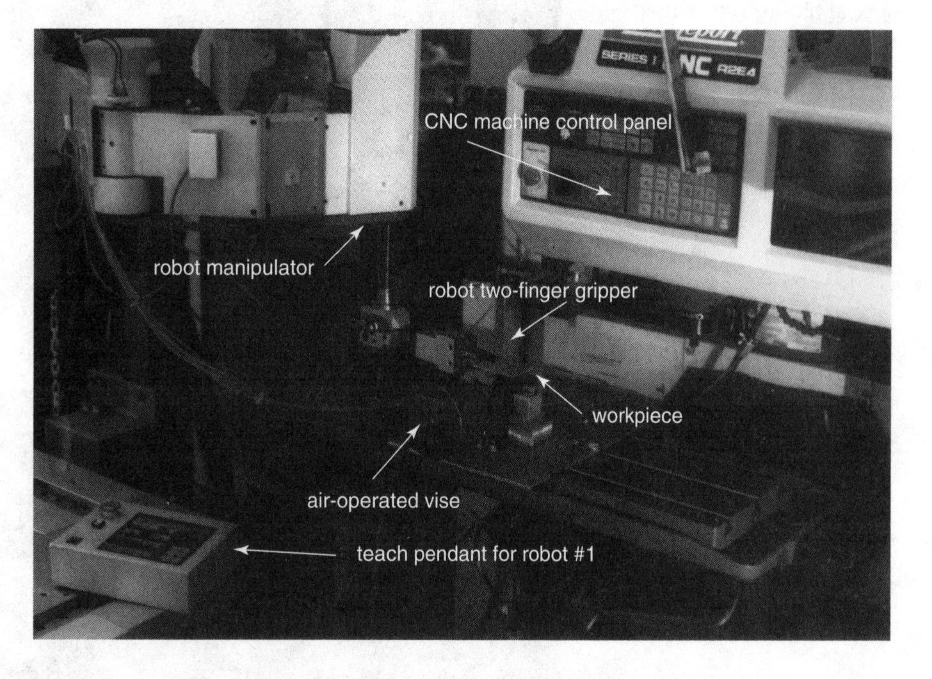

Figure 14-6
Robot 1 places a blank workpiece into an air-operated vise. The vise holds the workpiece stationary while the Bridgeport CNC milling machine operates.

As the milling operation takes place, the PLC monitors microswitch 3. This switch is located on the conveyor in front of robot 1. If the PLC detects that microswitch 3 has closed, this indicates that an empty pallet has arrived. It then operates the air solenoid at pallet stop 1 to stop the pallet.

At the end of its milling operation, the CNC milling machine signals the PLC. The PLC opens the air vise. It then issues a command to robot 1 to remove the part from the air vise and place it onto the empty pallet now at pallet stop 1. See Figure 14-4. The PLC next issues a command to the air solenoid at pallet stop 1, releasing the pallet holding the workpiece. The PLC then begins to monitor microswitch 4, located at pallet stop 2 (in front of robot 2).

Robot workstation 2

When the PLC detects that the pallet containing the workpiece has arrived, it sends a signal to the air solenoid at pallet stop 2. The pallet is captured. Next, the PLC turns on the edge-routing machine and issues a command to the robot controller to run its program. Robot 2 moves the workpiece from the pallet to the edge-routing machine, as shown in Figure 14-7A. A fixture located next to the routing machine, Figure 14-7B, allows the robot to rotate the workpiece so that all four edges can be machined.

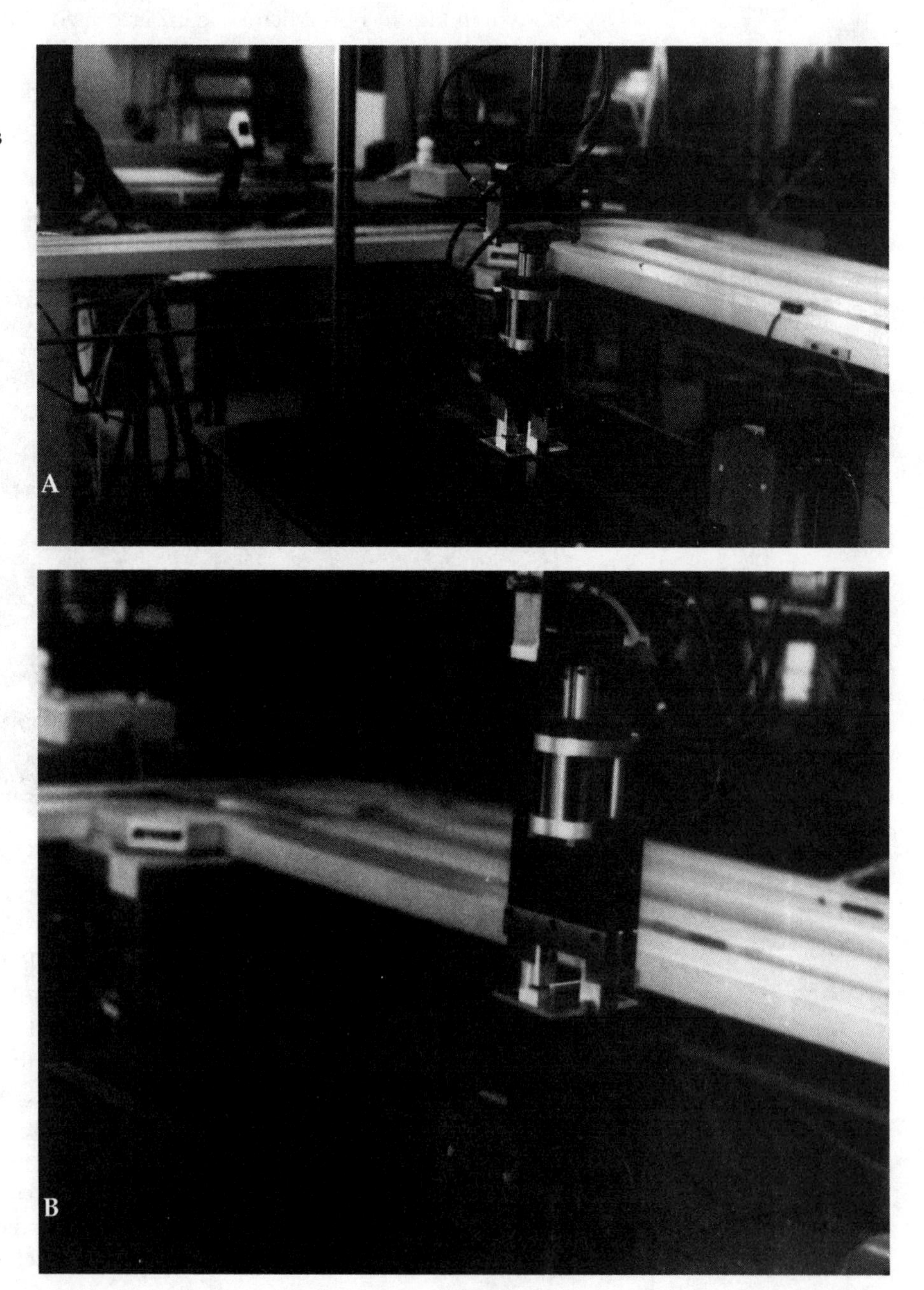

Figure 14-7
Edge routing operation. A—The edge routing machine places beveled edges on the workpiece. B—This special fixture allows the robot to rotate the workpiece so that all four edges can be machined.

When the routing operation is finished, robot 2 puts the part back onto the waiting pallet. Robot 2 then signals the PLC that routing is complete. The PLC sends a signal to the air solenoid at pallet stop 2, releasing the captured pallet. The workpiece on the pallet now has had a logo milled onto it and its edges beveled.

Robot workstation 3

The closing of microswitch 5 signals the PLC that a workpiece has arrived at pallet stop 3. Refer to Figure 14-4. It then issues a command to the air solenoid at pallet stop 3, capturing the moving pallet. The PLC turns on the air valve at the robot's cleaning station and directs the controller of robot 3 to run its program. The robot takes the workpiece from the captured pallet to the air-operated cleaning station, Figure 14-8. The robot passes the workpiece in front of a blast of air that removes chips. After placing the cleaned workpiece onto the waiting pallet, the controller signals the PLC that this operation is complete.

Figure 14-8
A workpiece is moved by robot 3 from the pallet at left to the cleaning station at right. A jet of air is used to remove chips left on the workpiece after machining.

Robot workstation 4

The PLC next sends a command to the air solenoid at pallet stop 3 to release the pallet. The PLC monitors microswitch 6 (located at workstation 4) to detect the arrival of the pallet. After issuing the command to capture the incoming pallet at workstation 4, the PLC then signals the controller at robot 4 to run its program. Robot 4 unloads the finished workpiece from the pallet and places it on the parts-removal conveyor, Figure 14-9. The part is then taken to the finished part bin. Figure 14-10 is a closeup view of a finished part.

Figure 14-9
The robot at station 4 unloads the finished workpieces onto a conveyor. This parts-removal conveyor moves the finished workpieces to a bin.

Figure 14-10
A closeup view of the finished workpiece prepared by the flexible manufacturing system.

Robots Outside the Factory

In the 1980s, the robotic industry was in high gear. Hundreds of companies were involved in manufacturing and installing industrial robotics systems. By the early 1990s, however, only a very small number of companies remained actively involved. The major growth for the 1990s and beyond seems to be taking place outside the industrial sector. This new area is referred to as ***service robotics***. Applications range from mobile robots used for building security patrols to totally automated robotic vehicles used for unmanned space exploration. Looking through a computerized business database gives a hint of the diverse applications that are currently being developed:

- Δ *Serbots* (service robots) are being used by Marriott hotels to clean floors.
- Δ Electric utility companies are considering using robots to work on transmission and distribution facilities that pose dangers for humans.
- Δ The University of Maryland is experimenting with a robotic "hired hand" for work on a dairy farm. The robot's arms adjust a cow's legs, wash the udders, and attach a milking machine.
- Δ Specialists at Carnegie-Mellon University are developing a 12-foot-tall, six-legged robot for NASA. The robot is named Ambler and is being designed to walk on Mars. This same type of robot could be used in logging and mining operations, for hazardous waste cleanup, and for emergency response.
- Δ Construction robots might be used to build high-rises.
- Δ Robotic couriers could have a role in healthcare facilities.
- Δ Robotic chefs could appear in the future. The robot in Figure 14-11 is transferring frozen hamburgers from a storage box onto a conveyor.
- Δ Robots are already being used for education, Figure 14-12, and entertainment. Museums use them to operate replicas of prehistoric animals.

Figure 14-11
This robot is used to transfer frozen hamburger patties from a box to a conveyor. (GE Fanuc)

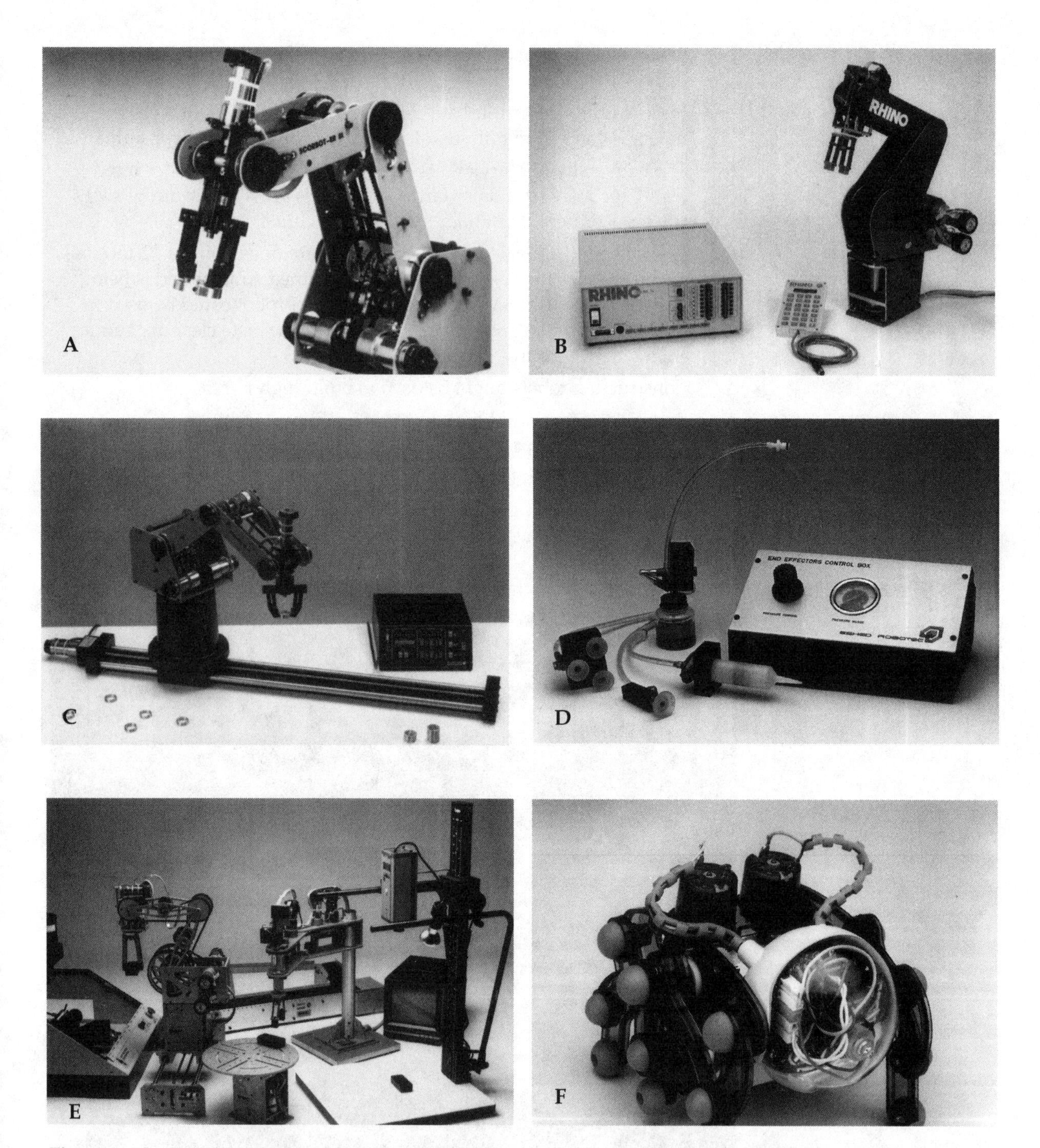

Figure 14-12 Educational robots. A—This robot is performing a simple pick-and-place operation. (Eshed Robotec, Inc.) B—This typical educational robot has a controller and teach pendant. (Rhino Robots, Inc.) C—A controller with a linear slide base provides an additional degree of freedom for the robot. (Eshed Robotec, Inc.) D—End-of-arm tooling available for one educational robot includes a spray-painting device, a syringe dispenser, a vacuum gripper with one suction cup, and a vacuum gripper with four suction cups. (Eshed Robotec, Inc.) E—This robotics workcell includes two different types of robots, one of them equipped with machine vision. (Rhino Robots, Inc.) F—A battery-powered robot kit that demonstrates the principles of optical sensing and walking motion. (OWI, Inc.)

- Δ *Telerobots*—robot arms and mobile robots operated by remote control—are candidates for work in law enforcement, the military, nuclear plants, and hazardous chemical sites.
- Δ NASA is developing robotic systems to automate assemblies in space.
- Δ ***Microbots***, the tiniest robots, may someday perform operations inside the human body. Doctors would guide their work by means of a computer. Imagine a robot submarine small enough to cruise through your veins and arteries!
- Δ Some people predict that by the year 2029, insect-sized robots will construct, alter, and clean everything from ranch houses to silk suits. One university professor is developing a robot "ant" that could perform as a tool or a tiny spy.

Several manufacturers have developed robots that are used primarily for education. These robots are smaller versions of those used in industry, and their operation is basically the same. Students who want to enter the field of robotics or related manufacturing fields can use them to learn the basic principles. Their main disadvantage is their expense.

Artificial Intelligence (AI) and Expert Systems

Is it possible to build machines that are genuinely intelligent, that seem humanlike in their problemsolving abilities? "Thinking machines" have long been the goal of scientists and experimenters. See Figure 14-13. However, most researchers agree that a truly humanlike thinking machine is a long way into the future. Some feel that such a machine can never exist. They ask the age-old question: "Can any mechanism really think?" They are convinced that a human being is more than a machine. This does not mean, however, that machines cannot be made more intelligent than they are currently.

Figure 14-13
Some historical milestones in artificial intelligence.

Date	Milestone
Middle Ages	Drawings for automatons (mechanical humans) appear.
1600-1700	Mechanical clocks with automatons increase in number and sophistication.
1940s	"Thinking machines" appear in the writings of scientists.
1950s	Research on neural networks increases.
1960s	Heuristic techniques find greater application.
1970s	Expert systems gain acceptance.
1980s	Machine learning research is on the increase and commercialization of AI begins.

Marvin Minsky, one of the fathers of the ***artificial intelligence (AI)*** field, defines it as "the science of making machines do things that would require intelligence if done by man." AI has been called the force behind the "second computer revolution." AI researchers have come from many scientific disciplines, including computer science, psychology, cognitive science, philosophy, biology, and engineering. AI research can be broadly classified into three major interrelated areas:

- Δ Robotics, which relates to machine vision, movement, and tactile sensing. The goal is to perform these operations in an intelligent, human-like manner.
- Δ Natural language processing, which looks at the "user interface" where human and machine meet. The goal is to develop machines that understand natural human speech.
- Δ *Expert systems,* programs that contain a knowledge base, acquired from human experts, which can be used to help nonexperts diagnose problems and make decisions.

Researchers at Carnegie-Mellon Robotics Institute in Pittsburgh are developing hardware and software using AI techniques to solve one of industry's most costly problems. This problem is the lengthy setup and process planning time needed to get machine tools ready for a new operation. With the help of AI, a new generation of intelligent machine tools will be able to run with very little human help. They will choose the correct tool, tool speed, cutting fluids, and optimum cutting strategies. They will then assemble and inspect the parts, maintaining correct tolerances.

The fastest-growing branch of AI is that of *expert systems.* An expert system contains a knowledge base, an inference engine, and a computer-human interface. Engineers get information from experts in a specific field, such as welding, and build that knowledge into the knowledge base of the system. Users who are not knowledgeable in that specific field can ask questions and receive the advice from the expert system.

There are a number of characteristics that separate expert systems from regular computer programs:

- Δ They reason like a human and can handle uncertainties.
- Δ They can explain why they have asked a specific question of the user.
- Δ They can tell the user how they reached a certain conclusion.
- Δ They can learn from experience.

Expert systems have come out of the laboratory and are being applied in industry and in the service and public sectors to solve a range of problems. The first expert system to be made available was known as MYCIN. It was a medical expert system used to diagnose diseases. Other applications include diagnosing equipment failures in cars, electronic gear, factory machines, and computers; optimizing system designs; helping sales personnel keep track of the thousands of parts required by some complex systems, and helping standardize tooling designs. A capital proposal package guides managers through bureaucratic forms for requisitioning goods. Another system makes it possible for sales people who are not engineers to design computer networks for their customers. Automotive personnel use AI to help detect motor problems before an actual breakdown occurs. A filler-metal selection system uses AI to recommend optimal filler materials for gas-metal arc (GMAW) and gas-tungsten arc (GTAW) welding of aluminum.

AI is the new frontier in the advancement of robotic and computer technology, with more and more practical applications becoming commercially available. Ultimately, AI could change our way of life.

Impacts on Society

There is little doubt that robots and other forms of computer-controlled machinery will continue to increase in number and sophistication. To date, few workers have been totally displaced by this technology. However, that will change. Eventually, the employment opportunities for semiskilled and unskilled workers will become fewer and fewer.

In a study of the effect of robotics on human resources, the authors reached this conclusion:

> *"The most remarkable thing about the job displacement and job creation impacts of industrial robots is the skill-twist that emerges so clearly when the jobs eliminated are compared to the jobs created. The jobs eliminated are semiskilled or unskilled, while the jobs created require significant technical background... this is the true meaning of the robotics revolution."*

Thus, the retraining of workers will be an important activity for every high technology company if it is to remain efficient and competitive. At the very least, computer literacy will become a prerequisite for employment for most workers in a majority of job fields.

Your Future in Robotics

Robotics and automated manufacturing offer careers that can be both exciting and challenging. A variety of educational opportunities are available to students and workers from diverse backgrounds.

Training is needed for those who will operate, maintain, or supervise robots. Training is also needed for the worker who is displaced by automation. Most companies attempt to retrain these workers for more satisfying jobs. In some cases, they may be retrained to program, install, or service robots. Although many of these workers have not been in a school environment for a long time, well-organized training programs offer a career boost for most.

Colleges and vocational-technical schools also offer programs in robotics. Some are focused toward operator training, while others develop supervisors for computer-integrated manufacturing. Operation and servicing is a major concern. Essential subjects include: robotic fundamentals, programming, electricity, basic electronics, hydraulics, pneumatics, digital electronics, microprocessors, programmable logic controllers, and machine vision systems. Basic courses in math, science, and communication skills are also strongly recommended.

Training is also provided by suppliers or robotic systems. This training is usually short and focuses on the operation and programming of the systems which are being installed. Maintenance is also covered by suppliers. Sometimes courses on special applications are offered to encourage the sale of products.

Important Terms

artificial intelligence (AI)
computer-integrated manufacturing (CIM)
expert systems
flexible manufacturing systems (FMS)
microbots
serbots
service robotics
telerobots

Review Questions

Write your answers on a separate sheet of paper.

1. Discuss several possible and practical uses for robots in the future.
2. For many decades, scientists, engineers, and plant managers have envisioned the "factory of the future." Write a detailed description of your vision of the factory of the future.
3. Define the term "flexible manufacturing system (FMS)." List some of its advantages.
4. Discuss the use of the programmable logic controller (PLC) in the flexible manufacturing system that was described in this chapter.
5. Define the term "artificial intelligence (AI)." Discuss its use in industry.
6. One of the fastest-growing areas of artificial intelligence is known as expert systems. What is an expert system? Describe how it is used in intelligent machine tools.
7. Describe the three main components of an expert system.
8. Name at least four ways in which an expert system is different from a normal computer program.
9. How do you think artificial intelligence and expert systems will impact society in the future?
10. Discuss the "skill-twist" that occurs when a manufacturing process is automated.

Glossary

A

Ac Synchronous Motor: Simple motor that contains no brushes, commutators, or slip rings. Commonly used in low-power applications; very reliable and has a long operating life.

Accumulator: Digital device that temporarily stores operands to be processed by the arithmetic logic unit (ALU). After processing, output of the ALU also appears in the accumulator.

Accuracy: Term used to express how close a robots end effector can be programmed to hit a desired point.

Acoustical Proximity Detector: Sensory device with a cylindrical, open-ended cavity that reacts to sound. A microphone may be used to detect a change in sound pressure and measure the distance of an object from the detector.

Actuator: A motor or valve that converts power into robot movement.

Address Register: The part of a microprocessor used to temporarily store the address of a location in memory.

Alternating Current (ac): Electrical energy produced when electrons flow first in one direction and then in the opposite direction. Each change in direction is called a cycle.

Analog Information: Information that varies continuously.

AND Gate: Logic circuit that has two or more inputs and one output; if all inputs are in the 1 state, then there is a 1 at the output.

Angular Actuator: A motor, the driving force of which provides angular rotation. Moves loads in an arc or circle.

Angstrom Å: Unit of light measurement. One angstrom unit is one tenth of a nanometer.

Anthropomorphic: A word used to describe a robot that is humanlike in form.

Arc Welding: A welding process that uses the heat of an electric arc to fuse two metals together along a joint.

Arithmetic Logic Unit (ALU): The part of a microprocessor that performs mathematical and logic operations.

Armature: An electromagnet placed between two permanent magnets in an electric motor and wound with electric wire. When a current is passed through the armature, it begins to rotate.

Artificial Intelligence (AI): The ability of a computer program to make decisions based on known information.

Asimov, Isaac: Science fiction author whose story "Runaround" established his three fundamental laws of robotics. The laws state that a robot (1) may not injure or allow a human being to come to harm; (2) must obey the orders given it by human beings as long as such orders do not conflict with the first law; and (3) must protect its own existence as long as such protection does not conflict with the first and second laws.

Assembler: A computer program that automatically translates lines of assembly language.

Assembly Language: A machine-oriented set of instructions for a specific machine's microprocessor.

Automated Guided Vehicle (AGV): Computer-controlled, battery-operated vehicle that follows an electronic guidepath in the floor.

Automatic Tool Changer: Adapter that provides rapid tool-changing capabilities with a minimum amount of time lost, enabling the robot to perform a wider range of assembly tasks and machining operations.

Automaton: A human-made object that moves automatically. The term was first used for clockworks introduced during the Middle Ages.

Avoidance Costs: Investment made in equipment for a new operation.

B

Bifilar Construction: Method of winding stator coils in a dc stepping motor. Two separate wires are wound into the coil slots at the same time.

Binary-Coded-Decimal Number System (BCD): System in which four binary digits are used to represent each decimal digit; used for writing large numbers in binary.

Binary Counter: Device used to count numerical information in binary form. In the counter, flip-flops are connected so that the Q output of the first circuit drives the trigger, or clock input, of the next circuit. Each flip-flop has a divide-by-two function.

Binary Logic Circuit: Computer circuit that makes logical decisions based upon input signals in binary form.

Binary Number System: A number system that has 2 as its base. Only the numbers 0 or 1 are used in the binary system.

Bistable: Term used to describe any electronic device that can assume one of two stable operational states.

Bit: A contraction of the letters *bi* from the word *binary* and *t* from *digit*; a single pulse of digital information.

Brushes: Carbon devices that rub against the commutator in an electric motor. When power passes through the brushes and commutator on its way to the armature, additional magnetic fields are created.

Bus: Distribution path in a microcomputer.

Byte: A measure of digital computer data; a group of 8 bits.

C

Capacitive Transducer: Device that measures a change in capacitance and used for such things as sensing fluid pressure.

Capek, Karel: Czech playwright (1890-1938) whose play *R.U.R, "Rossum's Universal Robots"* helped introduce the idea of robots to the popular imagination.

Capital Expenditure: The initial cost of an item paid for out of a company's working capital.

Cartesian Configuration: Geometric description of the arm movement of a robot along three intersecting perpendicular axes (straight lines). Movement along all three axes can start and stop simultaneously.

Central Processing Unit (CPU): The nerve center of a computer or programmable logic controller (PLC) system; a microprocessor is the basic element having address outputs and buses. It stores and handles data and monitors the status of input and output signals.

Centrifugal Pump: Pump that moves an indeterminate amount of fluid. Centrifugal force causes the fluid to move outward toward the wall of the housing. As a general rule, pumps of this type are only used for transferring large amounts of fluid at low pressure.

Closed Loop System: A system that allows for feedback that affects the output of the system.

Collet Gripper: Gripper used to pick and place cylindrical parts that are uniform in size; repeatability is an important characteristic.

Command Resolution: The closest distance between robots' movements.

Commutator: The part of a dc motor that switches current flow.

Comparator: In a closed-loop system, feedback from the controlled element goes to a comparator where it is compared to that of the reference source. A correction signal is developed by the comparator and sent to the control unit.

Compliance: The ability of a robot to tolerate misalignment of mating parts; prevents jamming, wedging, and galling of the parts; essential for assembly of close-fitting parts.

Compound-Wound Dc Motor: Motor that has two sets of field windings, one in series with the armature and one in parallel.

Computer-Integrated Manufacturing (CIM): The integration of engineering, production, marketing, and functions of a manufacturing enterprise in a computer-controlled operation.

Computer Vision: Computer sensing system used to sense spatial relationships.

Conditioning: Using an air filter with a condensation trap and drain to remove dirt and moisture from air that has been compressed for use in a pneumatic system.

Continuous-Path (CP) Motion: Motion of a robot manipulator that involves more points than point-to-point motion. The distances between points are thus extremely close, producing movement that appears smooth and continuous. Control of the path is of more concern than end-point positioning.

Control: Management of a robot by means of altering the flow of power and causing some kind of operational change in the system.

Controller: The part of a robot that coordinates all movements of the mechanical system. It is usually a microprocessor that receives input from the immediate environment through various kinds of sensors.

Cost Savings: Investment made for replacement of existing equipment.

Counter: Versatile logic device used to count a wide variety of objects in a number of different applications.

Counter Electromotive Force (CEMF): The voltage generated when the armature in a dc motor rotates.

Cycle: The change in direction of electricity. See alternating current (ac).

Cycle Timing: Complex timing system designed to provide some type of energizing action in an operational sequence; may include both interval and delay timing.

Cylindrical Configuration: Geometric description of a robot in which its range of motion assumes a cylindrical shape. The arm consists of two orthogonal slides mounted on a rotary axis. Reach is accomplished as the arm of the robot moves in and out. For vertical movement the carriage moves up and down on a stationary post, or the post can move up and down in the base of the robot.

Cylindrical Grip: Gripping movement used for grasping cylindrical objects.

D

Data Register: The part of a microprocessor that temporarily stores information applied to the data bus.

Dc Stepping Motor: Motor found in nearly all high-power servomechanisms. It is more efficient and develops more torque than the synchronous servomotor; used primarily to change electrical pulses into rotary motion.

Decade Counter: Counter used to change binary signals into a binary-coded decimal (BCD) form.

Decimal Point: The point of reference in number systems; used to determine place values.

Dedicated Equipment: Equipment designed to perform only one function.

Degrees of Freedom: Term used to describe a robot's dexterity, or freedom of motion. For each degree of freedom, a joint (or axis) is required.

Delay Timing: Timing in which a time delay occurs before the robot's load device actually becomes energized.

De-palletizing: Removing parts from a uniform series of positions; performed by pick-and-place robots.

Design for Manufacturability: Designing products to be assembled by automation.

Desiccant: A dry chemical substance designed to attract moisture.

Detector: The part of a sensing system designed to respond to energy from the source.

Detents: Two or more elements in an overload safety device that are held in position by spring-loaded mechanisms. They move from their original positions under excessive stress and cause shutdown of the robot.

Devol, Jr., George C.: Inventor who, in 1954, patented the first industrial robot.

Die Casting: The pumping of hot metals into closed dies to shape it. After the metal solidifies, the die is opened and the casting is removed.

Digital Electronics: The technology that controls robotic and other automated systems.

Digital Information: Information used in digital computers that occurs in separate chunks, as opposed to analog information. It can then be translated into binary code.

Digital System: An automatic system that processes digital information. Numeric instructions are supplied to the system by perforated paper tape, punched cards, magnetic tape, or variations in pressure, temperature, or electric current. The numeric signals are decoded and directed to specific machines or machine parts, which then perform the necessary operations.

Direct Current (dc): Electrical energy produced when electrons flow in only one direction.

Direct-Drive Electric Motor: High-torque motor that drives a robot arm directly without the use of reducer gears.

Direction Control Device: Device designed to start, stop, or reverse fluid flow without causing an appreciable change in pressure or flow rate; may be actuated by pressure, mechanical energy, electricity, or manual operation.

Displacement: The movement of a workpiece that a robot senses in reference to a fixed position or to the force required to move it from one position to another.

Dynamic Performance: Used to describe how fast a robot can accelerate, decelerate, and stop at a given point. Also called operational speed.

E

Eddy Current Proximity Sensor: Device that produces a magnetic field in a detector unit, which may be mounted in a probe. The magnetic field induces eddy currents into any conductive material that is near the probe.

Electric Drive: Any actuator driven by an electric motor.

Electric Motor: Motor that converts electrical energy into mechanical energy.

Electromagnetic Spectrum: The range of visible and invisible light; includes frequencies for radio, television, radar, infrared radiation, ultraviolet light, X-rays, and gamma rays.

Electromechanical Gripper: End effector that uses the magnetic field created by an electromagnet or permanent magnet to pick up an object. Also called a magnetic gripper.

Electromechanical System: A system that transfers power from one point to another through mechanical motion that is used to do work.

End Effector: A robot's gripper, or hand.

End-of-Arm Tooling: An end effector or other device attached to the wrist of the manipulator that can perform operations on a workpiece.

End-Stop: Limit switch.

Engelberger, Joseph F.: An engineer who teamed up with George C. Devol. He became president of Unimation, Inc., formed in 1958 to develop an industrial robot. He is considered to be the father of industrial robotics.

Error Detector: Device that receives data from both the input source and the output device. The output device is usually a synchro system that relays information back to the error detector for position comparison.

Error Signal: Feedback to a servo amplifier. Error signals are amplified by the servo amplifier and applied to the servo control valve along the appropriate manipulator axis.

Execute: Programmed information placed into memory that directs a microprocessor to perform an operation.

Expandable Gripper: Gripper used to clamp workpieces that have irregular shapes; uses a hollow rubber envelope that expands when pressurized, insuring an evenly distributed surface load.

Expert System: A computer program that uses artificial intelligence and a knowledge base acquired from experts to help nonexperts solve problems and make decisions.

F

Feedback: The interaction between the controlled element and the control unit in a closed-loop system.

Fetch: Programmed information that directs a microprocessor to get the next instruction from memory.

Field Winding: In dc motors, the coil wrapped around magnets to create a magnetic field.

Filter: Device that provides a finer grade of fluid conditioning than strainers. Typically, filters are made of some porous medium such as paper, felt, or fine-wire mesh.

Fixed-Sequence Robot: A manipulator that performs successive steps of a given operation repetitively, according to a predetermined sequence, condition, and position. Its instructions cannot be easily changed.

Flexible Automation: Machines that can be programmed to perform different tasks.

Flexible Manufacturing System (FMS): A type of manufacturing in which machines are able to make different products without retooling or similar changes. Numerically controlled tools, robots, and computer-controlled conveyors are usually used.

Flip-Flop: Memory device used in digital circuits that can be made to hold an output state even when the input is completely removed; can also change its output based on an appropriate input signal.

Flow Control Device: Device that alters the volume or flow rate at which fluid is delivered to the load of a fluid power system, determining its operational speed.

Flow Indicator: Device used to test flow rates from pumps and at the inlet and outlet ports of actuators.

Fluid Motor: Motor designed to convert the force of a moving fluid into rotary motion by means of vanes, gears, or pistons; similar to a pump in appearance and operation.

Fluid Power System: System designed to transfer power using air, oil, or a combination of air and oil.

Force: Any cause that tends to produce or modify motion; normally expressed in units of weight.

FRL Unit: In pneumatic systems, filter, regulator, and lubricator components placed together in a combination unit.

G

Gripper: An end effector that is designed to grasp an object and move it.

H

Hard Automation: Machinery that has been specifically designed and built to perform one particular task.

Heat Exchanger: Device used in some hydraulic systems to maintain the temperature of the fluid at a desired level. Heat exchangers may be forced-air fan units, water-jacket coolers, or gaseous cooling units.

Hexadecimal Number System: An arithmetical system used to process large numbers; the base of this system is 16.

Hierarchical Control: An arrangement in which a given level of robotic control is more elemental than the one above it and is dependent on the level above it for its instructions.

Hierarchical Control Programming: A method of programming in which control is partitioned into a number of different levels. Each level accepts commands from the next higher level and responds by generating simpler commands to the next lower level.

High-Level Language: A computer programming language that is closer to standard English and which must be translated into machine code by means of a program called a compiler.

Hook Movement: Hand movement used to pull drawers open or lift objects with bail-type handles.

Hydraulic System: Fluid power system generally consisting of an electric pump connected to a reservoir tank filled with fluid, some control valves, and a hydraulic actuator.

I

Image Acquisition: In machine vision, the process of illuminating a workpiece and digitally scanning its image.

Image Analysis: In machine vision, the process that involves gathering information from the image that will enable the robot to do a task. Analysis is performed by computer software.

Image Interpretation: In machine vision, the process using algorithms (special codes) by which the software describes and measures the features of the digital image, allowing the robot to make correct decisions about the work that needs to be done.

Image Preprocessing: In machine vision, the process of changing an analog signal into an equivalent digital signal using an analog-to-digital (AD) converter. These light intensity values are stored in memory where the digital image can then be analyzed and interpreted.

Indicator: A device that displays readings indicating operating conditions at various points throughout an automated system.

Inductive Transducer: Sensory device that has a stationary coil and a movable core. The movable core is connected to the object whose movement is to be measured.

Industrial Robot: A programmable, multifunction manipulator designed to move materials, parts, tools, or special devices through programmed motions for the performance of a variety of tasks.

Inertia: The property of a body which determines the amount of force needed to produce motion or its resistance to change.

Infrared Detector: Sensory device that picks up radiation in the infrared region of the electromagnetic spectrum.

Input-Output (I/O) Transfer: In digital electronics, an operation similar to the read/write operation. Code actuates an I/O port, which either receives data from the input or sends it to the output device.

Input Port: The connection through which a robot receives information in the form of digital data. This data is used by the robots computer program to initiate some action.

Instruction Decoder: In digital electronics, an application that examines a coded word and decides which operation is to be performed by the arithmetic logic unit.

Integrated Circuit (IC): A tiny silicon wafer containing thousands of transistors, resistors, and diodes.

Intelligent Robot: A robot that can detect changes in the work environment by means of sensory perception (visual and/or tactile), then, using its decision-making capability, proceeds with the appropriate operations.

Interface: The common point at which two or more systems communicate with each other.

Interfacing: The process that allows a robot to communicate and interact with other pieces of equipment.

Interrupt: A programming operation often used to improve efficiency by allowing a microprocessor to do other work while waiting for a signal from peripheral equipment such as keyboards, monitors, or printers.

Interval Timing: Timing that occurs after the robot's load has been energized; it may cause the load to remain energized only for a certain period.

Inverter: See NOT gate.

L

Laser: Acronym for light amplification by stimulated emission of radiation. A highly focused beam of light that can cut, carry messages, and perform other kinds of work.

Lateral Grip: Gripping movement used to grasp larger objects in a sideways motion.

Light Curtain: A programmable safety barrier consisting of photoelectric presence-sensing devices.

Light-Emitting Diode (LED): Small, lightweight opto-electronic device, easily used with digital and other miniaturized optical systems.

Light Pipes: A solid transparent plastic rod that can transmit light from one end to the other even if it is bent. *Also see Optical Fibers.*

Linear Actuator: Motor that provides motion along a straight line.

Load: The part or number of parts in a robotic system designed to produce work.

Logic Circuit: In digital electronics, the circuit that performs control functions.

Logic Gate: The individual components of a logic circuit that determine the path of information in the circuit.

Lubricator: Conditioning device, found only in pneumatic systems, that adds a small quantity of oil to the air after it leaves the regulator; makes valves and cylinders last longer and operate more efficiently.

M

Machine Vision System: A system built into automated workcells that gives the robot eyes and allows it to perform in a more intelligent manner by providing guidance, determining part orientation, measuring and inspecting parts, identifying images, etc.

Magnetic Gripper: A type of end effector that is magnetized for improved grasping ability. Also called an electro-mechanical gripper.

Manipulator: The part of a robot consisting of segments that may be jointed and that move about, allowing the robot to do work; a robot's "arm."

Manual Manipulator: A manipulator, or robot arm, worked by a human operator.

Manual Programming: Machine setup by an operator who adjusts the necessary end-stops, switches, cams, electric wires, or hoses to set up the programming sequence of the machine.

Manual Rate Control Box: A device used for robot programming having a knob and some switches that control the movement along each axis individually. The joints are moved one at a time until the robot is oriented in the desired location. The values of the position indicators are then stored in memory.

Mechanical Finger Gripper: The most common type of robot gripper having fingers that are used for grasping objects.

Mechanical Fuse: Overload protection device that consists of pins or tubes that break or buckle under extreme stress.

Memory: The ability of a computer to store information.

Microbot: A tiny robot designed for such tasks as operating inside the human body.

Microcomputer: Small computer built around a single integrated circuit; also called a PC.

Microprocessor Unit (MPU): The arithmetic logic unit and control section of a microcomputer that receives data and stores it for future processing.

Mini-PLC: Mini-programmable logic controller that can control simple machine operations and numerous manufacturing processes.

Mnemonics: An operation code in the form of symbols or words that is used in assembly language; easier to recognize and remember than binary code.

N

NAND Gate: Combination of a NOT logic gate and an AND logic gate. A NAND gate is an inverted AND gate.

Nanometer (nm): Unit of measurement for light wavelengths. A nanometer is 1×10^{-9}m.

NC (Numerically Controlled) Robot: A manipulator that is programmed by means of numerical data, using punched tapes, cards, or digital switches.

Nonpositive Displacement Pump: In a fluid system, a pump with no set amount of air or fluid that passes by the impeller blades during rotation. Flow depends upon the speed of the blades.

Nonprehensile Movements: Movements that include pushing, poking, punching, and hooking. They do not require any special dexterity.

Nonservo Robot: A nonintelligent robot. The U.S. defines a nonservo robot as the simplest type of robot.

NOR Gate: A combination of a NOT logic gate and an OR logic gate. A 1 is output only when both inputs are 0.

NOT Gate: A logic gate that has one input and one output. NOT gates are also called inverters. Output is opposite to input.

O

Occupational Safety and Health Act (OSHA): Federal act regulating health and safety in the workplace.

Octal Number System: A base-8 number system used to process large numbers.

Off-Line Programming: Programming of a robot by means of a computer to which the robot is not connected.

On-Line Programming: Programming of a robot by means of a computer at the robot's console.

Open-Loop System: A system in which no feedback mechanism is used to compare programmed positions to actual positions.

Operational Speed: The speed at which the robot can accelerate, decelerate, and stop at a given point. Also called dynamic performance.

Oppositional Grip: Gripping movement involving the use of the index finger and the thumb, which oppose one another.

Optical Fibers: Fibers made of glass or plastic; optical fibers can transmit light from one point to another. Also referred to as a light pipe.

Optical Proximity Sensor: Sensory device that measures the amount of light reflected from an object to determine its position; may respond to either visible or infrared light.

Opto-Electronic: Term that refers to the combination of optics and electronics.

OR Gate: A logic gate that has two or more inputs and one output. A 1 is output when both inputs are 1 or when either input is 1.

Output Port: A connection through which the robot controller sends digital data to peripheral equipment.

Overload Sensing: The sensing of obstructions or overload conditions so that the robot can be shut down before damage occurs.

P

Palletizing: Placement of parts in a uniform series of positions; performed by pick-and-place robots.

Palmar Grip: Gripping movement similar to that of a baby holding a bottle during feeding.

Parallel Port: A connecting point through which multiple bits of digital data can be transmitted very quickly.

Parallel Transmission: The method of delivering multiple bits of digital data at the same time, or in parallel with one another. Transmission is much faster than serial transmission but is generally cost prohibitive over long distances.

Pascal's Law: The principle which states that pressure applied to a confined fluid is transmitted undiminished throughout the fluid. It acts on all surfaces at right angles to those surfaces.

Paternoster: A device that holds components in place during assembly operations.

Payback Period: The time required to recover the amount expended for new equipment through savings in labor or material costs.

Payload: The maximum weight or mass of material a robot is capable of handling on a continuous basis; usually includes the weight of the end effector.

Period: The interval taken for a timing pulse to pass through a complete cycle from beginning to end.

Permanent-Magnet Dc Motor: Motor used when a low amount of torque is required. The dc power supply is connected directly to the rotor conductors through the brush-commutator assembly. The magnetic field is produced by permanent magnets mounted to the stator.

Personal Computer (PC): Any small desktop or portable computer built around a single integrated circuit or microprocessor.

Photoconductive Device: Device that changes in conductivity according to variations in light.

Photoemissive Device: Opto-electronic device that emits electrons in the presence of light. Phototubes are a type of photoemissive device.

Photovoltaic Device: Opto-electronic device that converts light energy into electrical energy; also called a solar cell.

Pick-and-Place Motion: The motion of a pick-and-place robot; points along each axis are comparatively few in number; the movement of the end effector follows a fixed pattern and generally only one axis of the robot moves at a time.

Piezoelectric Effect: The creation of electrical energy by applying pressure to a certain type of crystal; often used in microphones.

Pitch: The up-and-down movement of the robot's wrist; bend.

Pixels: Abbreviation for picture elements. Term used for the dots of light that form the image on a TV screen.

Place Value: The position of a digit with respect to a reference point called the decimal point. The largest digit that can be used in a specific place is determined by the base of the system.

Playback Robot: A manipulator that can store in memory and reproduce operations originally executed under human control.

Pneumatic System: Power system that uses air as its energy source. Its construction is similar to that of a hydraulic system.

Point-to-Point (PTP) Motion: The movement of the robot through a number of discrete points in space. A combination of axes is used to position the end effector at a desired point. The positions are recorded and stored in memory. The path of motion is a series of straight lines between the points.

Positive Displacement Pump: In a fluid system, a pump with a rather close clearance between the rotating member and the stationary components. A definite amount of fluid passes through the pump during each revolution.

Positive Logic: In digital electronics, a type of logic where the value of 0 is expressed as low-voltage or no voltage and the number 1 is used to indicate a voltage larger than 0.

Power: Term used to express the relationship between force and pressure (effort) and the length of time it takes to accomplish work.

Power Supply: The system that supplies the power required by a robot. The power supply may convert ac line voltage to dc voltage, or it may be a pump or compressor providing hydraulic or pneumatic power.

Prehensile Movements: Movements used to grasp an object with the aid of the thumb. Curled fingers and the opposing thumb provide the hand with the dexterity for these movements.

Preloaded Spring: Safety device used to prevent overload conditions. Excess stress causes the spring to release and the end effector breaks away from the work area.

Pressure: Term used to describe the amount of force applied to a specific area.

Pressure Indicator: Device having elements that physically change shape when different pressures are applied. The physical change moves an indicator on a scale or a stylus on a paper chart.

Pressure Regulator: Device used to create a balance between atmospheric air pressure and system line pressure. These regulators may be used in several places within a system.

Pressure Relief Valve: In a fluid system, a pressure control device used to dump the output of a positive displacement pump back into the reservoir when the pressure rises to a dangerous level; also serves as a safety device. In pneumatic systems, a device used to control smaller amounts of air.

Preventive Maintenance: The process of regularly checking equipment, cleaning and maintaining it, and replacing worn parts before breakdowns occur.

Program: A series of instructions, stored in the controller's memory, that controls robot movement.

Program Counter: A memory device that indicates the location in memory of the next instruction to be executed.

Programmable: Term used to describe a robot or other machine that can be given new instructions to meet new requirements.

Programmable Logic Controller (PLC): Simple control device designed specifically for industrial machines that performs logical operations compatible with traditional relay logic.

Proximity Sensor: Device that senses the absence or presence of an object within a certain region or that provides feedback about distance between it and something else, such as an end effector.

Q

Quality Assurance: Inspection system that ensures the production of the highest quality products.

R

Radial Traverse: Extension and retraction of the robot arm, creating in-and-out motion relative to the base.

Random Access Memory (RAM): A very fast type of read/write memory commonly found in computers. Any information stored in RAM is volatile and will be lost if power is interrupted.

Range Sensor: Device used to determine the precise distance from the sensor to an object.

Read-Only Memory (ROM): Memory in which data is permanently stored in a microcomputer. This data is not lost when the power source is turned off.

Read/Write Memory: Data stored in a microcomputer that can be read or written to (altered).

Reciprocating Pump: A pump that forces either air or oil from a chamber by the reciprocating action of a moving piston.

Rectification: The process of converting alternating current to direct current.

Reed Switch: Device that makes or breaks contact when exposed to a magnetic field.

Relay Contact: The basic programming element when using relay logic. This contact may be normally open (NO) or normally closed (NC).

Relay Ladder: A network of elements used in relay logic.

Relay Logic: An assembly language used to program programmable logic controllers (PLCs). A ladder diagram is entered into the system by means of a keyboard.

Remote-Center Compliance (RCC) Device: Device that fits in the wrist of the robot to give it compliance capability.

Repeatability: Term used to express how close the robot returns to its programmed positions time after time.

Resin-Core Solder: The type of solder that should be used for all electronic circuit work.

Resistance: Term used for the friction that occurs as fluid flows through the components of a power system.

Resistance Welding: Welding method in which current flows by means of electrodes through two metals, joining them at the point where they touch; also called spot welding.

Resistive Transducer: Sensory device that converts variations of resistance into electrical variations; often used to sense physical displacement.

Resolution: Term describing the smallest incremental movement a robot can make.

Revolute Configuration: Term used to describe the irregularly shaped work envelope of a robot with rotary joints.

Rework: The process of fixing parts that do not meet product specifications.

Robot: An automatic apparatus or device that performs functions ordinarily ascribed to human beings or that operates with what appears to be human intelligence; from the Czech word *robota*, meaning "forced labor."

Robotics: The use of robots.

Robotic Industries Association (RIA): Organization founded in 1975 to encourage development and use of robotics in America.

Roll: Swivel, or rotation, of the robot's wrist.

Rotary Actuator: Fluid power device designed to produce a limited amount of rotary motion in either direction (twisting or turning).

Rotary Electric Actuator: Electrical device that produces rotary motion and transmits it between locations without direct mechanical linkage.

Rotary-Gear Pump: A pump that contains two gears enclosed in a precision-machined housing. Rotary motion from the power source is applied to the drive gear.

Rotary Vane Pump: Pump having a series of sliding vanes placed in slots around the inside of the rotor. As the rotor turns, centrifugal force or spring action forces the vanes outward.

Rotational Traverse: Movement about a vertical axis. The side-to-side swivel of the robot's arm about its base.

Rotor: The rotating armature, shaft, and associated parts of a motor.

S

SCARA: Selective Compliance Assembly Robot Arm. Developed by Professor Makino of Yamanashi University, Japan. Its design allows it to be firmly yielding in horizontal motions and rigid in vertical motions.

Scrap: A part that cannot be fixed.

Sensing System: A system that responds to various forms of energy, such as light energy and electrical energy, and conveys information to a control unit.

Sensory Feedback: Information about its environment that affects how a robot responds or interacts.

Sequence Controller: A device that uses clock inputs to maintain the proper sequence of events required to perform a task.

Sequencing: The control of a load device in each of its stepping positions. Because a sequencer is basically a counting register, sequencing is widely used in automatic assembly operations and in production line control.

SERBOT: Abbreviation for service robot.

Serial Port: Connecting point through which a computer sends or receives digital data using the serial transmission method.

Serial Transmission: A method of delivering digital data over long distances. Data is delivered at a fairly slow rate because it must be transmitted one bit at a time.

Series-Wound Dc Motor: Motor in which the armature (rotor) and the field circuits are connected in a series arrangement.

Service Robot: Robot used outside a factory setting. Mobile service robots have the ability to move to the work area. Their command center can be located in the same building or miles away.

Service Robotics: The use of robots to perform services outside a factory setting.

Servomechanism: A special type of ac or dc motor that drives precision equipment in angular increments. Used in precise control devices or synchro systems that require increased torque.

Servomotor: A device used to achieve a precise degree of rotary motion. Two distinct types of servomotors are used today, the synchronous motor and stepping motor.

Servo Robot: A robot classified as intelligent or highly intelligent, determined by its level of awareness of its environment.

Servo System: Machine system that changes the position or speed of a mechanical object.

Shunt-Wound Dc Motor: Commonly used dc motor in which the field coils are connected in parallel with the armature (rotor).

Single-Phase Ac Motor: Motor that operates from a single-phase ac power source. The three basic types include universal motors, induction motors, and synchronous motors.

Single-Phase Induction Motor: Motor that has a solid, or squirrel cage, rotor. Large-diameter copper conductors are soldered at each end to a circular connecting plate. When current flows in the stator windings, a current is induced in the rotor.

Slip: The difference between the synchronous speed and the rotor speed of a motor.

Solder Sucker: Device used to properly remove solder from a defective component without damaging the printed circuit board.

Sound Sensing System: Sensory system that relies upon the piezoelectric effect to convert sound to electrical energy.

Spatial Resolution: The movement of the robot at the tool tip, taking into account command resolution and mechanical inaccuracy.

Speed Sensing: Methods of detecting speed, the results of which are used as inputs in robotic applications.

Spherical Configuration: Geometric description of the shape of a robot's work envelope, which resembles a sphere. A pivot point gives the robot its vertical movement. Reach is accomplished by means of a telescoping boom that extends and retracts. Rotary movement occurs around a rotary axis. Sometimes referred to as the polar configuration.

Spherical Grip: Gripping movement in which the fingers are used to hold round objects.

Spray Finishing: The application of a variety of paints, polyurethanes, or other protective coatings to the surface of a part or product.

Spread Movement: End-effector movement used to carry objects having holes or openings of some kind; the gripper's fingers spread out.

Squirrel Cage Rotor: The solid rotor in an induction motor.

Stadimetry: Process that determines the distance to an object based on the apparent size of a camera image.

Stator: The frame and other stationary components of a motor.

Strain Gauge: (1) Device that emits electrical signals based on the amount of pressure the mechanical fingers of a robot are exerting to lift an object. (2) Device used to sense a change in dimension as some material is subjected to stress.

Strainer: Device for removing impurities that contains a stainless steel screen having 60 to 200 wires per square inch; frequently placed in the reservoir filler opening, air breather, and pump inlet feed line of hydraulic systems.

Subroutine: A set of instructions within a computer program that has a beginning and an end.

Subsystem: A group of smaller systems, such as electromechanical and hydraulic systems that make up a robot as a whole.

Support Gripper: A type of end effector usually found on crane-type manipulators. The hook is the most common type. Some support grippers support the workpiece from the underside.

Switch Carpet: A sensing system consisting of a carpet that detects the presence of workers or objects when they apply pressure to switches embedded in it.

Synchronous Speed: The stator speed of a motor.

Synchro System: Motor-generator units connected together. With this type of system it is possible to achieve accurate control over great distances.

Synthesized System: Subsystems combined to make a new system.

System: An organization of parts that work together to form a unit.

T

Tachometer: A device used to measure speed.

Tactile Sensor: Sensory device that indicates the presence of an object by touch.

Task Level Programming: A user-friendly programming system for robots in which the goals of each task are specified rather than the motions required to achieve those goals. Instructions are input using English-like terms.

Teach Pendant: A device used to record movements into the robot's memory.

Teach Pendant Programming: Programming method in which the operator leads the robot through various positions. The end points are recorded into memory by pushing buttons on the teach pendant. The recorded points are used to generate the path the robot follows during operation.

Telerobot: A robot operated by remote control.

Thermistor: Temperature-sensitive resistor, the resistance of which increases with a decrease in temperature.

Thermocouple: Device that converts heat energy into electrical energy.

Thermo-Electric Sensor: Sensory device that produces a change in electrical output based on a change in temperature.

Three-Phase Ac Motor: Motor operated using three-phase ac power sources. The two basic types are induction motors and synchronous motors.

Three-Phase Induction Motor: Ac motor having a squirrel cage rotor. Three-phase voltage is applied to the stator so no external starting mechanisms are needed.

Three-Phase Synchronous Motor: Unique, specialized motor that delivers constant speed and can be used to correct power factors of three-phase systems.

Timing System: Mechanisms that turn a device on or off at a specific time or in step with an operating sequence.

Tool: End effector that performs a specific task, such as welding or painting.

Torque: The rotary motion produced by a motor; depends on the strength of the magnetic fields and the amount of current flowing through the motor's conductors. As the magnetic field or the current increases, the amount of torque also increases.

Touch-Sensitive Proximity Detector: Sensory device that operates on capacitance developed by a large conductive object; a conductive plate or rod is used to sense contact.

Trajectory: The resulting path of a robot's end effector as programmed in the controller.

Transducer: A device that converts one type of energy, such as light, heat, or mechanical energy, into electrical energy.

Transmission Path: The part of a system that provides a path for the transfer of energy.

Triangulation: Method of measurement based on measuring angles and the base line of a triangle.

Troubleshooting: Method of finding out why something doesn't work properly.

Truth Table: A table that shows combinations of inputs and the resulting outputs of a logic gate.

U

Ultraviolet Sensor: Sensory device that responds to electromagnetic radiation in the ultraviolet range.

Unimate: Name of the first industrial robot developed by Unimation and sold to General Motors in 1961. Unimate was used in a die-casting operation.

Universal Motor: Motor powered by either ac or dc sources.

V

Vacuum Gripper: End effector consisting of one or more suction cups made of rubber; extremely lightweight and simple in construction.

Variable-Sequence Robot: A manipulator that performs successive steps of a given operation repetitively, according to a predetermined sequence, condition, and position. Unlike those for the fixed-sequence robot, instructions for the variable sequence robot can easily be changed.

Vertical Traverse: Up-and-down motion of the robot arm.

Voice Recognition: Technology in which a robot is programmed to respond only to the vocal commands of an operator whose voice frequency has been recorded in memory.

W

Walk-through Programming: Programming method used for continuous-path playback robots. An experienced operator physically moves the end effector through the precise motion required by the task. While the robot moves along the desired path, points are recorded into memory for later playback.

WAVE: The first robot programming language, developed at the Stanford Artificial Intelligence Laboratory in 1973 for research purposes.

Weight: Gravitational force exerted on a body (or mass) by the earth.

Work: (1) That which occurs when energy is transformed. Heat, light, chemical action, sound, and mechanical motion are common forms of work. (2) The measure of what a system actually accomplishes when force and pressure (usually expressed in foot-pounds or newton-meters) are applied, taking into account the length of time it takes.

Work Envelope: The area within which a robot's end effector can reach.

X

X-Rays: Radiation produced by one band of frequencies in the electromagnetic spectrum.

Y

Yaw: Side-to-side movement of the robot's wrist.

These students are learning to program a robot using a joystick and teach pendant. (Ford Motor Co.)

Index

D

E

F

G

H

I

J

L

M

S